T0140126

Advances in Intelligent Systems and Computing

Volume 925

The series "Advances in Intelligent Systems and Computing" contains publications on theory, applications, and design methods of Intelligent Systems and Intelligent Computing. Virtually all disciplines such as engineering, natural sciences, computer and information science, ICT, economics, business, e-commerce, environment, healthcare, life science are covered. The list of topics spans all the areas of modern intelligent systems and computing such as: computational intelligence, soft computing including neural networks, fuzzy systems, evolutionary computing and the fusion of these paradigms, social intelligence, ambient intelligence, computational neuroscience, artificial life, virtual worlds and society, cognitive science and systems, Perception and Vision, DNA and immune based systems, self-organizing and adaptive systems, e-Learning and teaching, human-centered and human-centric computing, recommender systems, intelligent control, robotics and mechatronics including human-machine teaming, knowledge-based paradigms, learning paradigms, machine ethics, intelligent data analysis, knowledge management, intelligent agents, intelligent decision making and support, intelligent network security, trust management, interactive entertainment, Web intelligence and multimedia.

The publications within "Advances in Intelligent Systems and Computing" are primarily proceedings of important conferences, symposia and congresses. They cover significant recent developments in the field, both of a foundational and applicable character. An important characteristic feature of the series is the short publication time and world-wide distribution. This permits a rapid and broad dissemination of research results.

**** Indexing: The books of this series are submitted to ISI Proceedings, EI-Compendex, DBLP, SCOPUS, Google Scholar and Springerlink ****

More information about this series at http://www.springer.com/series/11156

Ewaryst Tkacz · Marek Gzik ·
Zbigniew Paszenda · Ewa Piętka
Editors

Innovations in Biomedical Engineering

Editors
Ewaryst Tkacz
Faculty of Biomedical Engineering,
Department of Biosensors and Biomedical Signals Processing
Silesian University of Technology
Zabrze, Poland

Marek Gzik
Faculty of Biomedical Engineering,
Department of Biosensors and Biomedical Signals Processing
Silesian University of Technology
Zabrze, Poland

Zbigniew Paszenda
Faculty of Biomedical Engineering,
Department of Biosensors and Biomedical Signals Processing
Silesian University of Technology
Zabrze, Poland

Ewa Piętka
Faculty of Biomedical Engineering,
Department of Biosensors and Biomedical Signals Processing
Silesian University of Technology
Zabrze, Poland

ISSN 2194-5357 ISSN 2194-5365 (electronic)
Advances in Intelligent Systems and Computing
ISBN 978-3-030-15471-4 ISBN 978-3-030-15472-1 (eBook)
https://doi.org/10.1007/978-3-030-15472-1

Library of Congress Control Number: 2019934096

This Springer imprint is published by the registered company Springer Nature Switzerland AG
The registered company address is: Gewerbestrasse 11, 6330 Cham, Switzerland

Preface

In several recent decades, it is easy to notice and observe an unusual progress of civilization development in many fields concerning achievements coming out from technical theories or more generally technical sciences. We can practice an extraordinary dynamics of the development of technological processes including different fields of daily life, which refers particularly to the problem of communicating. We are aspiring for disseminating of the view that the success in the particular action or activity is a consequence of the wisdom won over, collected, and appropriately processed. They are having in mind straight out about the coming into existence of the so called generally information society.

In such a context, the meeting of both experts and specialists dealing with the widely understood innovations in biomedical engineering give the opportunity of a new dimension associated with promoting something like state-of-the-art achievements. Because having the innovative approach as a pointer in today's world of changing attitudes and socioeconomic conditions can be perceived as one of the most important advantages. Consequently, the way from the universal globalization letting observe oneself of drastically changing surrounding world. Thanks to the development of new biotechnologies coming out from the rapid progress in biomedical sciences comprehending the contemporary needs of surrounding world, it may be said almost without any doubts that life without biomedical sciences would sooner or later stop existing.

At present, implementing the universal standardization of the transfer and processing of information is one of the most important issue what in the significant way influences for expanding the circle of Biomedical applications. It is undoubtedly a kind of challenge to put the proper weight into particular branches covered by biomedical engineering, and therefore, we decided to edit the book as four-part elaboration covering biomaterials, biomechanics, biomedical informatics, and last but not least biomedical signal processing. One should aspire to its permanent integration rather than the disintegration to progress in the context of technological development. Hence the constant observation and the appropriate problem analysis of biomedical sciences as well as checking the technologies development and their applications is picking the great importance up.

The monograph returned to the hands of readers being a result of meeting specialists dealing with above-mentioned issues should in the significant way contribute to the success in implementing consequences of human imagination into the social life. We therefore believe, being aware of a human weakness and an imperfection, that the monograph presenting a joint effort of the increasing numerically crowd of professionals and enthusiasts will influence the further technology development regarding generally understood biomedicine with constantly expanding spectrum of its socially useful applications.

The last part of this preface will be devoted to express our great thanks and appreciation to all the contributors of this book, which were listed in the special section as "contributors' list" and to persons who gave us an unusual help in the final editing process. Specials thanks are transferred to Szymon Sieciński for incredible engagement and help in creating the final version of this book.

October 2018

Ewaryst Tkacz
Marek Gzik
Zbigniew Paszenda
Ewa Pietka

Contents

Modelling and Simulations in Biomechanics

Engineering of Biomaterials

List of Contributors

Krzysztof Aniołek Faculty of Computer Science and Materials Science, Institute of Materials Science, University of Silesia in Katowice, Katowice, Poland

Paweł Badura Faculty of Biomedical Engineering, Silesian University of Technology, Zabrze, Poland

Grzegorz Bajor Department of Anatomy, School of Medicine in Katowice, Medical University of Silesia in Katowice, Katowice, Poland

Adrian Barylski Faculty of Computer Science and Materials Science, Institute of Materials Science, University of Silesia in Katowice, Katowice, Poland

Marcin Basiaga Department of Biomaterials and Medical Devices Engineering, Faculty of Biomedical Engineering, Silesian University of Technology, Zabrze, Poland

Andrzej Bieniek Department of Biomechatronics, Faculty of Biomedical Engineering, Silesian University of Technology, Gliwice, Poland

Maria Bieńkowska Faculty of Biomedical Engineering, Silesian University of Technology, Zabrze, Poland

Paweł Bożek Radpoint Sp. z o.o., Katowice, Poland

Marcin Bugdol Faculty of Biomedical Engineering, Silesian University of Technology, Zabrze, Poland

Miłosz Chrzan Department of Biomechatronics, Faculty of Biomedical Engineering, Silesian University of Technology, Gliwice, Poland

Szczepan Cofta Department of Pulmonology, Allergology and Respiratory Oncology, Poznań University of Medical Sciences, Poznań, Poland

Joanna Czajkowska Faculty of Biomedical Engineering, Silesian University of Technology, Zabrze, Poland

Miroslaw Czak Foundation of Cardiac Surgery Development, Zabrze, Poland

Adam Czaplicki Faculty of Physical Education and Sport in Biała Podlaska, Academy of Physical Education in Warsaw, Biała Podlaska, Poland

Małgorzata Czernek John Paul II Geriatric Hospital, Katowice, Poland

Jarosław Derejczyk John Paul II Geriatric Hospital, Katowice, Poland

Beata Dorzak Department of Anatomy, School of Medicine in Katowice, Medical University of Silesia in Katowice, Katowice, Poland

Paweł Drapikowski Institute of Control, Robotics and Information Engineering, Poznan University of Technology, Poznan, Poland

Lechosław B. Dworak Chair of Bionics, University of Arts, Poznan, Poland

Róża Dzierżak Faculty of Electrical Engineering and Computer Science, Lublin University of Technology, Lublin, Poland

Renata Ferduła Institute of Applied Mechanics, Faculty of Mechanical Engineering and Management, Poznań University of Technology, Poznań, Poland

Anna Filipowska Department of Biosensors and Processing of Biomedical Signals, Silesian University of Technology, Zabrze, Poland

Tomasz Flak Faculty of Computer Science and Materials Science, Institute of Materials Science, University of Silesia in Katowice, Katowice, Poland

Anna Fryzowicz Department of Biomechanics, Faculty of Physical Education, Sport and Rehabilitation, Poznań University of Physical Education, Poznań, Poland

Jadwiga Gabor Faculty of Computer Science and Materials Science, Institute of Materials Science, University of Silesia in Katowice, Katowice, Poland

Marta Galińska Faculty of Biomedical Engineering, Silesian University of Technology, Zabrze, Poland

Maciej Gawlikowski Foundation of Cardiac Surgery Development, Zabrze, Poland

Katarzyna Gieracka Science Club "SYNERGIA", Department of Biomaterials and Medical Devices Engineering, Faculty of Biomedical Engineering, Silesian University of Technology, Zabrze, Poland

Malgorzata Gonsior Foundation of Cardiac Surgery Development, Zabrze, Poland

Joanna Gorwa Department of Biomechanics, Faculty of Physical Education, Sport and Rehabilitation, Poznań University of Physical Education, Poznań, Poland

Jakub Krzysztof Grabski Institute of Applied Mechanics, Faculty of Mechanical Engineering and Management, Poznań University of Technology, Poznań, Poland

Adam Gramala Institute of Control, Robotics and Information Engineering, Poznan University of Technology, Poznan, Poland

Monika Grygorowicz Department of Spondyloorthopaedics and Biomechanics of the Spine, Wiktor Dega Orthopaedic and Rehabilitation Clinical Hospital, Poznan University of Medical Sciences, Poznan, Poland

Marek Gzik Department of Biomechatronics, Faculty of Biomedical Engineering, Silesian University of Technology, Zabrze, Poland

Eugeniusz Hadasik Faculty of Materials Engineering and Metallurgy, Silesian University of Technology, Katowice, Poland

Krzysztof Horoba Institute of Medical Technology and Equipment ITAM, Zabrze, Poland

Gabriela Imbir Institute of Metallurgy and Materials Science, Polish Academy of Sciences, Cracow, Poland

Małgorzata Aneta Janik Faculty of Computer Science and Material Science, University of Silesia in Katowice, Sosnowiec, Poland

Paweł Janik Faculty of Computer Science and Material Science, University of Silesia in Katowice, Sosnowiec, Poland

Janusz Jeżewski Institute of Medical Technology and Equipment ITAM, Zabrze, Poland

Katarzyna Jochymczyk-Woźniak Department of Biomechatronics, Faculty of Biomedical Engineering, Silesian University of Technology, Zabrze, Poland

Jacek Jurkojć Department of Biomechatronics, Faculty of Biomedical Engineering, Silesian University of Technology, Zabrze, Poland

Michał Jóźwiak Department of Biochemistry, First Faculty of Medicine, Medical University of Warsaw, Warsaw, Poland

Jarosław Kabaciński Department of Biomechanics, Faculty of Physical Education, Sport and Rehabilitation, Poznań University of Physical Education, Poznań, Poland

Marcin Kaczmarek Department of Biomaterials and Medical Devices Engineering, Faculty of Biomedical Engineering, Silesian University of Technology, Zabrze, Poland

Anita Kajzer Department of Biomaterials and Medical Devices Engineering, Faculty of Biomedical Engineering, Silesian University of Technology, Zabrze, Poland

Wojciech Kajzer Department of Biomaterials and Medical Devices Engineering, Faculty of Biomedical Engineering, Silesian University of Technology, Zabrze, Poland

Bogusław Kapelak Jagiellonian University, Medical College, John Paul II Hospital, Krakow, Poland

Wojciech Kapko John Paul II Geriatric Hospital, Katowice, Poland

Wojciech Kaspera Department and Clinical Division of Neurosurgery, Silesian University of Medicine in Katowice, Voivodeship Specialist Hospital no. 5, Sosnowiec, Poland

Jacek Kawa Radpoint Sp. z o.o., Katowice, Poland; Faculty of Biomedical Engineering, Silesian University of Technology, Zabrze, Poland

Paweł Kostka Department of Biosensors and Biomedical Signals Processing, Silesian University of Technology, Zabrze, Poland

Dominik Kowalczykowski Students Scientific "BIOKREATYWNI", Faculty of Biomedical Engineering, Silesian University of Technology, Zabrze, Poland

Michał Kręcichwost Faculty of Biomedical Engineering, Silesian University of Technology, Zabrze, Poland

Klaudia Kubik Faculty of Computer Science and Materials Science, Institute of Materials Science, University of Silesia in Katowice, Katowice, Poland

Dawid Kucharski Faculty of Mechanical Engineering and Management, Institute of Mechanical Technology, Division of Metrology and Measurement Systems, Poznan University of Technology, Poznan, Poland

Tomasz Kupka Institute of Medical Technology and Equipment ITAM, Zabrze, Poland

Przemyslaw Kurtyka Foundation of Cardiac Surgery Development, Zabrze, Poland

Przemysław Kurtyka Faculty of Biomedical Engineering, Department of Biomaterials and Medical Devices Engineering, Silesian University of Technology, Zabrze, Poland

Roman Kustosz Foundation of Cardiac Surgery Development, Zabrze, Poland

Damian Kusz Department of Orthopedics and Traumatology, School of Medicine in Katowice, Medical University of Silesia in Katowice, Katowice, Poland

Anna Kwaśniewska Department of Radiology, Medical University of Silesia in Katowice, Hospital SPSK M, Katowice, Poland

Andrzej Kępa Department of Radiology and Nuclear Medicine, Independent Public Clinical Hospital No. 4, Lublin, Poland

Sylwia Łagan Cracow University of Technology, Cracow, Poland

Marta Łężniak Faculty of Computer Science and Materials Science, Institute of Materials Science, University of Silesia in Katowice, Katowice, Poland

Jürgen M. Lackner Joanneum Research Forschungs-GmbH, Materials - Functional Surfaces, Leoben, Austria

Aleksander Lamża Department of Computer Biomedical Systems, University of Silesia, Institute of Computer Science, Sosnowiec, Poland

Monika Lewczuk Department of Biomaterials and Medical Devices Engineering, Faculty of Biomedical Engineering, Silesian University of Technology, Zabrze, Poland

Aneta Liber-Kneć Cracow University of Technology, Cracow, Poland

Roman Major Institute of Metallurgy and Materials Science, Polish Academy of Sciences, Cracow, Poland

Piotr Malesa Military University of Technology, Warsaw, Poland

Czesław Marcisz Department of Gerontology and Geriatric Nursing, School of Health Sciences, Medical University of Silesia, Katowice, Poland

Adam Matonia Institute of Medical Technology and Equipment ITAM, Zabrze, Poland

Marta Michalska Faculty of Mechanical Engineering and Management, Institute of Mechanical Technology, Division of Metrology and Measurement Systems, Poznan University of Technology, Poznan, Poland

Martyna Michałowska Institute of Applied Mechanics, Faculty of Mechanical Engineering and Management, Poznań University of Technology, Poznań, Poland

Robert Michnik Department of Biomechatronics, Faculty of Biomedical Engineering, Silesian University of Technology, Zabrze, Poland

Zuzanna Miodońska Faculty of Biomedical Engineering, Silesian University of Technology, Zabrze, Poland

Aldona Mzyk Institute of Metallurgy and Materials Science, Polish Academy of Sciences, Cracow, Poland

Katarzyna Nowakowska Department of Biomechatronics, Faculty of Biomedical Engineering, Silesian University of Technology, Zabrze, Poland

Hubert Okła Faculty of Computer Science and Materials Science, Institute of Materials Science, University of Silesia in Katowice, Katowice, Poland

Zbigniew Omiotek Faculty of Electrical Engineering and Computer Science, Lublin University of Technology, Lublin, Poland

Jerzy Pacholewicz Department of Cardiac, Vascular and Endovascular Surgery and Transplantology, Medical University of Silesia, Silesian Centre for Heart Diseases, Zabrze, Poland

Jarosław Paluch Department of Laryngology, School of Medicine in Katowice, Medical University of Silesia, Katowice, Poland

Patrycja Pastusiak Institute of Applied Mechanics, Faculty of Mechanical Engineering and Management, Poznan University of Technology, Poznań, Poland

Zbigniew Paszenda Faculty of Biomedical Engineering, Department of Biomaterials and Medical Devices Engineering, Silesian University of Technology, Zabrze, Poland

Mateusz Pawlik CABIOMEDE Sp. z o.o., Kielce, Poland

Michał Pielka Faculty of Computer Science and Material Science, University of Silesia in Katowice, Sosnowiec, Poland

Adam M. Pogorzała Hipolit Cegielski State College of Higher Education in Gniezno, Gniezno, Poland

Joanna Przondziono Faculty of Materials Engineering and Metallurgy, Silesian University of Technology, Katowice, Poland

Bartłomiej Pyciński Faculty of Biomedical Engineering, Silesian University of Technology, Zabrze, Poland; Radpoint Sp. z o.o., Katowice, Poland

Piotr Rasztabiga Faculty of Biomedical Engineering, Silesian University of Technology, Zabrze, Poland

Marek Świerczewski Military University of Technology, Warsaw, Poland

Jarosław Sacharuk Faculty of Physical Education and Sport in Biała Podlaska, Academy of Physical Education in Warsaw, Biała Podlaska, Poland

Marek Sanak Department of Medicine, Jagiellonian University Medical College, Cracow, Poland

Piotr Seiffert John Paul II Geriatric Hospital, Katowice, Poland

Jarosław Serafińczuk Faculty of Microsystem Electronics and Photonics, Department of Metrology, Micro and Nanostructures, Wroclaw University of Technology, Wroclaw, Poland

Szymon Sieciński Department of Biosensors and Biomedical Signals Processing, Silesian University of Technology, Zabrze, Poland

Hanna Sikora Department of Orthopedics and Traumatology, School of Medicine in Katowice, Medical University of Silesia in Katowice, Katowice, Poland

Piotr Siondalski Department of Cardiovascular Surgery, Medical University of Gdansk, Gdańsk, Poland

Michał Smoliński Radpoint Sp. z o.o., Katowice, Poland

Marta Sobkowiak Department of Biomechatronics, Faculty of Biomedical Engineering, Silesian University of Technology, Zabrze, Poland

Dominik Spinczyk Faculty of Biomedical Engineering, Silesian University of Technology, Zabrze, Poland

Mateusz Stasiewicz Faculty of Biomedical Engineering, Department of Biomechatronics, Silesian University of Technology, Gliwice, Poland

Paula Stępień Faculty of Biomedical Engineering, Silesian University of Technology, Zabrze, Poland

Andrzej Szymon Swinarew Faculty of Computer Science and Materials Science, Institute of Materials Science, University of Silesia in Katowice, Katowice, Poland

Janusz Szala Faculty of Materials Engineering and Metallurgy, Silesian University of Technology, Katowice, Poland

Marta Szczetyńska Institute of Applied Mechanics, Faculty of Mechanical Engineering and Management, Poznan University of Technology, Poznań, Poland

Janusz Szewczenko Department of Biomaterials and Medical Devices Engineering, Faculty of Biomedical Engineering, Silesian University of Technology, Zabrze, Poland

Paulina Szyszka Faculty of Physical Education and Sport in Biała Podlaska, Academy of Physical Education in Warsaw, Biała Podlaska, Poland

Grzegorz Sławiński Military University of Technology, Warsaw, Poland

Anna Tamulewicz Department of Biosensors and Biomedical Signal Processing, Faculty of Biomedical Engineering, Silesian University of Technology, Zabrze, Poland

Ewaryst Tkacz Department of Biosensors and Biomedical Signal Processing, Faculty of Biomedical Engineering, Silesian University of Technology, Zabrze, Poland

Klaudia Tokarska Faculty of Biomedical Engineering, Department of Biomaterials and Medical Devices Engineering, Silesian University of Technology, Zabrze, Poland

Tomasz Walczak Institute of Applied Mechanics, Faculty of Mechanical Engineering and Management, Poznań University of Technology, Poznań, Poland

Witold Walke Faculty of Biomedical Engineering, Silesian University of Technology, Zabrze, Poland

Łukasz Walusiak Department of Computer Biomedical Systems, University of Silesia, Institute of Computer Science, Sosnowiec, Poland;
Institute of Technology, Pedagogical University, Cracow, Poland

Jakub Wieczorek Faculty of Materials Engineering and Metallurgy, Silesian University of Technology, Katowice, Poland

Karol Wierzbicki Jagiellonian University, Medical College, John Paul II Hospital, Krakow, Poland

Agata Wijata Faculty of Biomedical Engineering, Silesian University of Technology, Zabrze, Poland

Piotr Wilczek Foundation for Cardiac Surgery Development, Zabrze, Poland

Robert Wilk Department of Orthopedics and Traumatology, School of Medicine in Katowice, Medical University of Silesia in Katowice, Katowice, Poland

Piotr Wodarski Department of Biomechatronics, Faculty of Biomedical Engineering, Silesian University of Technology, Gliwice, Poland

Wojciech Wolański Department of Biomechatronics, Faculty of Biomedical Engineering, Silesian University of Technology, Zabrze, Poland

Andre Woloshuk Weldon School of Biomedical Engineering, Purdue University, West Lafayette, IN, USA

Janusz Wróbel Institute of Medical Technology and Equipment ITAM, Zabrze, Poland

Zygmunt Wróbel Faculty of Computer Science and Material Science, University of Silesia in Katowice, Sosnowiec, Poland; Department of Computer Biomedical Systems, University of Silesia, Institute of Computer Science, Sosnowiec, Poland

Michal Zakliczynski Department of Cardiac, Vascular and Endovascular Surgery and Transplantology, Medical University of Silesia, Silesian Centre for Heart Diseases, Zabrze, Poland

Mikołaj Zimny Department and Clinical Division of Neurosurgery, Silesian University of Medicine in Katowice, Voivodeship Specialist Hospital no. 5, Students' Research Society, Sosnowiec, Poland

Anna Ziębowicz Faculty of Biomedical Engineering, Department of Biomaterials and Medical Devices Engineering, Silesian University of Technology, Zabrze, Poland

Aleksandra Zyśk Faculty of Biomedical Engineering, Silesian University of Technology, Zabrze, Poland

Informatics in Medicine

Automated Epidermis Segmentation in Ultrasound Skin Images

Joanna Czajkowska(✉) and Paweł Badura

Faculty of Biomedical Engineering, Silesian University of Technology,
Roosevelta 40, 41-800 Zabrze, Poland
joanna.czajkowska@polsl.pl

Abstract. The automated system for epidermis segmentation in ultrasound images of skin is described in this paper. The method consists of two main parts: US probe membrane segmentation and epidermis segmentation. The fuzzy C-means clustering is employed at the initial stage leading to probe membrane segmentation using fuzzy connectedness technique. Then, the upper (external) epidermis boundary is detected and adjusted using connectivity and line variability analysis. The lower (internal) boundary is obtained by shifting the upper edge by a constant vertical width determined adaptively during the experiments. The method is evaluated using a dataset of 13 US images of two registration depths. The validation relies on a ground truth delineations of the epidermis provided by two independent experts. The mean Hausdorff distances of 0.118 mm and 0.145 mm were obtained for the external and internal epidermis boundaries, respectively, with the mean Dice index for the epidermis region at 0.848.

Keywords: Skin imaging · Skin layers · High-resolution ultrasound · Image segmentation

1 Introduction

Fast development, availability, low price and no influence on the patient health underline the relevance of ultrasound (US) diagnosis and treatment. Since 1979 ultrasonography is employed in dermatology, to evaluate skin thickness [1,2], nowadays, the high frequency (>15 MHz) apparatus makes it possible to identify different skin layers and structures. The high-resolution US imaging of cutaneous lesions enables more accurate diagnosis providing the information on skin structure, which can be used in oncology and in combination with the histological image enables an analysis of dermatological diseases [1–4]. Despite numerous applications, the automated analysis of skin ultrasound as well as its segmentation is not widely explored in literature. Most of the proposed methods focus on the dermis segmentation and is based on active contour models [5,6]. The epidermis layer is omitted there. Moreover, as reported, the active contour approach is an efficient but time consuming technique. In [6] the authors claim, that

E. Tkacz et al. (Eds.): IBE 2018, AISC 925, pp. 3–11, 2019.
https://doi.org/10.1007/978-3-030-15472-1_1

the processing time is reduced, however in the final results the measured skin thickness is compared with manual segmentation instead of direct segmentation results.

Our application being a part of bigger software dedicated to skin wound treatment assessment targets in the automated epidermis segmentation. The developed methodology is insensitive to image artefacts and noise. The algorithm is divided into probe membrane and skin parts. The first one introduces clustering technique and fuzzy connectedness, whereas in the second one the clustering step is followed by the envelope analysis. The proposed method provides repeatable and reliable results, and the segmentation accuracy is evaluated using US scans previously delineated by two independent experts.

2 Materials and Methods

2.1 Materials

The database used during the study consists of 13 US images acquired by Linear B-Scan 18–22 MHz (Fig. 1). Each image has a size of 2048 × 1536 pixels with a 12 mm width and two different heights (depths): 8 mm (6 cases) and 16 mm (7 cases). For evaluation purposes, each image was subjected to two independent expert delineations of both (upper and lower) boundaries of the epidermis. Obtained contours as well as the epidermis region defined as a set of pixels between the contours (included) are used as a ground truth in validation experiments. The system is implemented using the Matlab software.

2.2 Methodology

The developed automated epidermis segmentation algorithm consists of two combined processing parts given in the block diagram (Fig. 2) as *Probe Membrane Segmentation* and *Epidermis Segmentation*, respectively. After a common clustering step the analysis is followed by the US probe membrane segmentation, whose results are then used to extract the epidermis.

Processing of the US image of skin layer starts with fuzzy C-means clustering (FCM) [7] step. The aim of the clustering method is to determine the internal structure of the image data. Despite some important constraints like sensitivity to image artifacts and noise, it is commonly used in medical image analysis [8]. Introducing 3 groups of interest, the FCM clustering algorithm makes it possible to divide the given image into: hipointensive area including ultrasound gel and air, hyperintensitive area including probe membrane as well as the considered epidermis, and the area including other structures. An exemplary US image and corresponding FCM clustering results are shown in Fig. 1 on the left and right side, respectively. Depending on the next step (probe membrane or epidermis segmentation) the cluster with the highest or lowest intensity level is chosen for further processing steps, respectively (see Fig. 2).

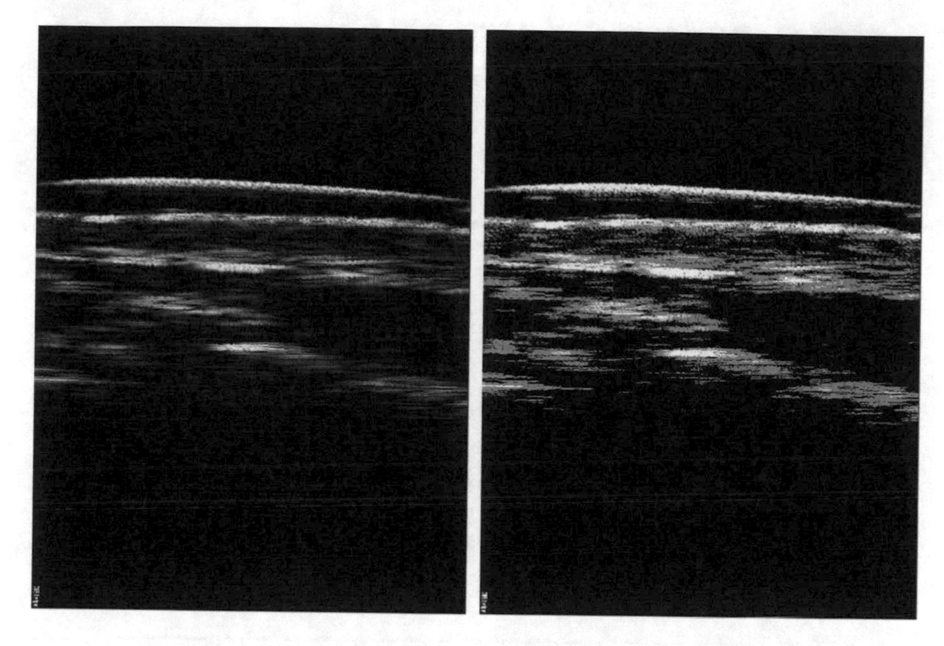

Fig. 1. An exemplary US image (left) and corresponding FCM clustering results (right)

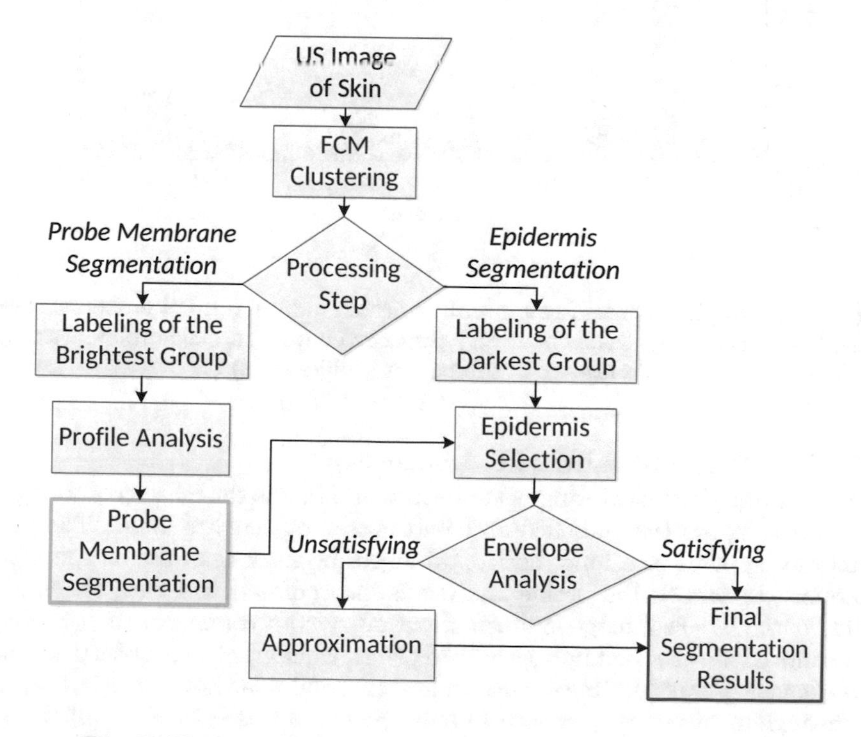

Fig. 2. Block diagram of the epidermis segmentation algorithm

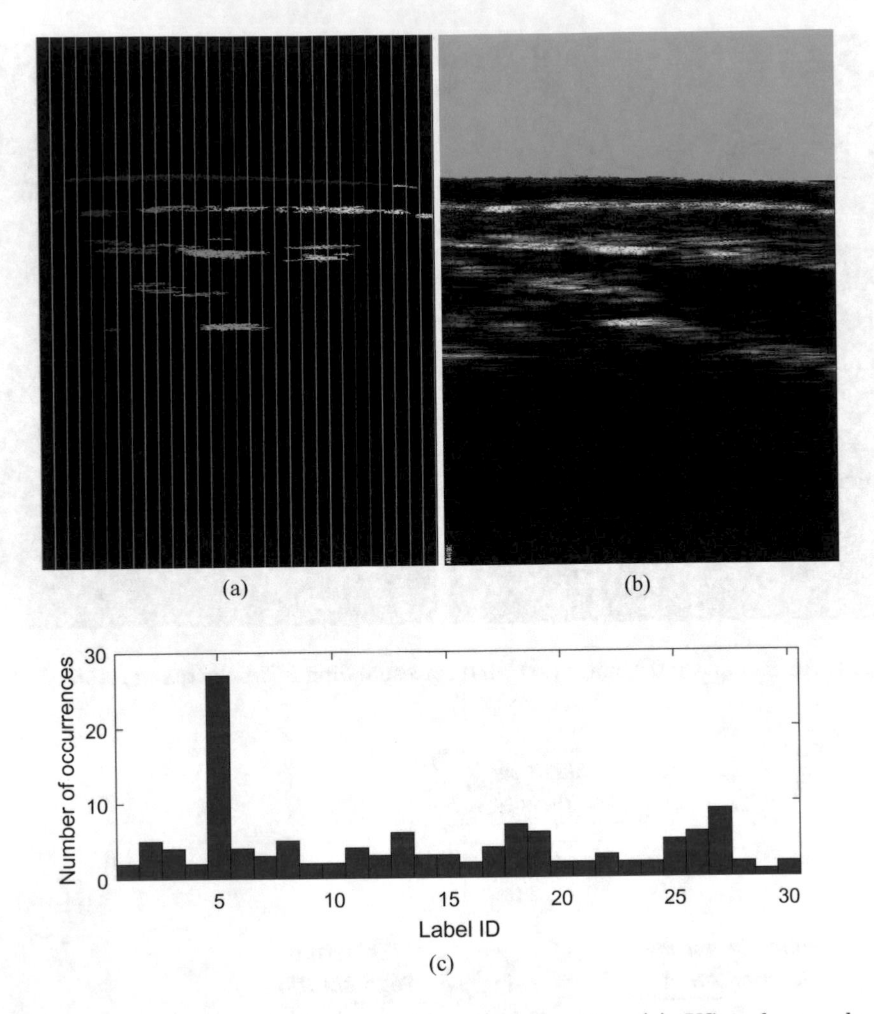

Fig. 3. An exemplary labelled image with indicated profiles (a), US probe membrane (green) and external epidermis boundary (lower red outline) segmentation results (b), and histogram of labels' presence in subsequent profiles in (a) (c)

2.2.1 US Probe Membrane Segmentation

The above described clustering results – in this case it is the mask containing the brightest area, are then labelled and 4-connected objects are found. The probe membrane appears as a long, horizontal object brighter than the surroundings. Therefore, to find it, the profile analysis is incorporated. For this, the vertical profile of the labelled image is taken about every 50th column of the US image. The width of the acquired images is 1536 pixels, what gives 30 profiles per image. The exemplary labelled image with indicated profiles is shown in Fig. 3(a) and the histogram of labels' presence in subsequent profiles – in Fig. 3(c). Due to the fact that in the US image the epidermis area looks similar to the probe membrane, two object labels most often appearing in the profile histogram are then treated as being a possible probe membrane. The object placed near to the upper image edge is chosen as the final probe membrane detection result.

To estimate the real probe membrane area, the already calculated mask is then used as the starting point set for segmentation algorithm. The introduced method is based on the fuzzy connectedness technique [9,10]. To exclude the probe membrane as well as the image part placed above it, from further analysis, the membrane area is horizontally extrapolated into the extreme columns. The final US probe membrane segmentation result with the image area above it is shown in Fig. 3(b) (green area).

2.2.2 Epidermis Segmentation

For this step, two previously obtained results are used. First, the further processing is based on the FCM clustering and second the probe membrane segmentation marks the region excluded from further analysis. The 4-connected objects within the darkest cluster are then extracted. The object located closest to the probe membrane is chosen as including the upper epidermis edge (Fig. 3(b), red-outlined area).

The first guess for the epidermis boundary is the lower edge of the already segmented object (Fig. 3(b)). However, we assume that the outer epidermis layer should not contain concavities and in the current version it is not always the case. Moreover, due to the image quality, in some cases the segmentation leads to false results shown in Fig. 4 left, in green. To overcome this problem, the vertical coordinate of the obtained line is analysed as the 1-dimensional signal. The signal envelope is estimated to define the external epidermis boundary (see Fig. 4). From upper and lower envelopes (blue and red lines in Fig. 4) the one with lower variance is chosen as the final result.

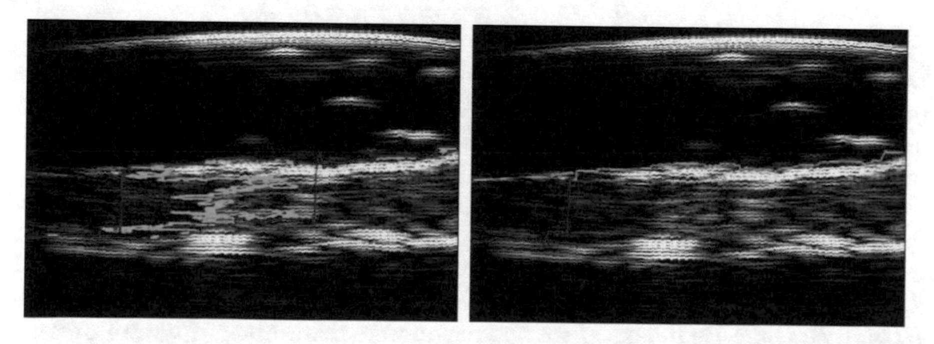

Fig. 4. False segmentation results (green) and its envelope (blue and red) – left, analysed envelope (red) and the upper epidermis boundary (green) – right

In case of segmentation inaccuracy the envelope variance is big and the maximum value of its derivative exceeds 0.015 mm. It can be observed in Fig. 4. In such case, further envelope analysis is required. For this, the envelope with lower variance is chosen (in Fig. 4 the blue one). Then, the automated detection of the envelope signal jump is applied. The incorrect signal samples to either left or right to the jump are replaced by the linear approximation of the neighbouring envelope samples. The obtained upper epidermis boundary is shown in Fig. 4 in green.

Finally, the lower (internal) epidermis boundary is detected by assuming constant width of the layer. For this, a submillimeter constant d_v is used to shift the external boundary towards the deeper layers of skin. An initial value of $d_v = 0.25$ mm was set, yet during the experiments using expert delineations (Sect. 3) it was subjected to a segmentation accuracy analysis yielding a robust value of 0.27 mm.

3 Results and Discussion

Evaluation of the epidermis segmentation algorithm relies on the Hausdorff distance *HD* [11] determined for each of the two epidermis boundaries separately. For given two finite sets of points (lines) $A = \{\mathbf{a}_1, \ldots, \mathbf{a}_n\}$ (**A**utomated segmentation) and $E = \{\mathbf{e}_1, \ldots, \mathbf{e}_m\}$ (**E**xpert delineation), the directed Hausdorff distance $HD(A, E)$ is defined as:

$$HD(A, E) = \max_{\mathbf{a}_i \in A} \min_{\mathbf{e}_j \in E} \|\mathbf{a}_i - \mathbf{e}_j\|, \tag{1}$$

where $\|\cdot\|$ is the Euclidean norm on the points of A and E in millimeters. Moreover, the spatial overlap of two epidermis regions: one segmented by the algorithm and the other delineated by an expert is assessed using the Dice index:

$$DI = \frac{2 \cdot TP}{2 \cdot TP + FN + FP}, \tag{2}$$

where TP, FN, FP denote the number of true positive, false negative, and false positive pixels, respectively. Since two different experts delineated both upper and lower boundaries of the epidermis in each of 13 cases, the validation is based on a total of $13 \cdot 2 = 26$ cases.

The mean Hausdorff distance $HD_u(A, E) = 0.118 \pm 0.048$ mm was obtained for the upper (external) boundary detection. Location of the lower (internal) boundary was assessed in the experiment with variable vertical epidermis width d_v. For each location of the lower boundary driven by d_v the Hausdorff distance $HD_l(A, E)$ and Dice index DI for the region were determined. Figure 5 presents the plots of both metrics as a function of d_v with optimum ($d_v = 0.27$ mm) marked in each case. With such vertical width the mean segmentation accuracy metrics reached: $HD_l(A, E) = 0.145 \pm 0.040$ mm and $DI = 0.848 \pm 0.044$.

The upper epidermis boundary segmentation produces satisfactory results in case of each image under consideration. With the mean value of $HD_u(a, E)$ at 0.118 mm the highest Hausdorff distance in individual case reaches 0.256 mm

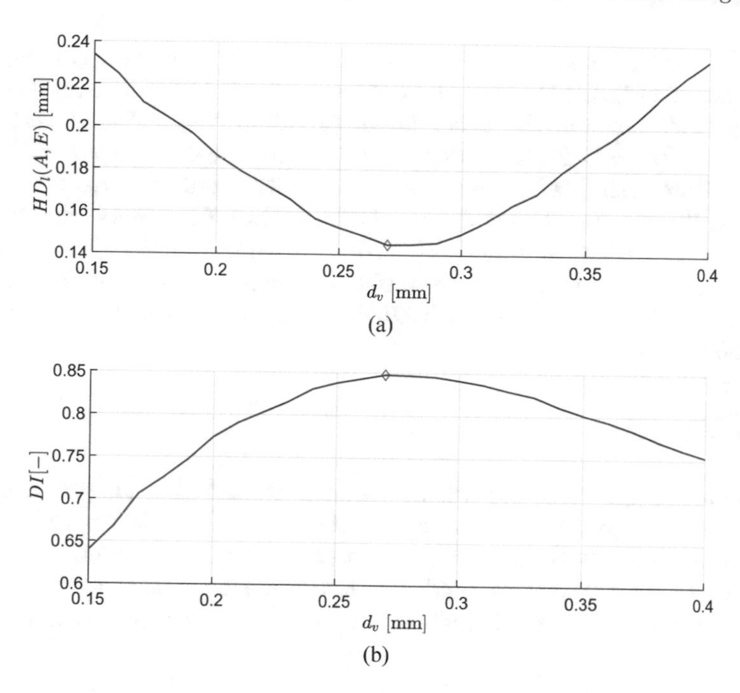

Fig. 5. Lower epidermis boundary detection results as a function of vertical width d_v: Hausdorff distance $HD_l(A, E)$ (a) and Dice index DI (b)

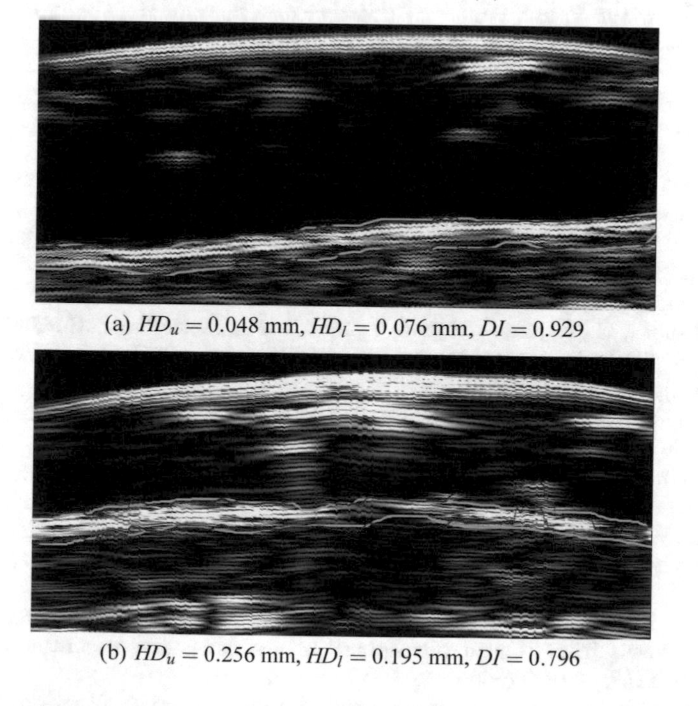

(a) $HD_u = 0.048$ mm, $HD_l = 0.076$ mm, $DI = 0.929$

(b) $HD_u = 0.256$ mm, $HD_l = 0.195$ mm, $DI = 0.796$

Fig. 6. Exemplary results (green line – expert delineation, red – segmented contour)

(below the vertical epidermis width d_v). This result, however, can be considered an outlier (Fig. 6(b)), since there are only 2 out of 26 Hausdorff distances exceeding 0.17 mm. On the other hand, the adaptively selected vertical epidermis width $d_v = 0.27$ mm used to determine the lower epidermis boundary yields Hausdorff distance exceeding 0.17 mm in only 6 cases with a maximum of 0.242 mm, also below d_v. Furthermore, in only 3 cases the Dice index falls below 0.8 with a minimum of 0.707. To value the reported results let's note, that the following metrics were obtained when comparing delineations of both experts: $HD_u(a, E) = 0.090 \pm 0.024$ mm, $HD_u(a, E) = 0.127 \pm 0.035$ mm, $DI = 0.860 \pm 0.047$.

4 Conclusion

The automated system for epidermis segmentation in US images of skin was described in this paper. The system is able to reliably detect the epidermis boundary end extract the region in a reasonable time. The method can stand for the first stage of a computer-aided diagnosis in skin treatment. The obtained results are comparable to this indicated by the expert and to the intra experts analysis.

Acknowledgements. This research was supported partially by the Polish National Science Centre (Narodowe Centrum Nauki) grant No. UMO-2016/21/B/ST7/02236 and partially by the Polish Ministry of Science and Silesian University of Technology statutory financial support No. BK-209/RIB1/2018.

References

1. Mandava, A., Ravuri, P., Konathan, R.: High-resolution ultrasound imaging of cutaneous lesions. Indian J. Radiol. Imaging **23**(3), 269–277 (2013)
2. de Oliveira Barcaui, E., Carvalho, A.C.P., Pineiro-Maceira, J., Barcaui, C.B., Moraes, H.: Study of the skin anatomy with high-frequency (22 MHz) ultrasonography and histological correlation. Radiol. Brasileira **48**, 324–329 (2015)
3. Ravichandra, G., Arjun, S., Ajmal, S., Manjunath, S., Ayshath, S.: High resolution ultrasonography in dermatology: a psoriasis experience. Indian J. Basic Appl. Med. Res. **5**(2), 121–125 (2016)
4. Pereyra, M., Dobigeon, N., Batatia, H., Tourneret, J.Y.: Segmentation of skin lesions in 2-D and 3-D ultrasound images using a spatially coherent generalized rayleigh mixture model. IEEE Trans. Med. Imaging **31**(8), 1509–1520 (2012)
5. Gao, Y., Tannenbaum, A., Chen, H., Torres, M., Yoshida, E., Yang, X., Wang, Y., Curran, W., Liu, T.: Automated skin segmentation in ultrasonic evaluation of skin toxicity in breast cancer radiotherapy. Ultrasound Med. Biol. **39**(11), 2166–2175 (2013)
6. Lagarde, J.-M., George, J., Soulcie, R., Black, D.: Automatic measurement of dermal thickness from B-scan ultrasound images using active contours. Skin Res. Technol. **11**(2), 79–90 (2005)
7. Bezdek, J.C.: Pattern Recognition with Fuzzy Objective Function Algorithms. Kluwer Academic Publishers Norwell, USA (1981). ISBN 0306406713

8. Bugdol, M., Czajkowska, J., Pietka, E.: A novel model-based approach to left ventricle segmentation. In: 2012 Computing in Cardiology, pp. 561–564, September 2012
9. Udupa, J.K., Samarasekera, S.: Fuzzy connectedness and object definition: theory, algorithms, and applications in image segmentation. Graph. Models Image Process. **58**(3), 246–261 (1996)
10. Badura, P., Kawa, J., Czajkowska, J., Rudzki, M., Pietka, E.: Fuzzy Connectedness in segmentation of medical images. A look at the pros and cons. In: International Conference on Fuzzy Computation Theory and Applications (FCTA 2011), pp. 486–492, October 2011
11. Huttenlocher, D.P., Klanderman, G.A., Rucklidge, W.: Comparing images using the hausdorff distance. IEEE Trans. Pattern Anal. Mach. Intell. **15**(9), 850–863 (1993)

Computer Analysis of Chest X-Ray Images to Highlight Pathological Objects

Łukasz Walusiak[1,2(✉)], Aleksander Lamża[1], and Zygmunt Wróbel[1]

[1] Department of Computer Biomedical Systems,
University of Silesia, Institute of Computer Science,
ul. Będzińska 39, 41-200 Sosnowiec, Poland
{lwalusiak,aleksander.lamza,zygmunt.wrobel}@us.edu.pl
[2] Institute of Technology, Pedagogical University,
ul. Podchorążych 2, 30-084 Cracow, Poland
lukasz.walusiak@up.krakow.pl

Abstract. New methods of analysis and processing the digital images create unique possibilities and prospects in modern medicine because the results of many medical examinations are in the form of images. The paper presents research aimed at creating methods supporting the diagnostic process related to lung diseases diagnosed with X-ray images, such as tuberculosis and pneumoconiosis. Due to the specificity of X-ray images, i.e. the occurrence of distortions in the image, possible poor quality of these photographs, these studies have their solid grounds. An original method was proposed, thanks to which it is possible to obtain an X-ray image with better diagnostic properties. Such results were obtained because the resultive image was transformed using methods such as filtration, original solution for histogram alignment as well as point transformations of the image, and determination of object boundaries.

Keywords: X-ray · Histogram · Segmentation · Object boundaries · Lungs · Tuberculosis

1 Introduction

Diagnosis establishment using X-ray images, in short called X-ray examination, is still the most popular method, especially in the case of chest diseases. Currently, these images have mainly a digital form. It is even the basic study in cases of detection and diagnosis of lung and pleural diseases. It is thanks to them that the physician or technician radiologist makes decisions about further diagnostics. However, in many cases the quality of these images may be lacking certain qualities to carry out diagnostics. Therefore, it is important to develop the analysis of digital medical images to improve diagnostic activities. The use of medical image analysis is possible not only for chest radiographs, but also for, for example, coronary arteries [1]. The doctor radiologist Dr. Marian Szwabowicz, who performed expert assessment of pathological changes revealed on the analyzed

E. Tkacz et al. (Eds.): IBE 2018, AISC 925, pp. 12–19, 2019.
https://doi.org/10.1007/978-3-030-15472-1_2

X-ray images, proved to be important in the course of research on radiographs of the chest. Also owing to this cooperation, it was possible to access real-life X-ray images of disease cases during or after treatment. The obtained results have an impact on the improvement of diagnostic activity, which was confirmed for disease entities such as tuberculosis and silicosis.

2 Research and Results

The research was carried out on 97 digital chest X-ray images. They were obtained thanks to cooperation with a physician radiologist Marian Szwabowicz and the units with whom he cooperates. These photos are devoid of personal data of patients. Methods and actions on the image were done in the Matlab application. X-ray images are monochromatic images whose distribution of shades of gray ranges from 0 to 255, which limits the perception of the elements contained therein. Most often, the image of the chest has a very diversified histogram (example - Fig. 1).

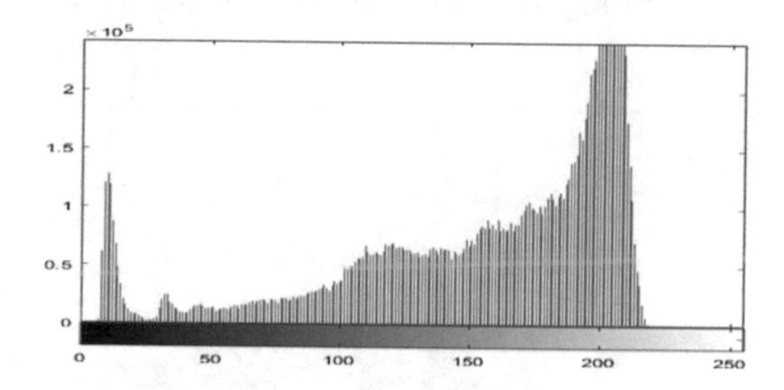

Fig. 1. Histogram for digital chest X-ray

The pixel distribution in such a photograph has a very irregular form. There is a large number of pixels, which in the gray scale can be described as almost black and a very large number of almost white pixels. It is influenced by the specifics of the operation of the X-ray apparatus based on the principle of registration of x-ray radiation not absorbed by the tissues. Dark places represent areas in the patient's body that have absorbed the least amount of X-rays. The greater the radiation absorption by a given tissue, the brighter the image. The main task of the research is to modify the image so that the resulting picture would be more clear and obvious for people analyzing it in terms of recognizing places that may indicate lesions. In addition, on such a modified image, a segmentation of pathological elements follows, recognized as tuberculous or pylic nodules.

3 Elements of Operation in the Image

The processing of the studied images consists of several stages, which are discussed in this subsection. Their order is important for the correctness of the result. The first action performed on the image is using a median filter. Its purpose is to eliminate noise that may eventually appear in the image. This filter does not introduce new values into the image, which means that the image after filtration does not require additional rescaling. However, more importantly, the median filter does not change the sharpness of the edges of individual objects, which is very important in terms of pathological changes that may be present in the images being examined. The next step is to obtain an image which, after median filtration, is subjected to adaptation and alignment of the histogram with the original quality regarding the shape of the histogram. By default, the method of adaptation and alignment of the histogram has three built-in properties that allow to obtain a flat, bell and curved histogram. Typically available properties did not allow to obtain such a good result image as expected. The change in the shape of the histogram added by the authors significantly affected the possibility of seeing (diagnosing by the doctor) pathological changes. This effect was obtained by the research method, checking the pixel properties, which enabled the determination of the main point of the value of 150, the limit of which is the histogram peak, and the pixel distribution for such a graph from values 150 to 250 contains significant amounts of pixels (Fig. 2).

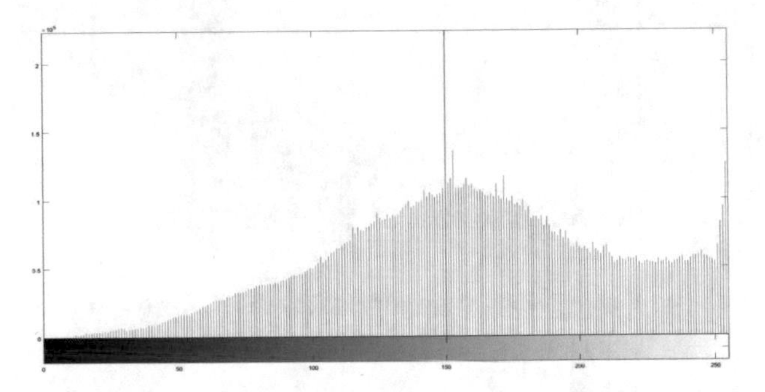

Fig. 2. Histogram of the image after applying the adaptation and alignment of the histogram with the author's quality regarding the shape of the histogram

Such a shift of pixel values in the author's method, relative to the results when using the basic properties (Fig. 3), made it possible to better determine pathological changes due to their specificity—tuberculous nodules [2,3] and changes associated with pneumoconiosis [4] (changes that look like a fine mesh).

The new property is characterized by a change in the value for the constant h, which affects the change in pixel values:

$$h = (2.8 * alpha)^2 \tag{1}$$

What affects the entire analyzed image and its structure:

$$temp = \sqrt{\left(\frac{1}{2}(-hconst)\log(1 - val)\right)} \tag{2}$$

Where: *alpha* - distribution parameter, specified as a nonnegative real scalar and *val* - the value of the quotient of the number of pixels and their values in relation to the maximum value.

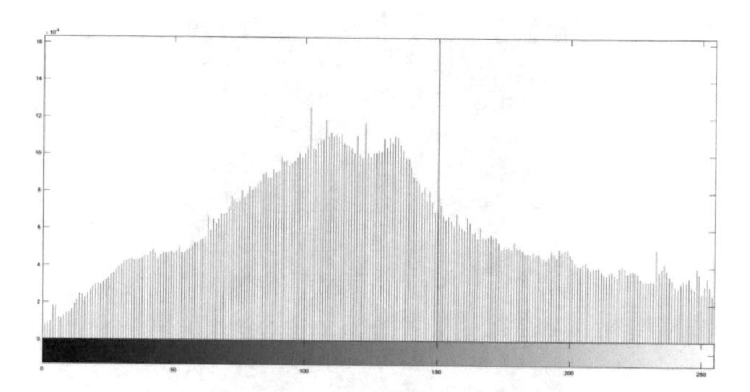

Fig. 3. Histogram of the X-ray image after applying the adaptation and alignment of the histogram with the "exponential" property

The next step is to determine nodular changes among objects appearing on the processed image. The pixel values for a given object are taken by determining the gray shade value of pixels pv from the position of a given pixel relative to the x, y axis. Then the given value is analyzed in the whole image by means of binary with two thresholds, where for a given image the value of the binarization threshold is the designated point for which pv $\pm$ 30 was assumed. This value was assumed by researching done on images. The use of only the pv value without its extension gave insufficient results.

$$L_{BIN}(m,n) = \begin{cases} 0 & L1(m,n) \leq (pv - 30) \\ 1 & (pv - 30) < L1(m,n) \leq (pv + 30) \\ 0 & L1(m,n) > (pv + 30) \end{cases} \quad (3)$$

The result of binarization with two thresholds for which the variable L was adopted is combined with the image after adaptation and alignment of the histogram with the original method, which was adopted as the variable AD. The result is an image containing the segmented elements in the image after the adaptation. However, such a result in the form of a combined image from both methods is not completely satisfactory. Therefore, on the binary image that was obtained after using the methods indicated in the previous steps, the next action, described below, was applied. The method consists in tracking the external boundaries of objects. This is done to accurately determine the objects obtained after the binarization and check their number. In the binary image, connections between objects are not clearly visible, especially if the objects are small and grouped in one part of the image. By tracking the boundaries of objects, it becomes possible to count them accurately (Fig. 4). This is very important in the course of patient treatmentias. It makes it possible to compare individual

X-ray results and to check whether there is a development of nodules or areas with fibrosis in selected areas of the lungs.

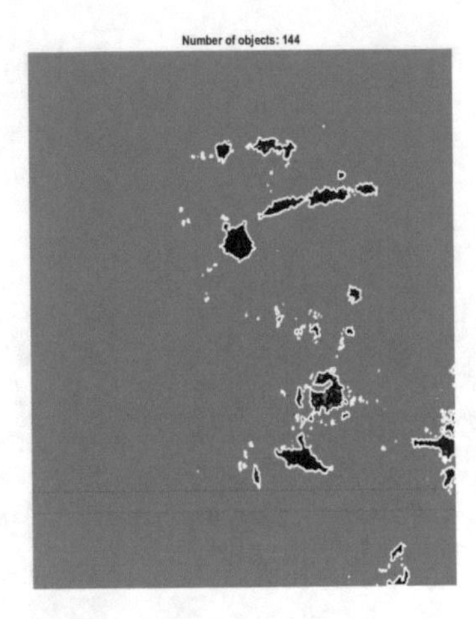

Fig. 4. Designation of objects by checking their boundaries, and counted number of objects present in the part of the lung

4 Results

The aim of the research was to obtain an image containing as much information as possible in the diagnostic process. The basic methods to improve the image quality, such as histogram alignment and the adaptation method with histogram alignment, did not allow to obtain a good result. In contrast, the use of original method for adaptation and alignment of the histogram has improved the quality of the image so much that the information obtained by processing the image affected the diagnostic action (Fig. 5).

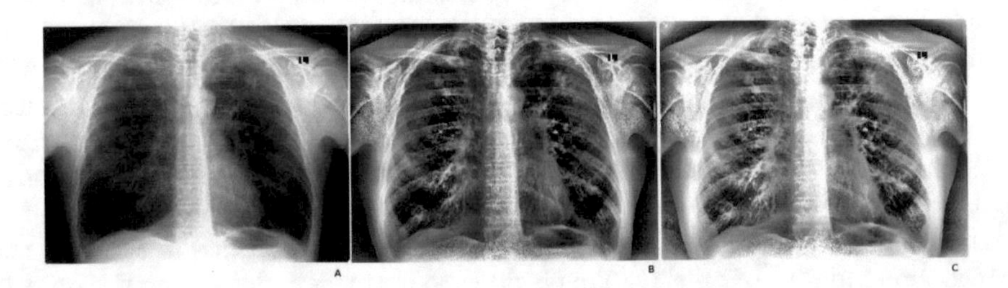

Fig. 5. The effect of various types of histogram alignment on the X-ray image; A- simple histogram alignment; B - Adaptation and alignment of the histogram with the "exponential" property; C - original quality in the process of adaptation and alignment of the histogram

Then, on this modified image, activities aimed at segmentation of pathological objects are carried out. In the current phase of research, the best results are obtained on the image slice. However, work is continued to achieve equally good results for the full picture. One of the elements is the semi-automatic collection of data about the object to be segmented. The studies have verified that nodular changes have similar properties regarding their grayness level. Therefore, when retrieving data about one pathological change, other objects with similar properties are checked. Then these objects are obtained in the binarization process (Fig. 6A).

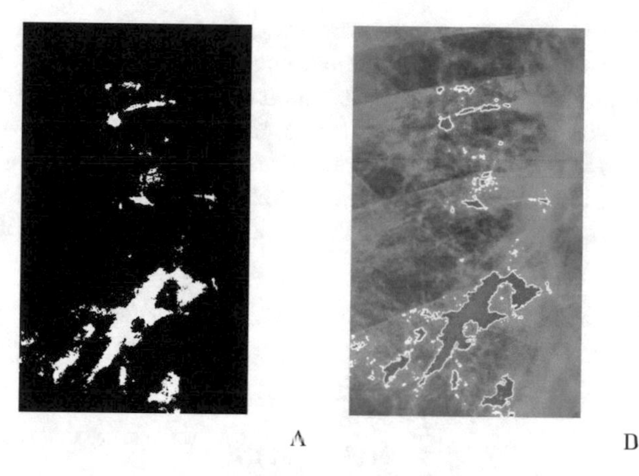

Fig. 6. A - segmented elements in the binarization process; B - segmented objects, along with the marking of their boundaries, imposed on the real lung fragment in the X-ray picture

For better recognition of the shape and determination of the number of objects after the binarization process, the action of checking the properties of the designated changes in the X-ray image was used. For this purpose, the object boundary tracking methods were used (Figs. 4 and 6B). As a result, more data was obtained, such as: the number of objects, shape of objects, information whether the building is homogeneous in terms of construction or consists of many smaller objects. Importantly, in comparison to existing commercial applications used to analyze X-ray images, the method presented in the article allows to extract a more precise number of objects. Image modifications are not limited only to the change of brightness, enlargement or image negation, change of contrast, which are most often offered by commercial solutions, but the analysis is more detailed.

These data will be used in the further stage of application development as elements of an expert system supporting the diagnostic process. Next, it was assumed that each fragment of the image under study is a coordinate system (Fig. 7).

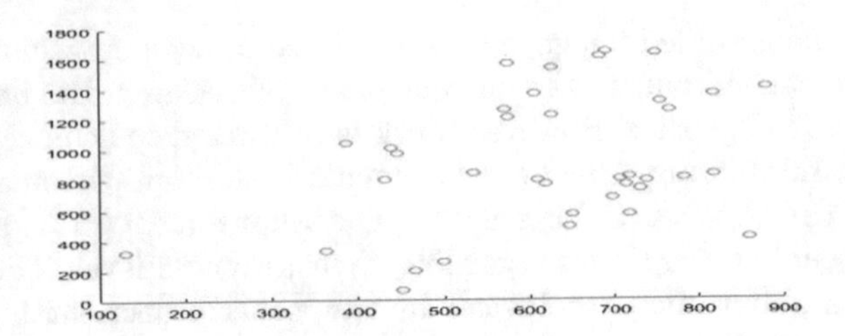

Fig. 7. Distribution of pathological changes dissected from the examined X-ray image

Data collected in this way serve to better define the disease classification. In the case of tuberculous lesions, the occurrence of nodules usually occurs in the upper and middle parts of the lung. It was assumed that the upper lung is all elements up to 1000 on the y-axis, and the middle part are values [1000–2000] on the y-axis. As can be seen on the above results, the nodules are mainly in the upper lung, which may indicate the development of tuberculosis. Detailed distribution of bumps is presented in the table below (Fig. 8) (Table 1).

Table 1. Part of the data on the distribution of nodules in the X-ray picture

	x axis	y axis
1	718	566
2	698	674
3	730	730
4	714	758
...	...	...

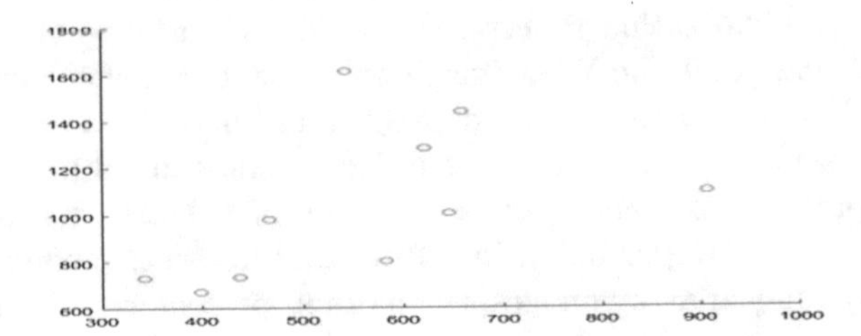

Fig. 8. The distribution of pathological changes in the image without the methods of adaptation and alignment of the histogram

5 Summary

X-ray images are still one of the basic diagnostic tools for many diseases, including chest diseases. The source of information are X-ray images often burdened with many imperfections that make it difficult to identify areas that indicate the existence and further progression of disease changes. The proposed solutions in the field of digital X-ray image processing facilitate diagnosis, increase the accuracy of the assessment and allow making quantitative and qualitative analyzes of lesions. Attention should also be paid to the utility of the method in a clinical setting, i.e. the software does not require excessive computing power, for these reasons it can be used with commonly available computers. In addition to the benefits of the field of medicine (improving diagnostics), it is also important to develop a new property for the matlab method of adaptation and alignment of the histogram, which can be used for other purposes than only digital chest X-ray images. This of course requires further research. The available literature regarding X-ray images shows that mainly attention was focused on the noise in these types of images [5,6] and object segmentation methods [7,8]. However, the discussed solutions are based on other concepts of image processing algorithms than those presented above. The presented solution, although it gives satisfactory results, which confirms the opinion of a specialist, is not a closed project.

References

1. Ogiela, M., Tadeusiewicz, R.: Image understanding methods in biomedical informatics and digital imaging. J. Biomed. Inf. **34**(6), 377–386 (2001)
2. Pruszyński, B.: Radiologia diagnostyka obrazowa Rtg, TK, USG. MR i radioizotopy. Wydawnictwo Lekarskie PZWL (2001)
3. Bilir, M., Sipahi, S., Mert, A., Yanardag, H., Ozaras, R., Aki, H., Karayel, T.: Sarcoidosis presenting with isolated right paratracheal mass on chest X-ray. Eur. J. Intern. Med. **15**(3), 198–199 (2004)
4. Gatey, C., Tattevin, P., Rioux, C., Ducot, B., Meyer, L., Bouvet, E.: Impact of early chest radiography and empirical antibiotherapy on delay in the diagnosis of pulmonary tuberculosis. Médecine et Maladies Infectieuses. **42**(3), 110–113 (2012)
5. Lu, C.T., Chen, M.Y., Shen, J.H., Wang, L.L., Hsu, C.C.: Removal of salt-and-pepper noise for X-ray bio-images using pixel-variation gain factors. Comput. Electric. Eng. **71**, 862–876 (2017)
6. Lee, M.S., Park, C.H., Kang, M.G.: Edge enhancement algorithm for low-dose X-ray fluoroscopic imaging. Comput. Methods Programs Biomed. **152**, 45–52 (2017)
7. Zhang, X., Jia, F., Luo, S., Liu, G., Hu, Q.: A marker-based watershed method for X-ray image segmentation. Comput. Methods Programs Biomed. **113**(3), 894–903 (2014)
8. Brown, M., Wilson, L., Doust, B., Gill, R., Sun, C.: Knowledge-based method for segmentation and analysis of lung boundaries in chest X-ray images. Comput. Med. Imaging Graph. **22**(6), 463–477 (1998)

Breast Lesion Segmentation Method Using Ultrasound Images

Agata Wijata(✉), Bartłomiej Pyciński, Marta Galińska, and Dominik Spinczyk

Faculty of Biomedical Engineering, Silesian University of Technology, Roosevelta 40, 41-800 Zabrze, Poland
Agata.Wijata@polsl.pl
http://ib.polsl.pl

Abstract. Breast cancer is one of the leading causes of death among women. A non-invasive ultrasound examination enables the location and type of lesion. The radiologist, based on the 2D ultrasound image, estimates the volume of the lesion and performs a core needle biopsy procedure. The lesion segmentation may support the diagnostic and therapeutic process. The purpose of this study is to develop a method of breast tumor segmentation using two-dimensional ultrasound images. The lesion mask is based on the fusion of several methods. The proposed method employs active contour models, gradient vector flow, region growing, thresholding, image gradient, watershed transform and morphological operations. The effectiveness of the method was checked with Dice, Jaccard and specificity coefficients. Median values were 0.915, 0.862 and 0.996, respectively.

Keywords: Breast ultrasound · Breast lesion · Ultrasound image segmentation · Image processing

1 Introduction

Breast cancer is the most common cancer type among women because 10% of all new cases of cancer are breast cancers [1]. Over the years, many methods of diagnosis and treatment of cancer have been developed, but the disease still has not been eliminated. Tumor size and site are important factors in assessing the clinical progression of the disease [2].

One of the commonly used imaging techniques in oncological diagnostics is ultrasonography. This method is safer for patient because of lack of x-ray radiation, which is present in mammography techniques. Using two-dimensional ultrasound images, it is possible to estimate the size and type of lesion. The core needle biopsy procedure is also performed under ultrasound control [3]. Breast tumors are seen in ultrasound images as areas whose echogenicity differs compared to surrounding tissues. This difference is dependent on lesion type. In addition to the tumors in the images, echo shadows are often seen. Echo shadows may blur tumor borders. This phenomenon significantly impedes an objective estimation of the lesion size or extraction sample for histopathological

E. Tkacz et al. (Eds.): IBE 2018, AISC 925, pp. 20–27, 2019.
https://doi.org/10.1007/978-3-030-15472-1_3

analysis. The development of a supporting segmentation method with objective and repeatable results, is a topic raised in the literature (Table 1).

Table 1. Summary results of ultrasound image segmentation methods, where DI – Dice Index, JI – Jaccard Index and SPE – specificity

Ref	Dim	Method	Measure	Mean	Median	Min	Max
[4]	2D	Median filter, region growing	DI	0.953	-	-	-
[5]	2D	Anisotropic diffusion, active contour	DI	-	0.810	0.640	0.870
[6]	3D	Median filter, multiscale blob detector, watershed transform, active contour	DI	-	0.854	0.766	0.926
			JI	-	0.745	0.620	0.863
			SPE	-	0.999	0.999	1.000
[7]	2D	Power watersheds	DI	0.810	-	0.560	0.930
[8]	2D	Swarm intelligence	DI (I set)	0.900	-	-	-
			DI (II set)	0.800	-	-	-
[9]	3D	Sobel edge detector, 2D watershed	JI	-	-	0.700	0.806
[10]	3D	Anisotropic diffusion, stick filter, Otsu threshold, Level-Set	SPE	0.925	-	-	-
[11]	2D	Gaussian smoothing, anisotropic diffusion, active contour	SPE	0.903	-	-	-
[12]	2D	Total-variation model, active contour	SPE	0.985	-	-	-

The purpose of this study is to propose a semi-automatic segmentation method of breast lesions using two-dimensional ultrasound images. Correct and objective methods of segmentation changes have a significant impact on the duration of the diagnostic process and repeatability of the results. The proposed method was verified on 16 sets of clinical data.

The paper is organized as follows. In Sect. 2 the materials and applied methods are described. Results and discussion are presented in Sect. 3. Finally, Sect. 4 presents conclusions and plans of application of proposed method in clinical practice.

2 Materials and Methods

2.1 Materials

The analysed dataset consists of 80 two-dimensional ultrasound images of 16 patients. The clinical images were recorded using The Imaging Source Video-to-USB 2.0 converter connected to the S-video output of Ultrasound Diagnostic Scanner Hitachi EZU-MT25-s1 and Hitachi EUP-L65 linear ultrasound transducer. Images were recorded by a radiologist during a routine breast biopsy procedure. The depth of the ultrasound beam was constant for each patient and the range was from 25 and 40 mm depending on the location of the lesion. The ultrasound resonant frequency of the scanner was related to the depth of ultrasound beam and it was from 6 to 14 MHz. Image resolution is dependent on the depth of the beam and ultrasound resonant frequency of the scanner. Motion artifacts of the video signal were deinterlaced using linear interpolation method and nearest rounding mode.

2.2 Methods

The workflow of the proposed method is shown in Fig. 1. This approach consists of three stages: image pre-processing, segmentation using the different methods and determination of output mask.

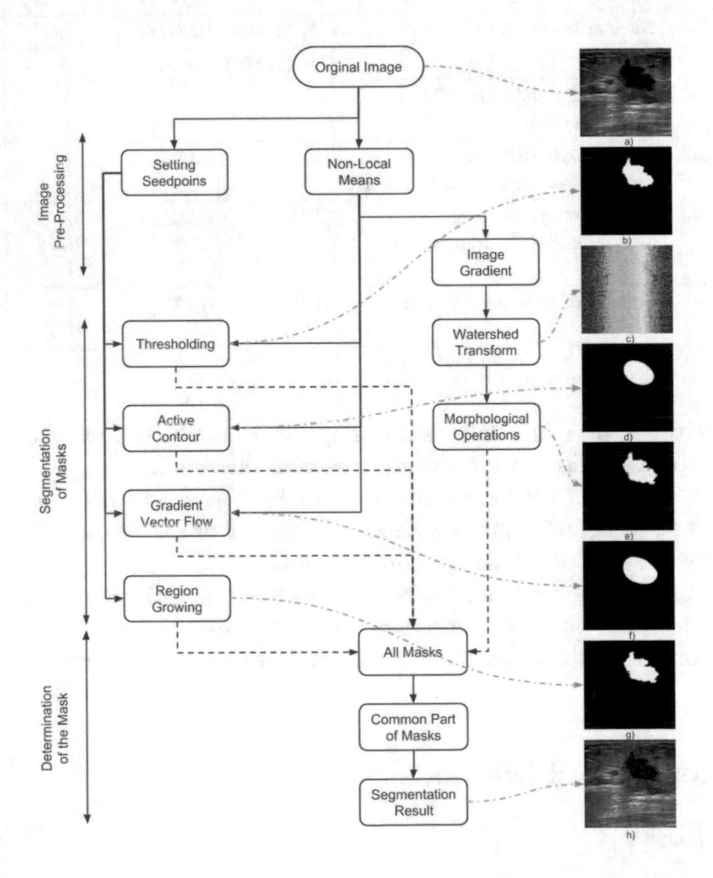

Fig. 1. Workflow of proposed method

First, the original image (Fig. 1a) is filtered using non-local means algorithm [13], which smoothes images without blurring edges. At least 4 seedpoints inside the lesion need to be marked (because this is number of points, which is needed to make closed, interpolated curve in active contour method; in case of 3 points, the curve is opened). Nearest interpolation is performed between seedpoints. The image gradient is calculated.

The mask segmentation stage consists of 5 segmentation methods, including one fully automatic and 4 semi-automatic methods which require starting points. The first mask is obtained using thresholding (Fig. 1b). Threshold value is the difference between the maximum value and the standard deviation of pixel intensity in region, which is indicated by seedpoints. Region growing algorithm [4], active contour models [14] and gradient vector flow [15] also are determined using seedpoints. Lesion masks are presented on Fig. 1d, f and g. The last mask is obtained using a watershed transform (Fig. 1c) based on the image gradient and morphological operations (Fig. 1e) [16].

The final stage is spelling the lesion mask, which is the common part of all binary masks received in the earlier stages. Final mask is the region of lesion that has been selected by at least 3 methods (Fig. 1h).

2.3 Quality Assessment

Segmentation result was compared with expert delineations (gold standard), which were prepared manually for each of the images. Examples of segmentations are presented on Figs. 3 and 4.

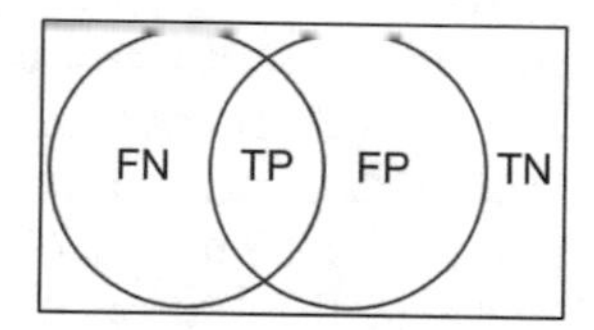

Fig. 2. Definition of pixel classification (red – gold standard, blue – segmentation result)

The similarity and diversity between the segmentation mask and gold standard were calculated using 3 measures:

- Dice Index (DI):

$$DI = \frac{2 \cdot TP}{2 \cdot TP + FP + FN} \tag{1}$$

- Jaccard Index (JI):

$$JI = \frac{DI}{2 - DI} \tag{2}$$

- Specificity (SPE):

$$SPE = \frac{TN}{TN + FP} \tag{3}$$

where: TP is number of True Positive, TN - True Negative, FP - False Positive and FN - False Negative pixels (Fig. 2).

3 Results and Discussion

Numerical results of segmentation are shown in Tables 2, 3 and 4.

Table 2. Dice Index for the proposed method (green – the best result, red – the worst result)

Method	Mean	Median	Std	Min	Max
Thresholding	0.4028	0.3374	0.3556	0.0000	0.9574
Region growing	0.6957	0.6812	0.1970	0.1443	0.9682
Active contour	0.7939	0.8347	0.1182	0.4365	0.9234
Gradient vector flow	0.7969	0.7964	0.0830	0.6532	0.9399
Watershed transform	0.8048	0.8693	0.1472	0.5215	0.9297
Proposed method	0.8886	0.9153	0.0519	0.7857	0.9488

The median values of Dice Index and Jaccard Index were 0.92 and 0.86, respectively. It shows that the method works well with clinical images. The best values of DI and JI were 0.94 and 0.90 and this case is presented in Fig. 3(a). The worst results of segmentation are shown on Fig. 4. In these cases, lesions have various shapes and blurred edges. The noise and artifacts caused distortion and border irregularity. The standard deviation of both indexes are small, which indicates that the results of this method are repeatable.

An independent comparison of the proposed method with the methods of which it is composed shows that adding masks improves the quality of segmentation. The median value ranged from 0.33 in case of thresholding to 0.87 for the watershed method. Comparing the method using minimum and maximum value shows the imperfections of these methods. The minimum value of the proposed method is 0.79 and unclear meaning. Specificity is an important metric in ultrasound, the median value of the proposed method was the second highest. Only the specificity of thresholding is higher than the proposed method, but the minimum value is smaller, which suggests that it works worse with difficult images.

Table 3. Jaccard Index for the proposed method (green – the best result, red – the worst result)

Method	Mean	Median	Std	Min	Max
Thresholding	0.3221	0.2029	0.3246	0.0000	0.9182
Region growing	0.5661	0.5165	0.2241	0.0777	0.9383
Active contour	0.6723	0.7163	0.1470	0.2792	0.8576
Gradient vector flow	0.6703	0.6618	0.1158	0.4850	0.8866
Watershed transform	0.6953	0.7689	0.1884	0.3528	0.9225
Proposed method	0.8483	0.8622	0.0417	0.7398	0.9025

Analysis of the DI value shows that the worst results were obtained for thresholding and region growing. The other three methods (active contour, gradient vector flow and watershed transform) have a similar effect based on median value analysis (Table 2). These values were in the range of 0.80 to 0.87, which means that segmentation results were worse than in the case of the proposed method (0.91). A similar tendency may be observed in the Table 3. According to Jaccard Index, the differences between the proposed method and its components are larger. The Dice Index indicates the sensitivity of the method while the Jaccard Index results also include the areas of oversegmentation (specificity), which for segmentation of lesion with irregular shapes, is undesirable. In the case of a pathological lesion in the breast based on ultrasound images, it is more beneficial to segment the mask, which lies in the area of the tumor, rather than showing its exact border.

Table 4. Specificity for the proposed method (green – the best result, red – the worst result)

Method	Mean	Median	Std	Min	Max
Thresholding	0.9874	0.9991	0.0252	0.8726	1.0000
Region growing	0.9231	0.9918	0.1235	0.5740	1.0000
Active contour	0.9777	0.9917	0.0292	0.8807	0.9999
Gradient vector flow	0.9584	0.9600	0.0231	0.8989	0.9930
Watershed transform	0.9761	0.9909	0.0277	0.9221	0.9999
Proposed method	0.9912	0.9963	0.0122	0.9488	1.0000

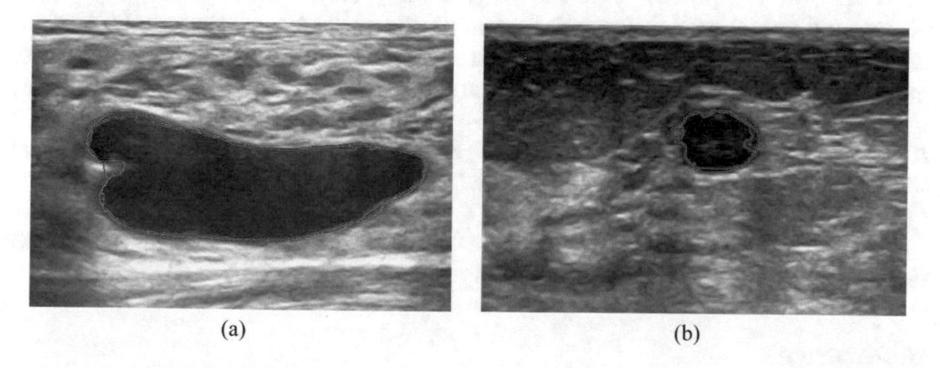

(a) (b)

Fig. 3. Delineation comparison for case No. 1: (a) the best case and (b) exemplary result of segmentation (red – gold standard, blue – segmentation result)

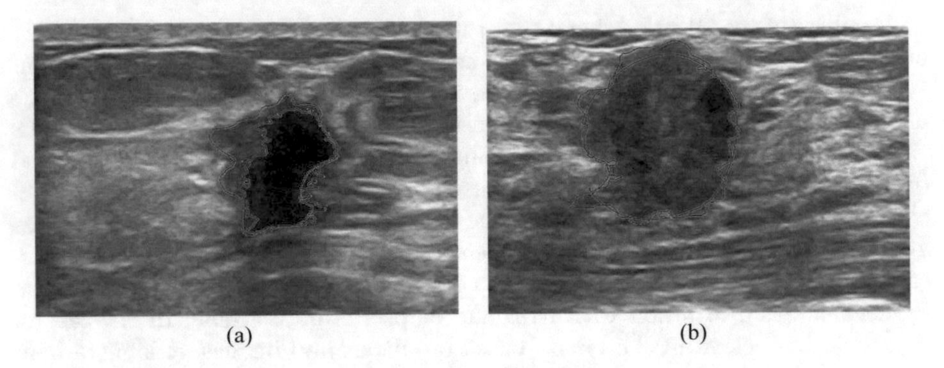

Fig. 4. Delineation comparison for case No. 1: (a) the worst case and (b) exemplary result of segmentation (red – gold standard, blue – segmentation result)

4 Conclusions

The method of lesion segmentation presented in the work accomplishes the task in three stages. The first is image pre-processing and filtration. The next step uses five different method segmentation and the last is determination of the result. In the best case, the Dice Index and Jaccard Index were 0.95 and 0.90, respectively. An important stage of ultrasound image analysis is filtering and noise reduction because the quality of the processed data determines the effectiveness of the segmentation. The quality of the ultrasound machines used during examinations is also an important factor.

The proposed method is the preliminary study on the effective method of lesion segmentation in two-dimensional ultrasound images. The next stage of work will be to automate the segmentation process by make it independent from the starting points. This may significantly affect the repeatability of the ultrasound examinations (especially the biopsy procedure). The method might be useful in clinical practice and segmentation results may result in a higher percentage of correctly diagnosed lesions and more effective oncology treatment.

Acknowledgements. This research was partially supported by the Polish National Centre for Research and Development (NCBR), grant no. STRATEGMED2 /267398/3/NCBR/2015.

The authors would also like to thank Andre Woloshuk for his English language corrections.

References

1. Ferlay, J., Hery, C., Autier, P., Sankaranarayanan, R.: Global burden of breast cancer. In: Breast Cancer Epidemiology, pp. 1–19 (2010)
2. Pflanzer, R., Hofmann, M., Shelke, A., Habib, A., Derwich, W., Schmitz-Rixen, T., Bernd, A., Kaufmann, R., Bereiter-Hahn, J.: Advanced 3D-sonographic imaging as a precise technique to evaluate tumor volume. Transl. Oncol. **7**(6), 681–686 (2014)

3. Czajkowska, J., Pyciński, B., Juszczyk, J., Pietka, E.: Biopsy needle tracking technique in US images. Comput. Med. Imaging Graph. **65**, 93–101 (2018)
4. Lee, L.K., Liew, S.C.: Breast ultrasound automated ROI segmentation with region growing. In: 2015 4th International Conference on Software Engineering and Computer Systems (ICSECS), pp. 177–182 (2015)
5. Galinska, M., Ogieglo, W., Wijata, A., Juszczyk, J., Czajkowska, J.: Breast cancer segmentation method in ultrasound images. In: Gzik, M., Tkacz, E., Paszenda, Z., Piętka, E. (eds.) Innovations in Biomedical Engineering, IBE 2017, Advances in Intelligent Systems and Computing, vol. 623, pp. 23–31 (2017)
6. Wieclawek, W., Rudzki, M., Wijata, A., Galinska, M.: Preliminary development of an automatic breast tumour segmentation algorithm from ultrasound volumetric images. In: Proceedings 6th International Conference Information Technology in Biomedicine, pp. 77–88 (2018)
7. Kollorz, E., Angelopoulou, E., Beck, M., Schmidt, D., Kuwert, T.: Using power watersheds to segment benign thyroid nodules in ultrasound image data. In: Bildverarbeitung für die Medizin, pp. 124–128 (2011)
8. Badura, P.: Virtual bacterium colony in 3D image segmentation. Comput. Med. Imaging Graph. **65**, 152–166 (2018)
9. Gu, P., Lee, W.M., Roubidoux, M.A., Yuan, J., Wang, X., Carson, P.L.: Automated 3D ultrasound image segmentation to aid breast cancer image interpretation. Ultrasonics **65**, 51–58 (2016)
10. Chang, R.F., Wu, W.J., Moon, W.K., Chen, D.R.: Automatic ultrasound segmentation and morphology based diagnosis of solid breast tumors. Breast Cancer Res. Treat. **89**(2), 179–185 (2005)
11. Minavathi, Murali, S., Dinesh, M.S.: Classification of mass in breast ultrasound images using image processing techniques. Int. J. Comput. Appl. **42**(10), 29–36 (2012)
12. Huang, Q., Yang, F., Liu, L., Li, X.: Automatic segmentation of breast lesions for interaction in ultrasonic computer-aided diagnosis. Inf. Sci. **314**, 293–310 (2015)
13. Buades, A., Coll, B., Morel, J.M.: A non-local algorithm for image denoising. In: Computer Vision and Pattern Recognition, pp. 60–65 (2005)
14. Kass, M., Witkin, A., Terzopoulos, D.: Snakes: active contour models. Int. J. Comput. Vis. **1**(4), 321–333 (1988)
15. Xu, C., Prince, J.: Gradient vector flow: a new external force for snakes. In: IEEE Conference on Computer Vision and Pattern Recognition, pp. 66–71 (1997)
16. Baraiya, N., Modi, H.: Comparative study of different methods for brain tumor extraction from MRI images using image processing. Indian J. Sci. Technol. **9**(4), 1–5 (2016)

Performance of Medical Image Transfer in High Bandwidth Networks

Bartłomiej Pyciński[1(✉)], Jacek Kawa[1], Paweł Bożek[1], Michał Smoliński[1], and Maria Bieńkowska[2]

[1] Radpoint Sp. z o.o., Kościuszki 173, 40-524 Katowice, Poland
{bartlomiej.pycinski,jacek.kawa,pawel.bozek,michal.smolinski}@radpoint.pl
[2] Faculty of Biomedical Engineering, Silesian University of Technology, Roosevelta 40, 41-800 Zabrze, Poland
maria.bienkowska@polsl.pl

Abstract. The paper summarizes testing the performance of transferring DICOM medical images. Experimental setup consists of several different workstations connected in gigabit network. The results state that legacy DICOM communication's efficiency is much worse than the transfer provided by novel technologies, especially DICOMweb. Hardware-related circumstances have also been discussed.

Keywords: DICOM · Performance · Networking · CAD · Testing environment

1 Introduction

A discovery of the X-rays in 19th century [10] changed the face of medicine. The Radiography became a crucial tool in assessment of fractures, diagnosis of tuberculosis etc. Further advances in the domain were brought by the invention of the ultrasound-based imaging. Next breakthrough is directly related to the invention of the computer. Advanced control, acquisition and processing abilities, enabled Computed Tomography, CT (first working model constructed by Sir Godfrey Hounsfield) [9] and Nuclear Magnetic Resonance Imaging (NMRI) [5] in the 1970s. The medical imaging irreversibly has become digital.

However, despite many advances in the field, at the beginning of the 1980s, the medical imaging devices from various manufactures were mostly incompatible in terms of storing or processing the imaging data. At the initiative of the joined groups of the American College of Radiology (ACR) and National Electrical Manufacturers Association (NEMA), a standardization in the field has started. The basic concepts embedded in two ACR-NEMA standards (1985 and 1988) [1] evolved into a DICOM v3 standard (Digital Imaging and Communication in Medicine version 3) in 1993 [6]. The DICOMv3 has been designed to stay open to the changes in the technology field. Basic definitions, network communication, encoding, etc. have been separated from each other and from

E. Tkacz et al. (Eds.): IBE 2018, AISC 925, pp. 28–35, 2019.
https://doi.org/10.1007/978-3-030-15472-1_4

the modality-dependent information. Every part of the standard is subjected to constant review and update procedure. The changes are backward-compatible (if possible) and even the *ancient* DICOM-conformant equipment can easily be integrated into modern DICOM network and communicate with the PACS (Picture Archive and Communication System) or a VNA (Vendor Neutral Archive – basically a next generation PACS, built over open components and able to store different kind of data).

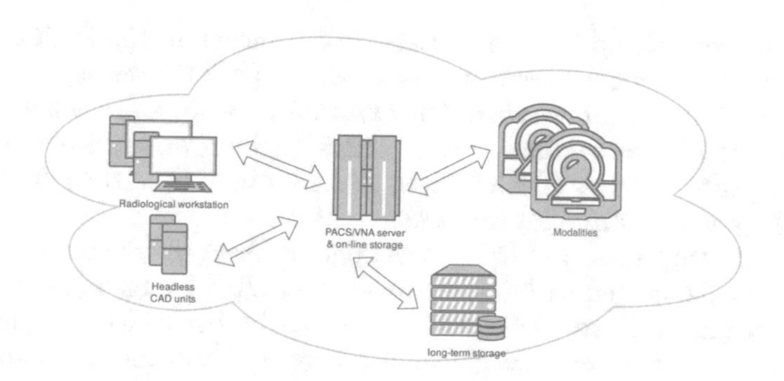

Fig. 1. Network in radiological laboratory

Modern DICOM network (Fig. 1) connects several kind of nodes [3]:

1. PACS or a VNA – a central storage and exchange node along with long-term backup,
2. modalities, i.e. acquisition equipment (CT scanner, NMRI machine, etc.),
3. radiological workstations,
4. peripheral devices: optical drives, radiological printers, etc.

Dedicated CAD (Computer Aided Diagnosis) modules are also present: either as PACS or radiological workstations plugins, or as separate units processing medical data. The main goal of CAD unit [7] is to automatically extract diagnostic information present in the image. Radiologist is able to use the results of the processing to speed up the diagnosis, detect hidden abnormalities or provide (otherwise often unfeasible) quantitative assessment (e.g. volumetric or spatial measurements of complex volumetric structures). CAD modules exist for brain tumours assessment [11], bone tumours [2], tissue segmentation [4], bone age assessment, multiple sclerosis or lung nodules detection, pneumothorax measurement [8], etc.

The CAD modules in radiological environment are typically either fully automated or semi-automated. In the latter case, some user input may be needed to facilitate the processing (e.g. pointing the lesion that needs to be analyzed). In the former case, such input is unnecessary, and the processing may be initiated as soon, as the data are received from the acquisition device. This feature is especially important for the time-consuming workflows, where waiting several hours (or days) for the user-initiated results may be infeasible. Fully automated

CAD modules are also easy to deploy into separate processing unit (e.g. headless personal computer or workstation-class computer with medically certified software).

A CAD unit must be compatible with the DICOM in the area of:

1. data encoding – supporting DICOM services on data objects storage,
2. data transfer – supporting DICOM services on transferring data objects through network.

There are several options for the network transfers in the DICOM, including: (1) binary protocol based on the DIMSE (DICOM message service element) employing TCP/IP services (DICOM PS3.7 2018b) and available from the announcement of the standard, and (2) REST (Representational State Transfer) HTTP-based services (DICOMweb, PS3.18 – part 18 of the DICOM 2018b standard) – a recent standard extension.

Due to security reason DICOM-networks are usually private: access is limited to terminals located within same network or VPN-enabled (Virtual Private Network) clients. Due to configuration complexity, the widely adopted binary DICOM protocol is rarely encrypted. It's easy to eavesdrop the transmission. DICOM nodes are traditionally identified by means of text-based identifiers and IP-addresses (only). Moreover, to receive an image, both sides needs to be able to initiate the connection to each other. Therefore, the image retrieval from outside the NAT (Network Address Translation) is not possible without dedicated NAT traversal options (making the VPN a natural solution for the outside access). Lastly, each processed DICOM request results in a multistep-DIMSE exchange (both sides negotiates the association, transfer syntax, etc.) which may severely impact the performance in high-latency networks.

DICOMweb option is much easier to configure and secure in modern network environment. The DICOMweb is based on the HTTP, a popular application Internet protocol. The communication follows the client-server schema. All the relevant options are present in the HTTP envelope. Standard web languages (e.g. JSON, JavaScript Object Notation) are used to define queries and answers. Security may be provided by enabling TLS (Transport Layer Security) transport. However, the DICOMweb is contemporary supported by only the minor subset of all available devices and software.

The goal of the paper is to analyze the performance of the high-bandwidth, low-latency CAD-PACS connection in terms of:

1. protocol (DICOM vs. DICOMweb),
2. encryption (no encryption vs. TLS-provided AES encryption).

The motivation is to select best option for the dedicated, autonomous CAD unit batch-processing the DICOM studies.

In the paper first, the network configuration as well as evaluated hardware configuration will be described. Next, the experiment details will be given. Finally, results will be provided and discussed.

2 Materials and Methods

2.1 Materials

For testing the transfer parameters, a dataset of DICOM images was prepared. The images were created artificially, using pydicom[1] and SimpleITK[2] toolkits. Each instance contained a grayscale image of 16-bit pixel's depth. The images' dimensions were 512 × 512 pix. Whole dataset consisted of 6000 images and it contained 3006 MB. The volume of the data matches a typical size of several CT or NRM studies, as CAD processing may be based upon a single study or several studies of a single patient.

The data were stored on a Linux based server with 16-core CPU Intel Xeon with working frequency of 2.4 GHz and 32 GB RAM DDR3. There were five barebone computers (BB) differing in hardware parameters, whose performance were also compared. All BB were manufactured by Qotom (Qotom Technology CO., LTD, China). They are denoted as BB1–BB5 in further part of the article and their hardware parameters are presented in Table 1 (please note, that BB4 and BB5 are exactly the same models). All BBs were hosting headless Ubuntu 16.04 version. The server and the BBs were connected in single 1 Gbps local Ethernet network.

Table 1. Comparison of barebone computers (BB1–BB5) used for testing. BB4 and BB5 are the same models

	BB1	BB2	BB3	BB4	BB5
Model name	Q220N-S02	Q210S-S02	Q220N-S02	Q190S-S02	Q190S-S02
CPU	Intel i5 3317U	Intel i3 3217U	Intel Celeron J1900	Intel Celeron J1900	Intel Celeron J1900
AES-NI	Yes	No	No	No	No
No of cores	4	4	4	4	4
Storage	32 GB SSD 2 TB SSHD	32 GB SSD	32 GB SSD 2 TB SSHD	32 GB SSD	32 GB SSD
RAM	4 GB	4 GB	4 GB	4 GB	4 GB

2.2 Methods

First, a network bandwidth and hardware components were evaluated separately. The tests included obtaining maximal performance of Ethernet interface, reading and writing streams of data to hard disks, as well as testing efficiency of consequent random I/O operations on the disks. Both solid state drives (SSD) and solid state hybrid drives (SSHD) were involved.

[1] https://pydicom.github.io.
[2] http://www.simpleitk.org.

Maximal speed of writing to disk was evaluated using standard GNU application dd. Maximal speed of reading from disk was yielded with hdparm tool. Another tool used for evaluating the performance of BBs was fio (flexible I/O tester). Bandwidth of Ethernet interface was tested with iperf.

Next tests concerned evaluation of encryption performance. Since ca. 2010 (1st generation Intel Core) processors have supported AES instruction set (AES-NI). Therefore the efficiency of applications that use cryptographic operations (e.g. TLS protocol) is expected to be significantly improved. In the experimental setup, only BB1 supports this instruction set (see Table 1).

DICOM communication was implemented by DCMTK toolkit[3]. An instance of dcmqrscp was running on each BB. Initially, their database had been pruned. During experiments, DICOM files were stored on BBs by sending the files from Xeon server with dcmstore utility. Repeatability of experiments was ensured by prepared Python scripts.

3 Experiments and Results

In this section, the performed experiments and the obtained results are shown. First two experiments show the limits of the network and storage. Next, the experiments concerning DICOM transfers are presented. The latter, directly related to the goal of the paper, are summarized in the Table 2. Moreover, for better readability, same data are presented in Fig. 2.

Table 2. Throughput comparison of all tested set-up. DICOM transmission has additionally been tested with two and three simultaneous connections; in these cases, the total throughput has been included. All values are given in MBps (Mega Bytes per second)

	BB-1	BB-2	BB-3	BB-4	BB-5
HTTP	96.90	96.77	82.27	96.12	90.80
TLS	96.30	96.37	57.21	68.88	63.56
DICOM (1 client)	36.78	33.24	25.49	25.21	25.40
DICOM (2 clients)	40.68	31.78	19.41	18.87	17.93
DICOM (3 clients)	29.62	21.56	13.21	13.08	12.56

3.1 Network Throughput

Measurements of separated components proved that Ethernet interfaces of all BBs can send or receive data with bitrate close to maximal 1 Gb Ethernet capacity. Observed data rates were between 929 and 941 Mbps (Mega bits per second), or 116 to 118 MBps (Mega Bytes per second).

[3] https://dicom.offis.de/dcmtk.php.en.

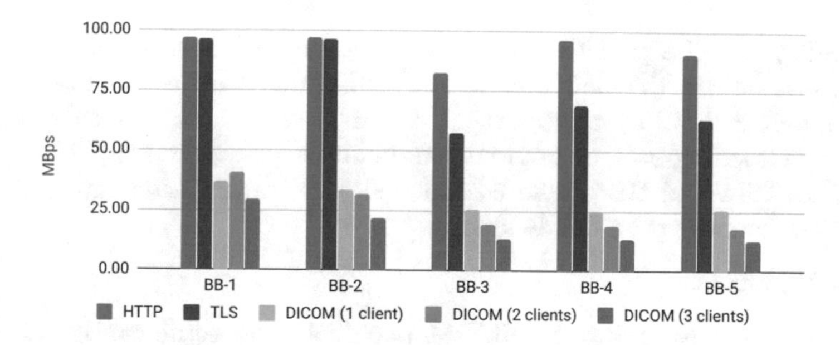

Fig. 2. Throughput comparison of all tested set-up. DICOM transmission has additionally been tested with two and three simultaneous connections; in these cases, the total throughput has been included. All values are given in MBps (Mega Bytes per second)

3.2 I/O Operations

I/O operations performance of both SSD and SSHD drives, as long as transfer in one direction of large amount of data (at least 500 MB) was concerned, exceeded significantly available Ethernet capacity, yielding 259 to 267 MBps (reading), 254 to 274 MBps (writing) for SSD, and 131 to 132 MBps (reading), 126 to 131 MBps (writing) for SSHD. However, when bidirectional tests of alternate input and output operations on SSD drives were performed with `fio` tool (random 75% of the whole I/O operations were reading, and the other were writing, block size was aligned to 4096 bytes, total amount of data was equal to 1 GB) differences between BBs were revealed. The performance decrease was the largest in BB4 and BB5 machines. Full results are presented in Table 3.

Table 3. Comparison of I/O bitrate when 1 GB of data in 4 kB blocks is transferred either in one way, or bidirectionally. All values are given in MBps (Mega Bytes per second)

	BB-1	BB-2	BB-3	BB-4	BB-5
Read only	157.03	149.78	118.38	53.20	59.62
Write only	168.32	166.23	114.63	75.18	105.70
Read (bidirectional operations)	96.53	95.82	79.70	35.06	41.27
Write (bidirectional operations)	32.41	32.17	26.76	11.77	13.84

3.3 DICOM Transfer

DICOMweb-like Mode

At this experiment, a DICOM files from the database were sent in (unencrypted) POST requests to the HTTP servers located on BBs, as in DICOMweb STOW-RS (Store Over the Web). The observed bitrates were consistently high for all BBs (between 82 and 96 MBps, see Table 2, first row).

Enabling the TLS-provided AES encryption revealed substantial differences between specific BBs (Table 2, second row). BB1 (equipped with AES-NI enabled CPU) as well as BB2 (not supporting AES-NI), were able to receive the same data rate, exceeding 99% of original bandwidths. Less powerful, Intel Celeron-based BB3, BB4 and BB5 featured much slower transmission rate (69–71% of the original, unencrypted bandwidth).

DICOM Protocol

Transfer of the dataset using DICOM protocol took significantly more time. Regardless of the number of connected clients working from the same Xeon machine, Intel Core based BB1 and BB2 devices received DICOM images with the highest bitrates. However, with the second and third parallel client connection, the total bandwidth of DICOM transmission decreased (Table 2, rows 3–5).

3.4 Final Observations

Two final remarks are as follows. Firstly, if SSHD device was used instead of SSD one, transfer of DICOM files decreased no more than by 10%. Secondly, system load average of BBs processors was monitoring during the experiments, and its value ranged between 0.5 and 1.5. As the number of CPU cores in all BBs was equal to 4, values of system load average (i.e. the average number of processes that were either in a runnable or uninterruptible state) indicated that there was no overwork of BBs during the experiments.

4 Discussion and Conclusion

The aim of this study was to select the best option for the dedicated, autonomous CAD unit batch-processing the DICOM studies. In this paper, the experiment which tested the transfer parameters of selected network and hardware configuration was presented. Experiments indicated that there are large bandwidth-related differences in high-performance local network, considering both communication protocol and hardware.

In general, the DICOM protocol was much slower than the DICOMweb, HTTP-based transmission. The reason is likely the design of the DICOM protocol and the DIMSE-based networking. The processing of a single high-level request (e.g. storying a single instance/image) results in an exchange of a number of low level messages. The round trip time (RTT) accumulates and the average performance drops. HTTP-based DICOMweb protocol processes same request in a single take (single HTTP request and response pair). Moreover, enabling parallel data transmission did not increase the performance. However, the effect might have been related to a processing-queue on the receiving side: if data from various clients were processed in single queue, there might be no visible gain in the overall processing speed.

Obtained results show, that, regardless of the used hardware, in the high-bandwidth radiological network, one may expect a much higher throughput using a modern DICOMweb standard extension: replacing DIMSE-based message exchange by HTTP has likely had a greatest impact on the overall performance in the tested environment. Furthermore, enabling encryption in such network may decrease performance (up to 30%), but the drop of throughput should be insignificant if both ends of connection employ hardware acceleration and/or have enough computational power to process data streams in real time (observed drop in performance was less than 1% in performed experiments).

It seem, that the best option in terms of PACS-CAD communication is to use the modern communication protocols and employ the hardware acceleration whenever possible. It may significantly change the perception of a CAD: make it working smoothly and process data faster and more smoothly.

Acknowledgment. This research was supported by the National Centre for Research and Development (NCBiR) grant no. POIR.01.01.01-00-0303/17-00.

References

1. Bidgood Jr., W., Horii, S.C.: Introduction to the ACR-NEMA DICOM standard. Radiographics **12**(2), 345–355 (1992)
2. Bugdol, M., Czajkowska, J.: Liver surface deformation model for minimally invasive surgery. In: 2011 Proceedings of the 18th International Conference Mixed Design of Integrated Circuits and Systems (MIXDES), pp. 52–55 (2011)
3. Huang, H.K.: PACS and Imaging Informatics. Wiley, Hoboken (2009)
4. Kawa, J., Juszczyk, J., Pyciński, B., Badura, P., Pietka, E.: Radiological atlas for patient specific model generation. In: Information Technologies in Biomedicine, Volume 4, vol. 284, pp. 69–84. Springer (2014). https://doi.org/10.1007/978-3-319-06596-0_7
5. Mansfield, P., Grannell, P.K.: "Diffraction" and microscopy in solids and liquids by NMR. Phys. Rev. B **12**, 3618–3634 (1975). https://doi.org/10.1103/PhysRevB.12.3618
6. NEMA PS3 / ISO 12052: Digital imaging and communications in medicine (DICOM) standard. National Electrical Manufacturers Association, Rosslyn. http://medical.nema.org/
7. Pietka, E., Kawa, J., Badura, P., Spinczyk, D.: Open architecture computer-aided diagnosis system. Expert Syst. **27**(1), 17–39 (2010)
8. Piętka, E., Kawa, J., Spinczyk, D., Badura, P., Więcławek, W., Czajkowska, J., Rudzki, M.: Role of radiologists in cad life-cycle. Eur. J. Radiol. **78**(2), 225–233 (2011)
9. Richmond, C.: Sir Godfrey Hounsfield. BMJ **329**(7467), 687 (2004)
10. Röntgen, W.C.: Ueber eine neue Art von Strahlen. Ann. der Physik **300**(1), 1–11 (1898). https://doi.org/10.1002/andp.18983000102
11. Szwarc, P., Kawa, J., Rudzki, M., Pietka, E.: Automatic brain tumour detection and neovasculature assessment with multiseries MRI analysis. Comput. Med. Imaging Graph. **46**, 178–190 (2015). https://doi.org/10.1016/j.compmedimag.2015.06.002

Assessment of Clinical Variables Importance with the Use of Neural Networks by the Example of Thyroid Blood Test Parameters

Martyna Michałowska[1(✉)], Tomasz Walczak[1], Jakub Krzysztof Grabski[1], and Monika Grygorowicz[2]

[1] Institute of Applied Mechanics, Faculty of Mechanical Engineering and Management, Poznan University of Technology, Jana Pawła II 24, 60-965 Poznań, Poland
martyna.michalowska@put.poznan.pl

[2] Department of Spondyloorthopaedics and Biomechanics of the Spine, Wiktor Dega Orthopaedic and Rehabilitation Clinical Hospital, Poznan University of Medical Sciences, 28 Czerwca 1956 135/147, 61-545 Poznan, Poland

Abstract. Screening blood tests for thyrometabolic status determination are difficult in interpretation because of many factors like age, sex or measurement method that influence their proper interpretation. To solve this problem machine learning techniques, like artificial neural networks (ANN) can be applied, but their application is very common belittled due to their very little explanatory insight to the relation between input parameters (e.g. test results) and model of the disease. In contrast to previous studies concerning application of neural networks in thyroid disease diagnosis, in this study the authors decided to focus on extraction of reliable dataset (with preserved proportion of diseased and health cases) and quantification of input parameters importance in neural network decisive process. The importance of the variables considered as the most significant in hypothyroidism detection was estimated based on two independent methods: connection weights method (according to the Garson's algorithm) and sensitivity analysis. The results show, that the most important factors in hypothyroidism detection are TSH, TT4, FTI and age, and the rejection of other analyzed in this study parameters (sex, T3, T4U) does not influence significantly the performance of the neural network model and its predictive power.

Keywords: Hypothyroidism · Neural network analysis · Parameter importance

1 Introduction

Tests that measure the concentration of thyroid hormones and their transport in blood are very popular among the medical tests used to assess thyroid metabolic

E. Tkacz et al. (Eds.): IBE 2018, AISC 925, pp. 36–46, 2019.
https://doi.org/10.1007/978-3-030-15472-1_5

activity, in order to diagnose thyroid disease [1]. They are usually performed by measurement of serum total circulating thyroid hormone levels (T4), serum thyrotropin stimulating hormone (TSH) concentrations, and thyroidal iodine-123 uptake, with the use of radioimmunoassay or isotopic methods. Thyroid activity is regulated by production of TSH and the anterior pituitary gland by thyrotropin-releasing hormone released by the hypothalamus. External iodine supply and mechanisms inside thyroid influence production of TSH hormone. Concentrated ingested iodine is then incorporated into the tissue-specific protein, thyroglobulin [2].

Usually, based on TSH and T4 measured level, there are three thyrometabolic statuses defined: hypothyroidism, euthyroidism and hyperthyroidism, with distinction to the subclinical and clinical status in regard to hypo- and hyperthyroidism. In simplified terms one can say, that euthyroidism defines proper functioning of the thyroid gland, while hyperthyroidism is caused by excessive production of thyroid hormones (e.g. TSH) and hypothyroidism is an effect of underactive thyroid gland [2]. These both pathological states have adverse effect on health. For example, retrospective study on pregnant women conducted by Arbib et al. [3] showed, that subclinical hypothyroidism is associated with an increased risk for preterm delivery, whereas subclinical hyperthyroidism is correlated with low birth weight and placental abruption.

In reference to the above, it might be suggested, that measurements of serum blood parameters are ideal as a screening methods for thyrometabolic status determination. However, normative range of values for different populations are significantly different due to the multiple factors like age, sex or pregnancy. Furthermore, measured hormone levels are sensitive to the measurement method and time of day the blood was taken. Big number of influential factors make analysis of serum blood tests difficult to interpret, but thanks to modern computational techniques, like neural networks it is possible to create nonlinear models that imitate complicated relations between disease and test results, in regard to many influential factors [4]. High efficiency of neural networks application in disease prediction or identity recognition issues was proven in many other studies [5–7].

In this study the authors used hypothyroid database from UCI Machine Learning Repository in order to evaluate the importance of particular serum blood parameters in diagnosing hypothyroidism. This database and similar databases concerning blood test parameters in thyroid disorders, were previously widely used in numerous of studies about application and development of machine learning techniques, but according to the best authors knowledge, none of these studies focused on clinical interpretation of gathered data and their importance in diagnosing hypothyroidism. For example, in study [8] Zhang et al. were considering an impact of sampling variability on classification to one of three groups (normal, under- or over-active thyroid), based on neural network posterior probability estimates. What is more, authors of this study used highly unbalanced dataset (502 diseased cases and 6613 euthyroid cases in total in test and training set), what in our opinion highly influenced achieved results. The performance of neural network in testing increased from 82.19% in hyperthyroid dataset to 98.16% in overall performance assessment due to much larger number of easy in prediction euthyroid cases. In discussion section Guo-qiang et al. emphasized, that there is a need to develop a variable selection method via neural networks to choose the best subset of variables from a large number of attributes what is one of purposes in our study. Next example are studies similar to [9] and [10] where authors investigated the influence of neural network structure on performance and it's level of generalization in thyroid disease diagnosis. None of these studies was focused on distinction of the most important blood test parameters in thyroid disease diagnosis, what is important in reference to minimalizing costs associated with diagnostic tests and treatment. What is more, in most studies conducted on dataset from UCI Machine Learning Repository the authors did not reject missing data. It means that single missing data was usually replaced with estimated value in order to preserve big quantity of dataset. Such a way of data supplementation may influence it's clinical interpretation. In contrast to that, in this study the authors focused on extraction of reliable dataset (with preserved proportion of diseased and health cases) and quantification of serum blood parameters importance in neural network decisive process.

2 Data

Data analyzed in this study has been obtained from UCI Machine Learning Repository donated by Garavan Institute in Sydney, Australia [11].

2.1 Data Selection and Characteristics

To provide the best view of data contained in the database and in order to avoid usage of estimated values of the missing data in the further analysis, the authors decided to restrict the data and present its basic statistics. All the records that missed the values of the key parameters were rejected in the following order: records with missing age (446), records of pregnant women at the time of the test (59), records with missing TSH level (419), records with missing T3 level (277). After this stage of data selection, there were no records with missing information about TT4, T4U or FTI levels. Due to the fact, that after selection only two records contained TBG level, the authors decided not to use this parameter in the further analysis. In this study the authors decided to reject not only the records with missing singular information of any input parameter, but also all the records of pregnant women. This decision was made based on large number of studies describing big influence of pregnancy on thyroid function. For example, in study conducted by the Soldin et al. [14], it was showed, that the levels of TSH and thyroxine are trimester-specific due to the fetal neurodevelopment of brain, heart and lungs. In reference to the above, the rejection of the data from pregnant women at the time of the test seems to be justified.

As the result the database with 1962 records, including 121 hypothyroid cases and 1841 euthyroid cases, was formed. Because of the fact, that the ratio of the distinguished dataset size to the not-distinguished dataset size influences significantly sensitivity and specificity of classifiers used to the assessment of the predictive tools (like neural networks), the authors decided to equalize the number of hypothyroid and euthyroid cases in the dataset. In reference to that, 121 (from the 1841 available) euthyroid cases were randomly chosen. Due to the fact, that age and sex were considered as factors having impact on interpretation of morphological test results (range of TSH, T3, TT4, T4U or FTI values considered as normative may vary due to the age or sex group), the euthyroid cases were chosen so as to not influence the mean age and percentage of the male sex in the analyzed probe (Table 1). More detailed input data characteristics, with distinction to the hypothyroid and euthyroid dataset is presented in the Table 2 and the overall process of data selection, from the UCI Machine Learning Repository to the dataset taken in neural network analysis, is presented in the Fig. 1.

Table 1. The influence of data selection on the mean age and percentage of male sex in analyzed dataset.

	Before selection	After selection (dataset used in ANN analysis)
Number of euthyroid cases	1841	121
Number of hypothyroid cases	121	121
Mean age ± SD	54.16 ± 19	55.61 ± 20
Percentage of male sex	33%	32%

Due to the big differences between values of TSH, TT4 and FTI in hypothyroid group compared to euthyroid group (Table 2), these parameters are expected to have big importance in neural network decision process. This hypothesis will be verified in the following part of this article concerning contribution of variables in neural network decisive process.

Table 2. Characteristic of input data chosen to the ANN analysis, with distinction to the hypothyroid and euthyroid data.

No. and name of feature	Hypothyroid data (121 cases including 25% of male sex cases)	Euthyroid data (121 cases including 40% of male sex cases)
1. Age	55.75 ± 19.84	55.47 ± 20.92
2. Male sex	-	-
3. TSH	60.26 ± 73.70	2.06 ± 4.79
4. T3	1.17 ± 0.93	1.87 ± 1.01
5. TT4 sex	36.69 ± 28.85	108.54 ± 36.84
6. T4U	1.05 ± 0.21	0.94 ± 0.19
7. FTI	34.65 ± 23.24	116.93 ± 36.27

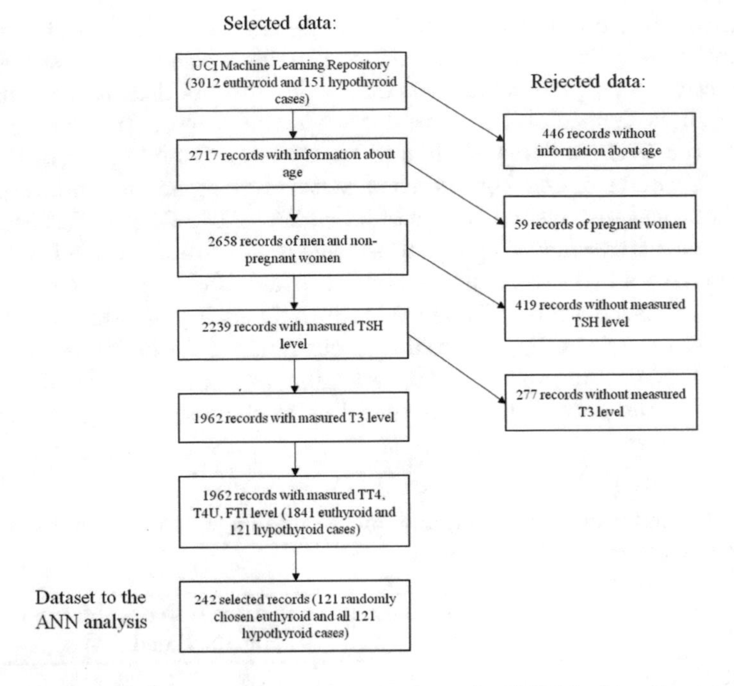

Fig. 1. Diagram of data selection to the study, from the UCI Machine Learning Repository to the dataset taken in neural network analysis.

3 Methods

3.1 Data Preprocessing

Because of the fact, that analyzed in this study neural networks use hyperbolic tangent activation function and to satisfy all the requirements connected with use of the connection weights method and Garson's algorithm (according to the study [13]), the non-binary analyzed variables (age, TSH, T3, TT4, FTI, T4U) were converted to the <0; 1> range of values.

3.2 Neural Network Design

In this study, task of the created neural networks was to distinguish people that suffer from hypothyroidism, from healthy people, based on the blood test results and information about age and gender. Converted information about age, TSH, T3, TT4, T4U, FTI levels and binary information about gender was used as input parameters of the Multi-Layer Perceptron (MLP) neural network with one hidden layer. The binary (categorical) input with information about gender was divided into two separate input neurons and similarly, the output with the binary information about the disease, was divided into two output neurons.

In the first step, to find the best architecture of the neural network (number of hidden neurons), the authors used Automated Network Search tool of TIBCO

Software Inc. (2017) Statistica (data analysis software system). The architecture of the sought neural network was restricted to MLP network type, with one hidden layer, 2 to 12 hidden neurons and hyperbolic tangent (Tanh) activation function in hidden and output layer. The software was set to train 5000 neural networks with randomly chosen initial weights. The performance of the neural networks was assessed with the use of Sum of Squares (SOS) error function. The dataset was classically divided into three randomly chosen subsets: training (70% samples of the dataset), test (15%) and validation set (15%) and BFGS algorithm with 200 cycles was set as training algorithm. As the result the Automated Network Search tool indicated neural network with 4 neurons in hidden layer (MLP 8-4-2) as the network with the highest performance. The chosen architecture with all analyzed input parameters is presented in the Fig. 2 and the basic information about chosen neural network and its performance is presented in Table 3 (neural network no. 1).

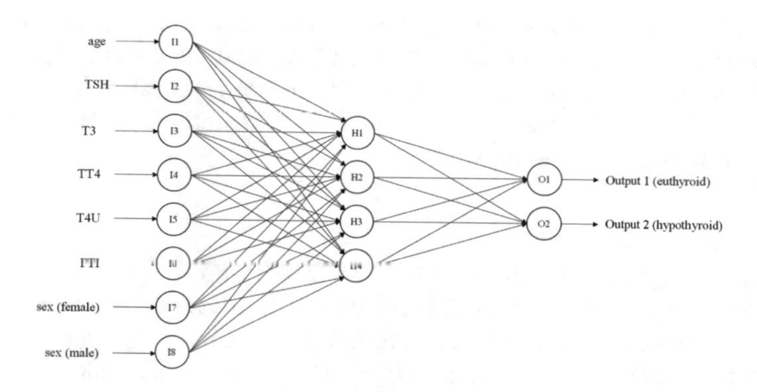

Fig. 2. Graphical representation of MLP 8-4-2 neural network.

Table 3. Performance for the MLP neural network chosen by Automated Neural Network Search Tool of Statistica software (NN no. 1) and performance achieved after looking for the best weights of the neural network connections, chosen by Custom neural network tool (no. 2).

No.	Net architecture	Training perf.	Test perf.	Validation perf.	Correct predictions
1	MLP 8-4-2	98.24%	100%	97.22%	98.34% (96.69% in euthyroid and 100% in hypothyroid samples)
2	MLP 8-4-2	100%	97.22%	100%	99.59% (100% in euthyroid and 99.17% in hypothyroid samples)

In order to decrease the influence of random selection of connection weights, and to obtain the best set of weights to the further analysis, the authors decided to repeat the process of neural network training, based on architecture selected in the previous step. This time Custom neural network tool of Statistica software was selected as strategy for creating predictive models and all the restrictions for the neural network architecture was preserved, as it was described in the first step. The software was set to train 500 MLP neural networks with 8 input, 4 hidden and 2 output neurons, with randomly chosen initial weights at the start point of training, and the network with the highest training, test and validation performance was selected as basic model to asses the contribution of variables in predicting hypothyroidism.

3.3 Determining the Contribution of Variables in Predicting Hypothyroidism

To determine the relations between an individual input parameter of the designed neural network and dependent output which inform about disease diagnosis, the authors decided to estimate importance of each input variable (age, sex, TSH, T3, TT4, T4U or FTI) with the use of two independent methods: Garson's algorithm and sensitivity analysis.

3.3.1 Garson's Algorithm

This easy in application method was described in study [13] as "Connection weights method", where it was validated on one database and proven as method equivalent to the other methods, known in literature, focusing on extracting the contribution of independent variables in neural networks. This method sums the products of connection weight between the input neuron of assessed variable and hidden neuron, and connection weight between hidden neuron and the output neuron, according to the formula [13]:

$$R_{ij} = \sum_{k=1}^{H} W_{ik} \cdot W_{kj}, \tag{1}$$

where:

- R_{ij} - relative importance of the input variable i in reference to output neuron j,
- H - number of neurons in the hidden layer, $H = 1, \ldots, k$,
- W_{ik} - connection weight between input neuron i and the hidden neuron k,
- W_{kj} - connection weight between hidden neuron k and the output neuron j.

These products are the basis of calculated variable importance in the neural network, what is described in details in [12]. To calculate relative importance of the parameters analyzed in this study, the authors adjusted above mentioned formula (1) to the neural network architecture considered as the best fit to diagnose hypothyroidisms (MLP 8-4-2 with 99.59% of correct % predictions), so the R_{ij} was estimated separately for output neuron indicating hypothyroid and euthyroid case. Calculations were performed with the use of script created in Matlab software.

3.3.2 Sensitivity Analysis

To compare results achieved from Garson's algorithm, sensitivity analysis in Statistica software was performed. According to the information provided by the software manufacturer, in sensitivity analysis the importance of the individual variable is estimated based on quotient of the neural network error obtained for the dataset without one (analyzed) variable to the error obtained with the full set of variables, so in this type of assessment bigger error for one variable (in comparison to the other analyzed variables) indicates higher influence of this variable on neural network decisive process. If the quotient of errors for one of the variables is less or equal 1, then removal of this variable does not influence or even improve the performance of the neural network.

Sensitivity was calculated for all input parameters selected to the study, simultaneously for the training, validation and test samples.

4 Results

As it was described in the previous section, created neural network models achieved very high performance in predicting hypothyroidism (98.34% and 99.59% what is shown in Table 3) and the importance of the parameters used

Table 4. Axon connection weights between input and hidden layer for the neural network MLP 8-4-2 with the best performance of the training, validation and test set and calculated for this network relative importance of the parameters being analyzed.

Predictor variable, i	No. of hidden neuron k and the connection weight between parameter i and hidden neuron k, W_{ik}				Relative importance of predictor variable	Error calculated from sensitivity analysis
	1	2	3	4		
1. Age	6.43	−0.57	4.19	−13.54	11.80%	3.43
2. TSH	70.90	−41.81	59.65	−5.70	48.85%	15.50
3. T3	5.52	−3.98	5.80	−1.31	4.92%	2.21
4. TT4	−15.31	12.68	−13.07	6.88	15.62%	8.70
5. T4U	4.12	−1.43	2.44	−0.32	2.24%	2.11
6. FTI	−8.92	11.17	−11.85	−8.48	14.46%	5.29
7. Female sex	−1.36	−1.59	−0.18	0.29	1.08%	3.25
8. Male sex	0.03	−1.39	0.51	−0.61	1.02%	3.25

as input values of the neural networks was assessed with the use of Garson's algorithm and sensitivity analysis, calculated on model with the highest performance in training, test and validation set. Both applied methods congruently indicated, that the biggest impact on correct prediction have four parameters in following order: TSH, TT4, FTI and age (Table 4).

In the last step the authors created the neural network model with only four the most relevant input parameters (TSH, TT4, FTI and age), according analysis presented in Table 4. It showed that the reduction of input parameters into the most significant four parameters (with previous applied model architecture), only slightly influenced the neural network model performance (decrease in performance from 100% to 96.69% in euthyroid test set) (Table 5).

Table 5. Comparison of performance achieved for neural network with all (n = 8) and reduced (n = 4) number of input parameters.

No.	Net architecture	Training perf.	Test perf.	Validation perf.	Correct predictions
1	MLP 8-4-2	100%	97.22%	100%	99.59% (100% in euthyroid and 99.17 in hypothyroid samples)
2	MLP 4-4-2	97.65%	97.22%	100%	97.93% (96.69% in euthyroid and 99.17 in hypothyroid samples)

5 Conclusions

The results of calculations presented in this study showed, that three serum blood parameters, that is TSH, TT4, FTI levels in reference to age, have the biggest influence on correct diagnosis of hypothyroidism. The analysis was performed on real dataset, however the data were obtained from machine learning repository, so the results should be confirmed in prospective, controlled study with large dataset. However, methodology used in this study (restrictive choice of dataset, searching for the best neural network model with the best set of connection weights, estimation of input parameter importance in decisive process) can be used in the future studies concerning application of neural networks as diagnostic or predictive model. Apart from this fact, it has been proved, that neural network models can be very efficient tools (even 99.59% of accuracy) in helping physicians to diagnose hypothyroidism. The big advantage of presented method is fact, that it was clearly indicated which of the analyzed input parameters were necessary to build diagnostic model with high effectiveness (the most significant occurred to be TSH level, then TT4, FTI and age). Created neural network models were not anymore treated like a "black box" without knowledge about its internal structure. In medical applications it is very important, because of at least two reasons: it helps to reduce the number of diagnostic measurements or tests required to diagnose a patient (less input parameters to provide) and it may serves as a lead in planning the treatment strategy. So knowledge about input parameters important from the point of view of classification task, might be helpful in reduction of time and costs in healthcare, thanks to reduction of necessary medical diagnostic procedures and in creation of disease models in order to provide better treatment strategy.

Acknoweledgements. This study was conducted with the use of database from UCI Machine Learning Repository donated by Garavan Institute in Sydney, Australia [11].

Presented research results were funded with the grant 02/21/DSMK/3529 and 02/21/DSPB/3513 allocated by the Ministry of Science and Higher Education in Poland.

References

1. Franklyn, J., Shephard, M.: Evaluation of thyroid function in health and disease. In: De Groot, L., Chrousos, G., Dungan, K., et al. (eds.) Endotext, South Dartmouth (MA) (2000)
2. Pellitteri, P.K., Ing, S., Jameson, B.: Disorders of the thyroid gland. In: Flint, P., Haughey, B.H., Lund, V., et al. (eds.) Cummings Otolaryngology, 6th edn, pp. 1884.e3–1900.e3. Elsevier Inc. (2015)
3. Arbib, N., Hadar, E., Sneh-Arbib, O., et al.: First trimester thyroid stimulating hormone as an independent risk factor for adverse pregnancy outcome. J. Matern Neonatal Med. **30**, 2174–2178 (2017)
4. Tadeusiewicz, R., Korbicz, J., Rutkowski, L., Duch, W.: Sieci neuronowe w inzynierii biomedycznej. Tom 9. ang. Neural Networks in Biomedical Engineering, vol. 9. Akademicka Oficyna Wydawnicza EXIT, Warszawa (2013)
5. Michałowska, M., Walczak, T., Grabski, J.K., Grygorowicz, M.: Artificial neural networks in knee injury risk evaluation among professional football players. In: AIP Conference Proceedings, Lublin, p. 70002 (2018)
6. Walczak, T., Grabski, J., Grajewska, M., Michałowska, M.: Application of artificial neural networks in man's gait recognition. In: Advances in Mechanics: Theoretical, Computational and Interdisciplinary Issues, pp. 591–594. CRC Press (2016)
7. Grabski, J.K., Walczak, T., Michałowska, M., Cieslak, M.: Gender recognition using artificial neural networks and data coming from force plates. In: Gzik, et al., M. (eds.) Innovations in Biomedical Engineering, IBE 2017, Advances in Intelligent Systems and Computing, vol. 623, pp. 53–60. Springer, Cham (2018)
8. Zhang, G., Berardi, V.L.: An investigation of neural networks in thyroid function diagnosis. Health Care Manag. Sci. **1**, 29–37 (1998)
9. Temurtas, F.: A comparative study on thyroid disease diagnosis using neural networks. Expert Syst. Appl. **36**, 944–949 (2009)
10. Ozyilmz, L., Yildirim, T.: Diagnosis of thyroid disease using artificial neural network methods. In: Proceedings of the 9th International Conference Neural Information Process, vol. 4, pp. 2033–2036 (2002)
11. Dua, D., Karra Taniskidou, E.: UCI Machine Learning Repository. University of California, School of Information and Computer Science, Irvine (2017). http://archive.ics.uci.edu/ml. Accessed 30 Mar 2018

12. Olden, J.D., Jackson, D.A.: Illuminating the "black box": a randomization approach for understanding variable contributions in artificial neural networks. Ecol. Model. **154**, 135–150 (2002)
13. de Oña, J., Garrido, C.: Extracting the contribution of independent variables in neural network models: a new approach to handle instability. Neural Comput. Appl. **25**, 859–869 (2014)
14. Soldin, O.P., Soldin, D., Sastoque, M.: Gestation-specific thyroxine and thyroid stimulating hormone levels in the United States and worldwide. Ther. Drug Monit. **29**, 553–559 (2007)

Mobile Test of Manual Dexterity in the Diagnostics of Frail Geriatric Patients – Pilot Study

Piotr Seiffert[1](✉), Jacek Kawa[2], Czesław Marcisz[3], Paula Stępień[2], Małgorzata Czernek[1], Marcin Bugdol[2], Wojciech Kapko[1], and Jarosław Derejczyk[1]

[1] John Paul II Geriatric Hospital, Morawa 31, 40-001 Katowice, Poland
piotrseiffert@gmail.com,
{malgorzata.czernek,jaroslaw.derejczyk}@emc-sa.pl
[2] Faculty of Biomedical Engineering, Silesian University of Technology, Roosevelta 40, 41-800 Zabrze, Poland
{jacek.kawa,paula.stepien,marcin.bugdol}@polsl.pl
[3] Department of Gerontology and Geriatric Nursing, School of Health Sciences, Medical University of Silesia, Ziołowa 45/47, 40-635 Katowice, Poland
klinwewtychy@poczta.onet.pl

Abstract. Mortality rate increases exponentially with people's age. They gather age-related diseases and become 'frail', i.e. increasingly vulnerable to various stressors. The gold standard to evaluate the severity of upper limb motor symptoms is to use the UPDRS-part III (motor examination) yet it is proven that simple tests of manual dexterity can identify persons at high risk for neurodegenerative diseases.

The aim of this study is to evaluate a novel, objective, observer-independent and assistance-independent tablet device-based method which could be helpful in the upper limb function diagnostics designed for telemedicine technology and frailty assessment. The test employs a tablet and a fixed-height obstacle. The patient is required to move his/her hand above the obstacle and touch the round fields displayed on the tablet. The test is performed for 30 s, as fast as possible, using pointing finger of the dominating hand. Several parameters are registered using the tablet.

In the study, the test is evaluated in the group inpatients of geriatric hospital separated into Frail (14 patients) and Control (14 patients) groups. The patients in the Control group are featuring (Mann-Whitney U-test, p-value < 0.05) more correct touches (23.4 ± 8.2 vs. 17.1 ± 12.4) than the members of the Frail group. The reaction time and mean time between touches in Control group is shorter than in the Frail group (respectively Mann-Whitney U-test and Student's t-test; p-value < 0.05).

Keywords: Telegeriatrics · Frailty · Mobile assessment · Manual dexterity test

E. Tkacz et al. (Eds.): IBE 2018, AISC 925, pp. 47–54, 2019.
https://doi.org/10.1007/978-3-030-15472-1_6

1 Introduction

Mortality rate increases exponentially with people's age [6] but individual persons do not die from becoming old itself. They gather age-related diseases and become increasingly vulnerable to various external and internal stressors which may result in death [10]. To describe a state of such vulnerability the 'frailty' term was established and has become one of the major problems in contemporary geriatrics. Frailty is associated with the greater risk of disability, institutionalization and mortality [16]. The etiology of this state is multifactorial and includes frail immune system [13] or lower brain dopamine level [12]. Frailty phenotype can be diagnosed with the use of Fried Criteria (weight loss, handgrip strength, gait speed, exhaustion, physical activity) [5].

Functional status disorders of the geriatric patient depend, among other things, on physical condition and can be easily detected using popular scales such as the Activities of Daily Living scale by Katz [8]. Mobility of the older individuals is fundamental to active aging and maintaining fine functional status [15] and is assessed on Geriatric Wards by Comprehensive Geriatric Assessment [3]. Limb motor skills are also at highest importance in maintaining independence and their alterations can be observed in many diseases associated with aging, for example in Parkinson's disease [11]. The gold standard to evaluate the severity of upper limb motor symptoms is the UPDRS-part III (motor examination) which is time-consuming and requires considerable clinical experience [14]. To assess functions of the upper limbs in other diseases specialized tests were also introduced. Dexterity of the hands can be evaluated and compared for example by Box and Block Test [9] or Nine Hole Peg Test [7], but usually experienced staff assistance and specialized tools are obligatory for the comprehensive diagnosis. Still, it is proven that simple tests of manual dexterity such as Purdue Pegboard Test can identify persons at high risk of neurodegenerative diseases in community [1].

A simple test which can be easily performed in domestic conditions and does not require assistance of the professionals is needed to increase the availability of the upper limb function diagnostics. The aim of this study was to develop and evaluate an objective, observer-independent and assistance-independent tablet device-based method which could be helpful in the upper limb function diagnostics designed for telemedicine technology and frailty assessment. This paper presents the development and evaluation process of the Mobile Test of Manual Dexterity with the special interest in frail geriatric patients diagnostics.

The proposed test is based on the design of Derejczyk et al. [2]. In their approach, two round fields are presented to the patient, separated by a wall-like obstacle with regulated height. The patient must touch both fields alternately, as fast as possible during 30 s. The supervising person evaluates the test by counting the number of correct (left/right field was touched after the right/left one and the hand was moved above the obstacle) and incorrect iterations.

In our study, the tablet device is used to display the fields and count the correct and incorrect touches as well as to calculate additional parameters (reaction time, pause between touches). The XPS (extruded polystyrene) foam obstacle placed on the tablet surface has the fixed-height of 15 cm.

2 Materials and Methods

2.1 Material

A group of 14 frail and 14 control (Table 1) geriatric inpatients (age > 60 years old) was included in the study (23 women, 5 men). The exclusion criteria were: previous diagnosis/treatment of PD (Parkinson's Disease), acute diseases or aggravation of chronic diseases, uncorrectable sight or hearing deficits, Mini-Mental Scale Examination [4] result above 20 points, neuroleptic treatment.

Frailty criteria [5] were assessed by experts. Each patient was evaluated either as Frail (≥3 pts) or Control (less than 3 pts).

To all patients the goal of the study was explained and they gave their consent. The work was a part of the 'Study of selected geriatric syndromes' and was approved by the Bioethics Committee of the Academy of Physical Education in Katowice.

Table 1. Control and Frail groups. Parameters statistically different in both groups are marked using double asterisk ** (Mann-Whitney U-test, p-value < 0.05) or single asterisk * (Student's t-test, p-value < 0.05)

Parameter	Control			Frail		
	Mean ± Std	Min	Max	Mean ± Std	Min	Max
Age (yrs)**	79 ± 7	70	97	83 ± 7	62	93
Weight (kg)*	77.9 ± 14.5	60	107	66.9 ± 16.9	37	102
Height (cm)	156.8 ± 6.9	145	168	155.6 ± 8.4	144	169
BMI (kg/cm^2)	32.0 ± 7.2	23.7	44.2	27.6 ± 6.6	17.1	38.9
Force (kgf)	17.1 ± 6.5	2.0	28.0	17.0 ± 7.6	2.0	32.0
Speed of walk (m/s)**	1.1 ± 0.3	0.7	1.7	0.6 ± 0.3	0.3	1.2
Fried (pts)**	1.4 ± 0.8	0	2	3.6 ± 0.6	3	5
MMSE (pts)**	27.2 ± 1.9	22	29	25.6 ± 2.2	22	29

2.2 Methods

In this section, the Manual Dexterity Test is first presented. Next, the evaluation procedure is given.

2.2.1 Manual Dexterity Test

The Manual Dexterity Tester has been designed as a system of two complementary elements: (1) physical obstacle, superimposed over a tablet screen and (2) software application, displaying two fields on the tablet screen and acquiring the measurements.

The wall-like obstacle is placed over the tablet in the middle of the longer side of the screen (separating the screen into two halves). The height of the obstacle is 15 cm from the surface of the tablet screen.

The round, star-like, orange fields of 5 cm diameter are displayed by the application at a fixed distance (center-to-center) of 15 cm.

During the test the patient is required to take a firm sit with his/hers dominate hand laying freely on the tablet screen. The tablet itself should be placed on a table, desk or other hard surface, at typical height comfortable for writing. Once the test application runs, the countdown starts and a signal to start the test is given: the pointing finger of a dominating hand should be used to touch the round fields visible on the tablet screen, alternately. Between touches, the hand should be moved over the obstacle. The actual test takes 30 s.

The touch is assessed as correct by the application when it is made within the displayed field and is either a first touch in the test or was preceded by a touch (either correct or incorrect) made on the other side of the tablet. Time between the end of counting down and the first touch (either correct or incorrect) is measured as a reaction time. Mean time between touches (either correct or incorrect) is also registered.

Mobile Test of Manual Dexterity is performed in a sitting position to stabilize the patient body and ensure selective movement of only upper limb and also to avoid patient fatigue or falls as this is a high risk group. The movement which is enforced by the protocol, requires the involvement of a large group of muscles. Otherwise, in case of the lack of the obstacle, the induced motion could involve only a small amount of muscle (mainly of the shoulder girdle). What is more, the precise move of the finger is required to correctly perform the Mobile Test of Manual Dexterity.

Brain activation patterns during movements of the upper limb (especially the fingers) involve a huge part of the whole body motor cortex representation in the brain. Authors believe that the information gathered using the presented examination protocol reflects the activation of vast parts of the central nervous system.

2.2.2 Evaluation Procedure

In this study, a Lenovo Yoga 2 1051L tablet with the Windows 10 OS was used. The tablet was horizontally placed over the custom-made, XPS (extruded polystyrene) foam basis. The obstacle was 15 cm high (over the surface of the screen), 2 cm thick, and fixed to the basis (Fig. 1).

During the procedure, the patient was asked to sit firmly on the chair, behind the desk. He/she was told to keep the dominant hand free (as for writing) and find a comfortable position. Next, the test application was started by the supervisor and the patient was given the test instructions regarding available time (30 s) and the procedure (two orange fields visible on the table screen should be touched alternately as fast as possible, using pointing finger of the dominant hand). To each patient a one-time, 10 s long training session was given.

The registered datasets were analyzed statistically. First, the distribution of the parameters was analyzed using Shapiro-Wilk test. Normally distributed features with comparable variances (two-sample F-test) were further analyzed employing Student t-test. Remaining parameters were analyzed using Mann-Whitney U-test.

Fig. 1. Tablet with the XPS obstacle

3 Results and Discussion

During the evaluation, 28 datasets were correctly acquired. Each set contained four basic parameters: reaction time (s), number of correct (OK) and incorrect touches (NOK), mean pause between touches (s). Summary of the results for each group (Frail and Control) is presented in the Table 2 and in the Fig. 2.

Three of four registered parameters were statistically different in both analyzed groups. The patients in the Control group featured more correct touches (23.4 ± 8.2 vs. 17.1 ± 12.4) in the 30 s test window than in the Frail group. The number of incorrect touches did not differentiate the groups (Mann-Whitney U-test, $p > 0.05$).

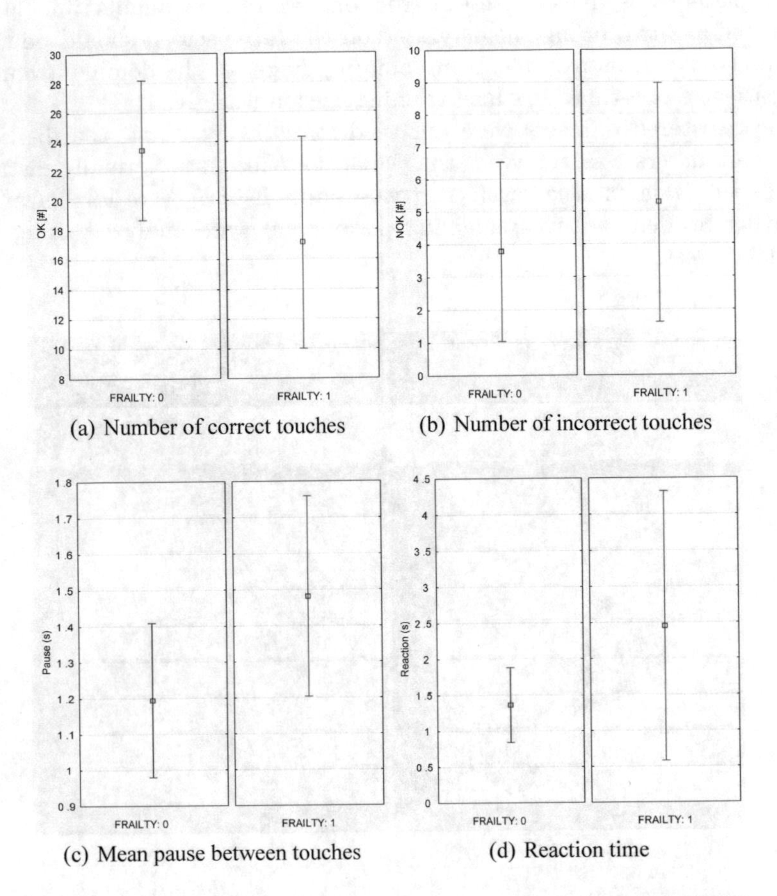

(a) Number of correct touches

(b) Number of incorrect touches

(c) Mean pause between touches

(d) Reaction time

Fig. 2. Results summary including mean ± confidence interval (95% confidence level)

Table 2. Results for Control and Frail groups. Parameters statistically different in both groups are marked using double asterisk ** (Mann-Whitney U-test, p-value < 0.05) or single asterisk * (Student's t-test, p-value < 0.05). *OK* and *NOK* stand for the number of correct and incorrect touches, respectively. *Pause* denotes mean pause between touches, and *Reaction* is the reaction time (in seconds)

Parameter	Control					Frail					p-value
	Mean ± Std	Med	Iqr	Min	Max	Mean ± Std	Med	Iqr	Min	Max	
Pause (s)*	1.19 ± 0.37	1.16	0.46	0.61	2.08	1.48 ± 0.48	1.66	0.67	0.61	2.13	0.045
OK (#)**	23.4 ± 8.2	23.5	15.0	11	39	17.1 ± 12.4	14.5	8.0	3	41	0.046
NOK (#)	3.8 ± 4.8	2.5	6.0	0	18	5.3 ± 6.4	2.0	8.0	0	17	0.426
Reaction (s)**	1.36 ± 0.9	1.0	1.5	0.3	2.8	2.5 ± 3.2	1.5	1.6	0.2	12.6	0.323

Moreover, the reaction time was significantly shorter in the Control group (1.36 ± 0.9 s) than in the Frail group (2.5 ± 3.2 s) as was the mean pause between touches (1.19 ± 0.37 s vs. 1.48 ± 0.48 s).

The obtained results confirm the assumptions that the Mobile Test of Manual Dexterity results differ depending on the frailty status. The observation probably reflects the fact that the frailty phenotype is assessed based inter alia on the physical condition of the patient and includes criteria of lowered grip strength or slow walking speed. Our study encourages attempts to replace previously used methods by simple tests that can be carried out under telemedicine conditions, such as Mobile Test of Manual Dexterity. What is more, one of the biggest advantages of our test is the ability to perform it also by patients who are not able to walk (what is often observed in the geriatric population).

The main limitation of the study is the small diversity of the subjects' overall physical conditions in both analyzed groups. More specifically, the participants assessed as Frail and Control, were all inpatients of the geriatric hospital, supposedly during diagnostic or undergoing treatment. Moreover, the specificity of the findings may be limited: the elongated reaction time or lower number of correct touches may not be restricted to the frail patients only.

4 Conclusion

In this study a novel test of manual dexterity is presented. The test employs a tablet and a fixed-height obstacle. The patient is required to move his/her hand above the obstacle and touch the round fields displayed on the tablet. The test is performed for 30 s, as fast as possible, using pointing finger of the dominating hand. During the test, a number of parameters is acquired, including: number of correct and incorrect touches, reaction time and mean time between touches.

Three of four parameters (number of correct touches, reaction time and mean time between pauses) differentiate the Frail and the Control group.

More research is needed to assess the specificity of the test and confirm the results in a larger group.

References

1. Darweesh, S., Wolters, F.J., Hofman, A., Stricker, B., Koudstaal, P.J., Ikram, M.A.: Simple test of manual dexterity can identify persons at high risk for neurodegenerative diseases in the community. Alzheimer's Dement. **12**(7), P585 (2016). https://doi.org/10.1016/j.jalz.2016.06.1148
2. Derejczyk, J., Płoska, M., Kawa, J.: Tester do kontroli sprawności psychofizycznej człowieka (Psychophysical Activity Tester) (2016). https://grab.uprp.pl/sites/WynalazkiWzoryUzytkowe/Opisy/Patenty%20i%20Wzory%20uytkowe/228963_B1.pdf. PL Patent 228963

3. Ellis, G., Gardner, M., Tsiachristas, A., Langhorne, P., Burke, O., Harwood, R.H., Conroy, S.P., Kircher, T., Somme, D., Saltvedt, I., Wald, H., O'Neill, D., Robinson, D., Shepperd, S.: Comprehensive geriatric assessment for older adults admitted to hospital. Cochrane Database Syst. Rev. (2017). https://doi.org/10.1002/14651858.cd006211.pub3
4. Folstein, M.F., Folstein, S.E., McHugh, P.R.: Mini-mental state. J. Psychiatr. Res. **12**(3), 189–198 (1975). https://doi.org/10.1016/0022-3956(75)90026-6
5. Fried, L.P., Tangen, C.M., Walston, J., Newman, A.B., Hirsch, C., Gottdiener, J., Seeman, T., Tracy, R., Kop, W.J., Burke, G., McBurnie, M.A.: Frailty in older adults: evidence for a phenotype. J. Gerontol.: Ser. A **56**(3), M146 (2001). https://doi.org/10.1093/gerona/56.3.M146
6. Gompertz, B.: On the nature of the function expressive of the law of human mortality, and on a new mode of determining the value of life contingencies. Philos. Trans. R. Soc. Lond. **115**, 513–583 (1825). https://doi.org/10.1098/rstl.1825.0026
7. Grice, K.O., Vogel, K.A., Le, V., Mitchell, A.W., Muniz, S., Vollmer, M.A.: Adult norms for a commercially available nine hole peg test for finger dexterity. Am. J. Occup. Ther.: Off. Publ. Am. Occup. Ther. Assoc. **57**(5), 570–3 (2003)
8. Hartigan, I.: A comparative review of the Katz ADL and the Barthel Index in assessing the activities of daily living of older people. Int. J. Older People Nurs. **2**(3), 204–212 (2007). https://doi.org/10.1111/j.1748-3743.2007.00074.x
9. Mathiowetz, V., Volland, G., Kashman, N., Weber, K.: Adult norms for the box and block test of manual dexterity. Am. J. Occup. Ther. **39**(6), 386–391 (1985). https://doi.org/10.5014/ajot.39.6.386
10. Mitnitski, A.B., Rutenberg, A.D., Farrell, S., Rockwood, K.: Aging, frailty and complex networks. Biogerontology **18**(4), 433–446 (2017). https://doi.org/10.1007/s10522-017-9684-x
11. Postuma, R.B., Berg, D., Stern, M., Poewe, W., Olanow, C.W., Oertel, W., Obeso, J., Marek, K., Litvan, I., Lang, A.E., Halliday, G., Goetz, C.G., Gasser, T., Dubois, B., Chan, P., Bloem, B.R., Adler, C.H., Deuschl, G.: MDS clinical diagnostic criteria for Parkinson's disease. Mov. Disord. **30**(12), 1591–1601 (2015). https://doi.org/10.1002/mds.26424
12. Seiffert, P., Derejczyk, J., Kawa, J., Marcisz, C., Czernek, M., Szymszal, J., Kapko, W., Bugdol, M., Torbus, A., Stępień-Wyrobiec, O.: Frailty phenotype and the role of levodopa challenge test in geriatric inpatients with mild Parkinsonian signs. Biogerontology **18**(4), 641–650 (2017). https://doi.org/10.1007/s10522-017-9716-6
13. Semba, R.D., Nicklett, E.J., Ferrucci, L.: Does accumulation of advanced glycation end products contribute to the aging phenotype? J. Gerontol. Ser. A: Biol. Sci. Med. Sci. **65A**(9), 963–975 (2010). https://doi.org/10.1093/gerona/glq074
14. Tavares, A.L.T., Jefferis, G.S., Koop, M., Hill, B.C., Hastie, T., Heit, G., Bronte-Stewart, H.M.: Quantitative measurements of alternating finger tapping in Parkinson's disease correlate with UPDRS motor disability and reveal the improvement in fine motor control from medication and deep brain stimulation. Mov. Disord. **20**(10), 1286–1298 (2005). https://doi.org/10.1002/mds.20556
15. Webber, S.C., Porter, M.M., Menec, V.H.: Mobility in older adults: a comprehensive framework. Gerontologist **50**(4), 443–450 (2010). https://doi.org/10.1093/geront/gnq013
16. Wen, Y.C., Chen, L.K., Hsiao, F.Y.: Predicting mortality and hospitalization of older adults by the multimorbidity frailty index. PLoS One **12**(11), e0187825 (2017). https://doi.org/10.1371/journal.pone.0187825

The Influence of the Normalisation of Spinal CT Images on the Significance of Textural Features in the Identification of Defects in the Spongy Tissue Structure

Róża Dzierżak[1(✉)], Zbigniew Omiotek[1], Ewaryst Tkacz[2], and Andrzej Kępa[3]

[1] Faculty of Electrical Engineering and Computer Science, Lublin University of Technology, Lublin, Poland
rozadzierzak@gmail.com

[2] Faculty of Biomedical Engineering, Silesian University of Technology, Zabrze, Poland

[3] Department of Radiology and Nuclear Medicine, Independent Public Clinical Hospital No. 4, Lublin, Poland

Abstract. The aim of the study was to determine the effect of normalisation of spinal CT images on the accuracy of automatic recognition of defects in the spongy tissue structure of the vertebrae on the thoracolumbar region. Feature descriptors were based on the grey-levels histogram, gradient matrix, run-length matrix, coocurrence matrix, autoregression model and wavelet transform. Six methods of feature selection were used: Fisher coefficient, minimisation of classification error probability and average correlation coefficients between chosen features, mutual information, Spearman correlation, heuristic identification of noisy variables, linear stepwise regression. Selection results were used to build 6 popular classifiers. The following values of individual classification quality factors were obtained (before normalisation/after normalisation): general accuracy of classification - 90%/82%, classification sensitivity - 89%/85%, classification specificity - 96%/82%, positive predictive value - 95%/95%, negative predictive value - 89%/84%. For the applied set of textural features, as well as the methods of selection and classification, image normalisation significantly worsened the accuracy of the automatic diagnosis of osteoporosis based on CT images of the spine. Therefore, it is necessary to use this operation with caution so as not to remove from the processed images information significant from the point of view of the purpose of the research.

Keywords: Osteoporosis · CT images · Image normalisation · Feature selection · Classification

1 Introduction

Bone tissue is an active tissue and is subject to the remodelling process throughout human life. The condition for its proper course is the balance between

E. Tkacz et al. (Eds.): IBE 2018, AISC 925, pp. 55–66, 2019.
https://doi.org/10.1007/978-3-030-15472-1_7

the excitation of osteoblast osteogenic activity and osteoclastogenesis and bone resorption. The disruption of this process leads to the loss of bone mass and the weakening of the spatial structure of the bone referred to as osteoporosis. This disease is metabolic, causing changes in the microarchitecture of the bone tissue consisting in the reduction of the number of bone beams in the spongy structure and a high risk of fractures. Although the overall bone mass of the entire skeleton decreases with age, some of its structures are particularly vulnerable. Highly risk areas are vertebrae, especially around the thoraco-lumbar section [1].

A commonly used diagnosis of osteoporosis is a densitometric examination involving the measurement of bone density. The X-ray absorptiometry method is used to study the entire skeleton or selected sites, particularly vulnerable to fractures. The bone mineral content determines the amount of minerals at the site of measurement, which divided by the surface area gives the bone mineral density (BMD). Therefore, this technique provides information on the correct mineral density of the entire study area, without the exact identification of the section where the defect occurs. The index corresponding to the standard deviation from the peak bone mass serves to qualify the patient as healthy or ill. Considering that it is a constant classification value of the indicator, there is a risk that it is not 100% adequate to each bone segment examined [2].

Due to doubts related to the correct and precise diagnostics of bone defects, new diagnostic solutions are still sought [3–6]. An opportunity to create an effective system is to use the spinal CT image results. Images from contemporary tomographs, in comparison with other medical imaging techniques, are characterised by high accuracy, high resolution and contrast. Thanks to the good quality of the obtained results, it is possible to analyse the texture of tissues [7]. Its aim is to find a set of parameters, called textural features, each of which is a numerical measure of a specific texture property [8]. To obtain the best information about the tissue under investigation, it is necessary to select appropriate pre-processing activities. This article presents the impact of the image normalisation process on changes in its significant textural features.

2 Materials and Methods

2.1 Material

The CT images of the thoraco-lumbar region from 13 healthy patients and 11 with osteoporosis were used for the analysis. From the series of images, cross-sections were selected on which the inside of the vertebra together with the spongy being is visible (Fig. 1).

The size of the separated samples was chosen so as to maximise the surface of the texture containing the potential information in the image of a vertebra with a cross-section. As a result, 168 samples with dimensions of 50×50 pixels were obtained, where 84 samples presented normal tissue structure and 84 samples showed defects in spongy tissue.

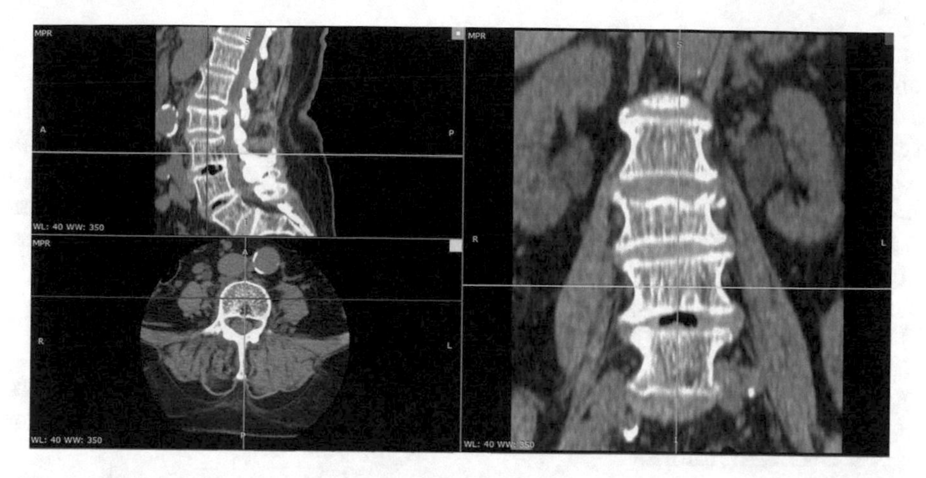

Fig. 1. Alignment of the axis in the centre of one of the vertebra (picture in three views).

2.2 Pre-processing

When analysing the texture of tissues, the pre-processing procedures should be limited in order not to distort the natural features of the images. Therefore, only the conversion from RGB mode to eight bit grey scale (two hundred and fifty six shades) was performed. The next step was to normalise the brightness range of the obtained samples with linear correction. The results are presented below (Fig. 2).

	Sample before image normalisation	Sample after image normalisation
Healthy tissue		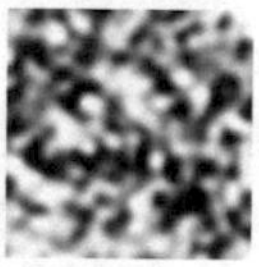
Tissue with cavities		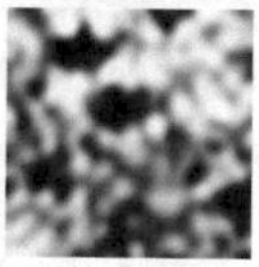

Fig. 2. Structure of spongy tissue before and after image normalisation.

2.3 Image Analysis

Image analysis was made with the MaZda program (version 4.6), developed at the Institute of Electronics of the Łódź University of Technology, whose author is Piotr M. Szczypiński [9]. This program has been made available free of charge on the Internet for scientific purposes. It allows to analyse grey cardboard images and determine the numerical values of image features. A detailed description of these features can be found in [10–13], as well as in the MaZda documentation. The advantage of the program is that in addition to the statistical approach to image analysis, it also uses a mathematical model (autoregression model) and a transformational approach (wavelet transform). The set of features has been obtained on the basis of:

- Histogram (9 features): histogram's mean, histogram's variance, histogram's skewness, histogram's kurtosis, percentiles 1%, 10%, 50%, 90% and 99%.
- Gradient (5 features): absolute gradient mean, absolute gradient variance, absolute gradient skewness, absolute gradient kurtosis, percentage of pixels with nonzero gradient.
- Run length matrix (5 features $\times$ 4 various directions): run length nonuniformity, grey level nonuniformity, long run emphasis, short run emphasis, fraction of image in runs.
- Coocurrence matrix (11 features $\times$ 4 various directions $\times$ 5 between-pixels distances) angular second moment, contrast, correlation, sum of squares, inverse difference moment, sum average, sum variance, sum entropy, entropy, difference variance, difference entropy.
- Autoregressive model (5 features): parameters 1, 2, 3, 4, standard deviation.
- Haar wavelet (24 features): wavelet energy (features are computed at 6 scales within 4 frequency bands LL, LH, HL, and HH).

The textural features obtained by the above methods have been used by the authors in previous studies, where they allowed for high classification accuracy of the thyroid ultrasound images [14].

2.4 Feature Selection

The results of image analysis were subjected to the process of reducing the dimension of feature space. The following selection methods were used [15–19]:

- Fisher coefficient (FC);
- minimisation of classification error probability and average correlation coefficients between chosen features (POE+ACC);
- mutual information (MI);
- Spearman correlation between chosen features and decision variable (SC);
- heuristic identification of noisy variables (HINoV);
- linear stepwise regression (LSR).

2.5 Classification

The complete set of data contained 168 observations, of which 84 were in the healthy category and 84 in the sick category. From this set, a training part (56 cases) and a test part (28 cases) were randomly selected. Thus, the training set was 2/3, and the test set was 1/3 of the full set. The training and testing procedure for classifiers was repeated 10 times. As the final result of the testing process, the average value from all tests was taken. This approach required the preparation of 10 training sets and 10 test sets for each data set. Determination of these sets was based on draws with returning. Six popular supervised learning

Table 1. Feature selection results.

Method	Not normalised images				Normalised images			
	Feature name	Fisher coeff.	Feature name	Fisher coeff.	Feature name	Fisher coeff.	Feature name	Fisher coeff.
FC	Perc.01%	3.2420	S(3,0)SumAverg	2.5486	S(0,5)SumAverg	0.5214	S(5,5)SumAverg	0.3893
	Perc.10%	2.9733	S(0,1)SumAverg	2.5422	S(0,4)SumAverg	0.5154	S(3,0)Entropy	0.3824
	Mean	2.5760	S(4,0)SumAverg	2.5408	S(0,3)SumAverg	0.4728	S(1,0)Entropy	0.3774
	S(1,0)SumAverg	2.5664	S(5,0)SumAverg	2.5340	S(5,0)Entropy	0.4369	S(4,0)Entropy	0.3722
	S(2,0)SumAverg	2.5567	S(1,-1)SumAverg	2.5303	S(0,2)SumAverg	0.3990	S(4,4)SumAverg	0.3709
	Feature name	POE+ACC coeff.	Feature name	POE+ACC coeff.	Feature name	POE+ACC coeff.	Feature name	POE+ACC coeff.
POE+ ACC	S(2,0)InvDfMom	0.4571	S(5,5)Entropy	0.5068	S(5,5)SumOfSqs	0.4731	S(0,5)SumAverg	0.5341
	Teta2	0.4824	S(1,0)Entropy	0.5072	S(4,4)SumEntrp	0.4866	S(1,-1)DifEntrp	0.5348
	S(5,0)SumAverg	0.4860	S(4,4)Entropy	0.5099	S(4,4)Entropy	0.5202	S(5,0)SumAverg	0.5380
	WavEnLL_s-1	0.4871	S(1,0)SumEntrp	0.5204	Perc.50%	0.5225	135dr_LngREmph	0.5416
	Mean	0.5043	WavEnLL_s-2	0.5952	S(0,3)Entropy	0.5252	Teta2	0.8869
	Feature name	Mutual inf. coeff.	Feature name	Mutual inf. coeff.	Feature name	Mutual inf. coeff.	Feature name	Mutual inf. coeff.
MI	WavEnLL_s-2	0.4937	S(0,1)SumAverg	0.4689	Teta2	0.1535	S(0,1)Correlat	0.1251
	S(1,1)SumAverg	0.4692	S(3,0)SumAverg	0.4686	S(0,1)DifEntrp	0.1352	S(3,3)SumOfSqs	0.1231
	S(1,-1)SumAverg	0.4692	Mean	0.4683	S(1,1)SumAverg	0.1345	S(2,-2)SumAverg	0.1224
	S(0,2)SumAverg	0.4692	S(1,0)SumAverg	0.4683	S(3,-3)SumAverg	0.1286	S(0,3)InvDfMom	0.1216
	S(0,3)SumAverg	0.4692	S(2,0)SumAverg	0.4683	S(4,-4)SumAverg	0.1262	S(0,5)SumAverg	0.1184
	Feature name	Spearman corr. coeff.	Feature name	Spearman corr. coeff.	Feature name	Spearman corr. coeff.	Feature name	Spearman corr. coeff.
SC	Perc.01%	-0.664066	S(2,0)SumAverg	-0.612954	S(1,0)AngScMom	-0.363675	S(1,0)Entropy	0.300463
	Perc.10%	-0.648436	S(4,0)SumAverg	-0.612709	S(2,0)AngScMom	-0.324153	S(2,0)Entropy	0.295063
	WavEnLL_s-2	-0.618355	S(3,0)SumAverg	-0.612709	S(0,4)SumAverg	0.317647	S(3,0)Entropy	0.292362
	Mean	-0.613936	WavEnLL_s-1	-0.612463	S(5,0)Entropy	0.311264	S(4,0)Entropy	0.292117
	S(1,0)SumAverg	-0.612954	S(0,1)SumAverg	-0.612218	S(0,5)SumAverg	0.307091	S(0,2)SumAverg	0.284631
	Feature name	Corrected Rand index	Feature name	Corrected Rand index	Feature name	Corrected Rand index	Feature name	Corrected Rand index
HINoV	S(4,4)Entropy	98.1598	S(2,-2)SumOfSqs	96.5498	S(0,4)DifEntrp	62.9893	S(0,3)DifEntrp	59.7709
	S(5,-5)Entropy	97.0083	S(0,1)SumOfSqs	96.4397	S(1,1)DifEntrp	61.8615	S(0,3)Correlat	59.6221
	S(4,4)AngScMom	96.9330	S(2,2)SumOfSqs	96.4305	S(2,2)DifEntrp	61.7884	S(3,0)Correlat	58.9342
	S(3,3)Entropy	96.8439	S(0,3)SumOfSqs	96.4305	GrMean	61.1595	S(2,2)DifVarnc	58.8955
	S(3,-3)Entropy	96.7139	S(3,3)SumOfSqs	96.4305	GrVariance	60.6692	S(2,2)Contrast	58.7767
	Feature name	p-value	Feature name	p-value	Feature name	p-value	Feature name	p-value
LSR	S(0,1)InvDfMom	5.1987e-09	S(4,4)InvDfMom	0.0483	S(1,0)DifVarnc	2.1316e-11	S(2,-2)DifVarnc	1.0875e-05
	Mean	7.0798e-07	S(5,5)InvDfMom	0.0736	S(1,1)SumEntrp	9.7742e-07	S(0,3)DifVarnc	8.4893e-05
	S(0,3)SumAverg	2.5439e-06			S(1,1)DifEntrp	1.4112e-06	S(0,3)InvDfMom	0.0939
	Teta2	4.9077e-06			S(0,5)SumAverg	3.3761e-06		

The symbols of the feature selection methods: FC - Fisher coefficient; POE+ACC - minimisation of classification error probability and average correlation coefficients between chosen features; MI - mutual information; SC - Spearman correlation between chosen features and decision variable; HINoV - heuristic identification of noisy variables; LSR - linear stepwise regression.

Table 2. Classification results.

Feature selection method	Not normalised images						Normalised images					
	Classifier	ACC	TPR	TNR	PPV	NPV	Classifier	ACC	TPR	TNR	PPV	NPV
FC	LDA	83.21	81.43	85.00	85.12	82.64	LDA	63.39	67.14	59.64	62.85	65.90
	QDA	82.32	**85.00**	79.64	81.19	**84.91**	QDA	**65.54**	68.57	62.50	64.76	66.74
	NBC	78.75	78.21	79.29	80.54	78.91	NBC	65.00	**78.93**	51.07	61.92	**71.58**
	DT	85.00	76.43	93.57	93.00	80.44	DT	61.43	67.14	55.71	61.70	62.75
	9-NN	**85.54**	75.36	**95.71**	**95.13**	80.23	3-NN	65.18	53.57	**76.79**	**70.11**	62.35
	RF	82.68	79.29	86.07	85.64	81.29	RF	63.21	66.43	60.00	62.65	64.63
POE+ACC	LDA	85.71	80.36	91.07	90.49	82.47	LDA	68.04	72.14	63.93	67.76	69.56
	QDA	89.11	**89.29**	88.93	89.12	**89.50**	QDA	**71.25**	**79.64**	62.86	68.55	**75.60**
	NBC	86.07	81.79	90.36	89.77	83.50	NBC	66.25	76.07	56.43	63.81	70.62
	DT	83.75	76.79	90.71	90.99	80.12	DT	63.57	61.43	65.71	64.27	64.65
	5-NN	**90.54**	85.71	**95.36**	**95.29**	87.29	5-NN	69.29	57.86	**80.71**	**75.44**	65.96
	RF	88.04	82.50	93.57	93.20	84.66	RF	70.18	70.00	70.36	70.74	70.82
MI	LDA	83.93	77.50	90.36	88.98	80.32	LDA	70.36	76.07	64.64	68.81	73.35
	QDA	**84.29**	**80.71**	87.86	86.91	**82.26**	QDA	**72.50**	69.64	**75.36**	**73.87**	71.58
	NBC	78.75	75.36	82.14	81.62	77.09	NBC	66.07	**82.14**	50.00	62.28	74.76
	DT	81.79	68.21	95.36	93.74	75.33	DT	66.43	60.00	72.86	69.68	65.29
	9-NN	81.43	66.79	**96.07**	**94.87**	74.72	9-NN	71.07	71.07	71.07	71.18	71.98
	RF	79.11	76.79	81.43	81.37	78.44	RF	72.32	75.71	68.93	71.20	**75.83**
SC	LDA	85.36	83.21	87.50	87.24	83.99	LDA	62.14	63.57	60.71	62.11	63.03
	QDA	**88.04**	**85.36**	90.71	90.29	**86.23**	QDA	**65.00**	75.71	54.29	63.18	68.92
	NBC	80.71	80.36	81.07	81.08	80.66	NBC	62.68	**80.36**	45.00	59.45	**70.05**
	DT	83.39	75.36	91.43	91.16	79.34	DT	59.29	70.71	47.86	58.95	62.09
	6-NN	85.36	77.14	**93.57**	**92.76**	80.53	1-NN	62.68	64.64	60.71	62.21	63.91
	RF	84.82	81.43	88.21	87.64	82.89	RF	64.11	63.93	**64.29**	**64.53**	64.50
HINoV	LDA	**67.14**	65.36	**68.93**	**68.18**	**66.94**	LDA	69.11	69.64	68.57	69.25	69.63
	QDA	60.89	**68.93**	52.86	60.13	63.63	QDA	**71.79**	**75.36**	68.21	**70.52**	**73.85**
	NBC	63.93	55.36	72.50	66.74	62.29	NBC	52.86	47.14	58.57	54.53	51.73
	DT	58.21	56.79	59.64	60.84	57.97	DT	56.25	59.29	53.21	56.62	57.04
	3-NN	60.71	62.86	58.57	60.59	61.13	1-NN	69.64	68.57	**70.71**	70.12	70.17
	RF	57.68	61.07	54.29	57.71	58.06	RF	65.18	64.29	66.07	65.41	65.38
LSR	LDA	87.68	83.93	91.43	90.85	85.22	LDA	**81.61**	81.43	81.79	**82.29**	81.80
	QDA	**90.00**	87.50	92.50	92.14	88.34	QDA	79.46	**85.36**	73.57	76.60	**83.73**
	NBC	80.89	74.29	87.50	85.72	77.42	NBC	68.75	77.14	60.36	66.77	73.08
	DT	84.29	75.36	**93.21**	**92.87**	79.74	DT	63.93	62.50	65.36	64.72	63.87
	1-NN	89.11	**88.57**	89.64	89.87	**88.97**	4-NN	74.29	66.43	**82.14**	79.11	71.22
	RF	86.255	79.64	92.86	91.75	82.35	RF	73.04	72.14	73.93	73.23	73.09

The meaning of the classifier symbols used in Table 2 is as follows: LDA, QDA - linear and quadratic discriminant analysis; NBC - naive Bayes classifier; DT - decision tree; *K*-NN - *K*-nearest neighbours; RF - random forests. The symbols of the classification quality indices: ACC - overall classification accuracy; TPR - classification sensitivity; TNR - classification specificity; PPV - positive predictive value; NPV - negative predictive value. All numerical values are expressed in %.

methods [20–25] were used: linear and quadratic discriminant analysis (QDA, LDA), naive Bayes classifier (NBC), decision tree (DT), K-nearest neighbours (K-NN) random forests (RF). To assess the accuracy of classifiers, the following were used: overall classification accuracy (ACC), true positive rate (TPR, classification sensitivity), true negative rate (TNR, classification specificity), positive predictive value (PPV) and negative predictive value (NPV).

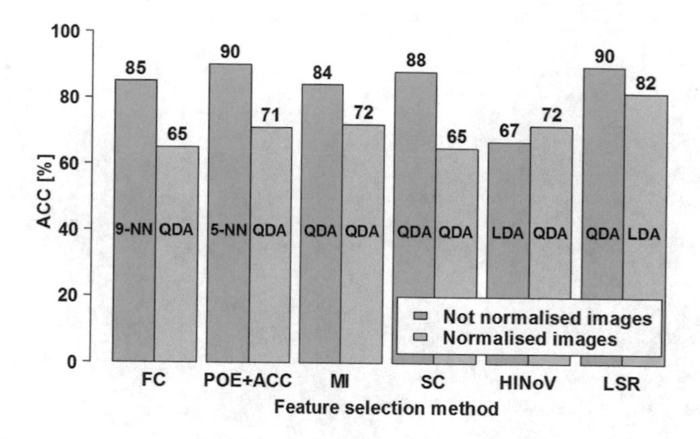

Fig. 3. Overall classification accuracy for individual feature selection methods.

3 Results

As a result of the image analysis, 290 descriptors of textural features were obtained. In the further part of the research, the dimension of the space of features was reduced. For this purpose, 6 selection methods were used, which are characterised in Subsect. 2.4. Table 1 presents the result obtained by each of the methods for the images before and after normalisation. Each of the first five methods of selection drew 10 features occupying the initial place in the ranking made according to its own coefficient. The sixth method, i.e. linear stepwise regression, returned 6 features for images before normalisation and 7 features for images after this operation. The data presented in Table 1 show that sets of features obtained by the same selection method are different for images before and after image normalisation. This is due to the fact that the operation mentioned changes the brightness of pixels, thus changing the results of the analysis.

The sets of features selected in the selection process were evaluated using six popular supervised classification methods. These were: linear and quadratic discriminant analysis, naive Bayes classifier, decision tree, K-nearest neighbours and random forests. The data sets needed for the construction and testing of classifiers were prepared as described in Sect. 2.5 Classification factors have been used to assess the quality of classification, which are often used in medical research (ACC, TPR, TNR, PPV, NPV). Detailed classification results are given in Table 2. For each method of feature selection, the bold font indicates the largest value of the appropriate classification quality factor. This allowed to indicate the classifier, which for the given selection method proved to be the best.

The best classification results in Table 2 are shown in bar charts (Figs. 3, 4 and 5). They present the values of individual indicators of the quality of classification, depending on the applied method of selection of features and image form (before and after normalisation). In addition, there is a symbol of the classification

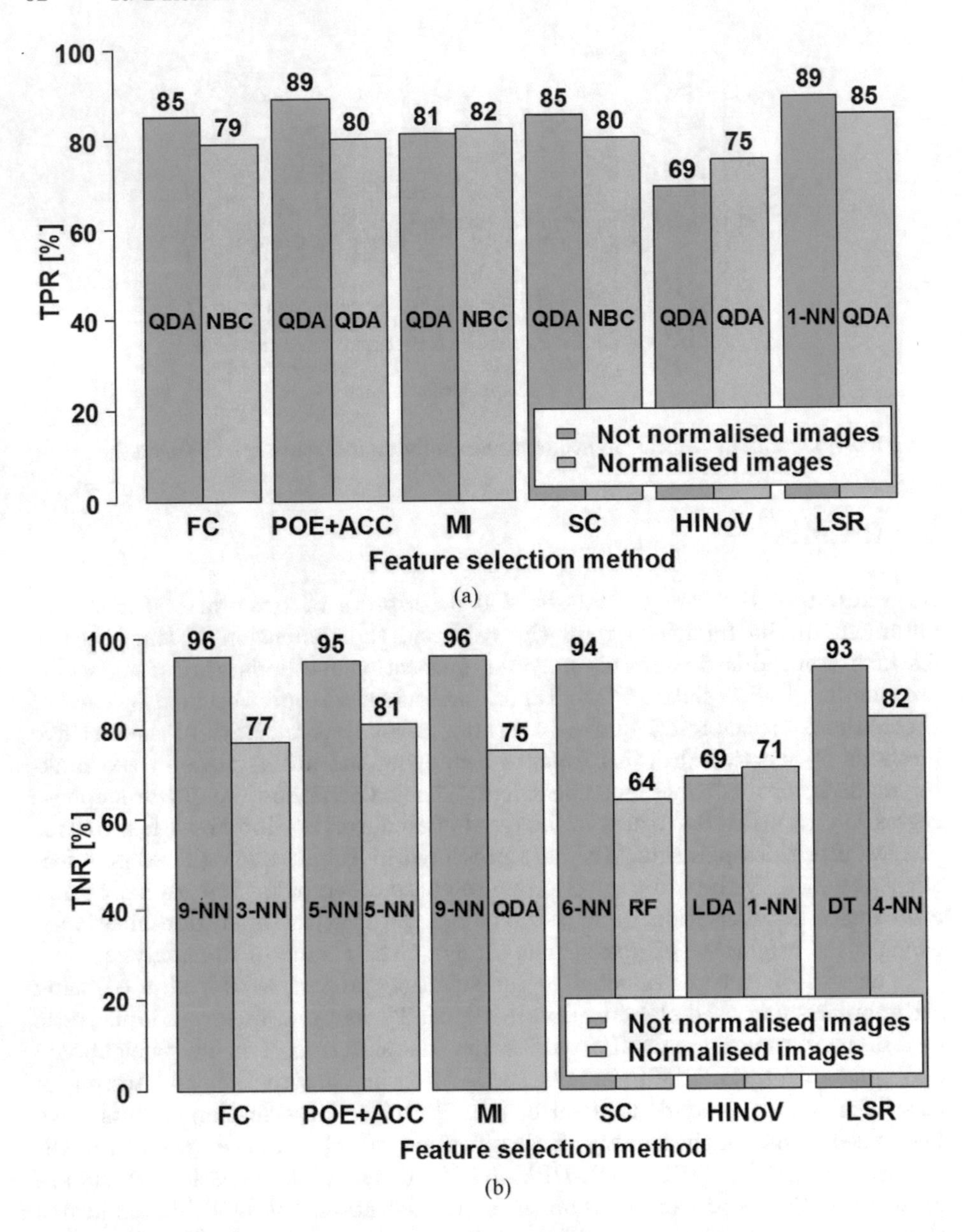

Fig. 4. Classification sensitivity (a) and specificity (b) for individual feature selection methods.

method in each bar graph, by means of which a given result was obtained. It should be noted that in almost all cases, a higher value of individual quality indicators was obtained for images without normalisation of brightness.

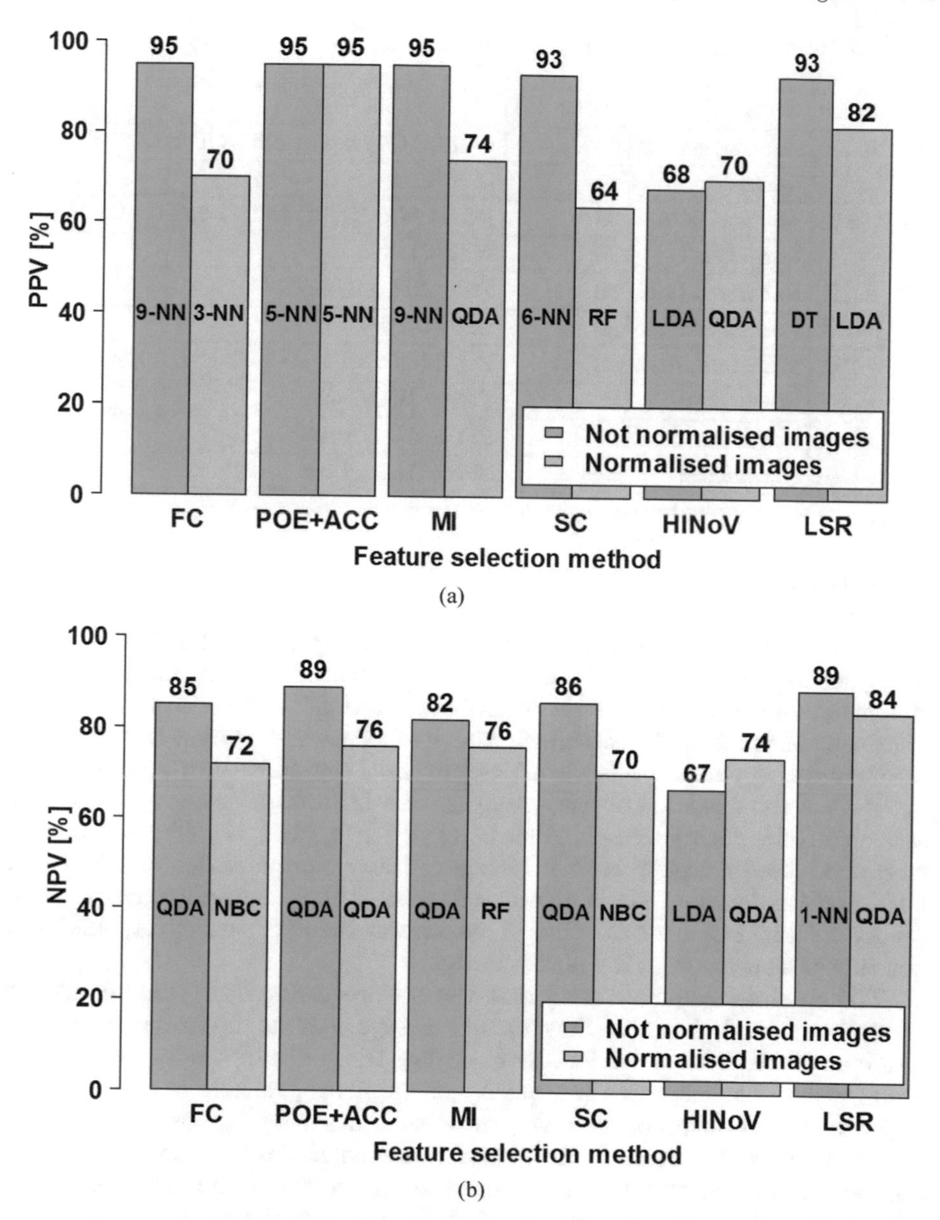

Fig. 5. Positive predictive value (a) and negative predictive value (b) for individual feature selection methods.

Table 3 gives a summary of the best results for individual classification quality factors. In almost all cases, greater accuracy was obtained for images without normalisation of brightness. The exception is the PPV coefficient, the value of which (95%) turned out to be the same for both types of images.

Table 3. A summary of the best classification results shown in Figs. 3, 4 and 5.

Index	Image	Value (%)	Method/Classifier
ACC	Not normalised	90	POE+ACC/5-NN, LSR/QDA
	Normalised	82	LSR/LDA
TPR	Not normalised	89	POE+ACC/QDA, LSR/1-NN
	Normalised	85	LSR/QDA
TNR	Not normalised	96	FC/9-NN, MI/9-NN
	Normalised	82	LSR/4-NN
PPV	Not normalised	95	FC/9-NN, POE+ACC/5-NN, MI/9-NN
	Normalised	95	POE+ACC/5-NN
NPV	Not normalised	89	POE+ACC/5-NN
	Normalised	84	LSR/QDA

4 Conclusions

Conducting the normalisation of the examined images caused a change in the brightness of individual pixels. As a result, the textural features of the images changed, as demonstrated by the results of the analysis. The sets of descriptors, obtained as a result of the chosen selection methods, were different for the images before and after the normalisation. The individual sets of features were evaluated in terms of the accuracy of the automatic classification of images into healthy and sick (osteoporotic) cases. Almost all of the applied classification quality factors achieved a higher value for images without normalisation of brightness. Differences in deterioration in accuracy, as a result of normalisation, ranged from 4% for TPR to 14% for ACC. The exception was the PPV coefficient, which did not change as a result of the normalisation.

The obtained results showed that the pre-processing operations should be carefully selected, because they do not always lead to improvement of the intended effects. Normalisation, by extending the range of shades to the maximum acceptable, improves the quality of the image, making it easier to distinguish the details on it. However, as research has shown, information that is important from the point of view of texture recognition may be lost. For the applied feature descriptors and classification methods, normalisation of spinal CT images led to the deterioration of the accuracy of the automatic diagnosis of osteoporosis. Therefore, this operation should not be recommended for use in the context of the present Research.

References

1. Downey, P.A., Siegel, M.I.: Bone biology and the clinical implications for osteoporosis. Phys. Ther. **86**, 77–91 (2006)
2. Marcus, R., Feldman, D., Dempster, D., Luckey, M., Cauley, J.: Osteoporosis, 4th edn. Academic Press (2013)
3. Reshmalakshmi, C., Sasikumar, M.: Trabecular bone quality metric from X-ray images for osteoporosis detection. In: 2017 International Conference on Intelligent Computing, Instrumentation and Control Technologies (ICICICT), pp. 1694–1697 (2017)
4. Nasser, Y., Hassouni, M., Brahim, A., Toumi, H., Lespessailles, E., Jennane, R.: Diagnosis of osteoporosis disease from bone X-ray images with stacked sparse autoencoder and SVM classifier. In: 2017 International Conference on Advanced Technologies for Signal and Image Processing (ATSIP), pp. 1–5 (2017)
5. Reshmalakshmi, C., Sasikumar, M.: Fuzzy inference system for osteoporosis detection. In: 2016 IEEE Global Humanitarian Technology Conference (GHTC), pp. 675–681 (2016)
6. Tejaswini, E., Vaishnavi, P., Sunitha, R.: Detection and prediction of osteoporosis using impulse response technique and artificial neural network. In: 2016 International Conference on Advances in Computing, Communications and Informatics (ICACCI), pp. 1571–1575 (2016)
7. Shahabaz, Somwanshi, D.K., Yadav, A.K., Roy, R.: Medical images texture analysis: a review. In: 2017 International Conference on Computer, Communications and Electronics (Comptelix), pp. 436–441 (2017)
8. Strzelecki, M., Materka, A.: Tekstura obrazów biomedycznych. Wydawnictwo Naukowe PWN (2017)
9. MaZda. www.eletel.p.lodz.pl/programy/cost/progr_mazda.html. Accessed 6 May 2018
10. Haralick, R.: Statistical and structural approaches to texture. Proc. IEEE **67**(5), 786–804 (1979)
11. Haralick, R., Shanmugam, K., Dinstein, I.: Textural features for image classification. IEEE Trans. Syst. Man Cybern. **3**(6), 610–621 (1973)
12. Hu, Y., Dennis, T.: Textured image segmentation by context enhanced clustering. IEE Proc.-Vis. Image Sig. Process. **141**(6), 413–421 (1994)
13. Lerski, R., Straughan, K., Shad, L., et al.: MR image texture analysis - an approach to tissue characterization. Magn. Reson. Imaging **11**, 873–887 (1993)
14. Omiotek, Z.: Improvement of the classification quality in detection of Hashimoto's disease with a combined classifier approach. Proc. Inst. Mech. Eng. Part H: J. Eng. Med. **231**(8), 774–782 (2017)
15. Shurmann, J.: Pattern Classification. Wiley, Hoboken (1996)
16. Dash, M., Liu, H.: Feature selection for classification. Intell. Data Anal. **1**(3), 131–156 (1997)
17. Tourassi, G.D., Frederick, E.D., Markey, M.K., Floyd, C.E.: Application of the mutual information criterion for feature selection in computer-aided diagnosis. Med. Phys. **28**(12), 2394–2402 (2001)
18. Carmone, F.J., Kara, A., Maxwell, S.: HINoV: a new method to improve market segment definition by identifying noisy rariables. J. Mark. Res. **36**, 501–509 (1999)
19. Omiotek, Z., Burda, A.: Feature selection methods in image-based screening for the detection of Hashimoto's thyroiditis in first-contact hospitals. Barometr Regionalny **14**(2), 187–196 (2016)

20. Breiman, L., Friedman, J., Olshen, R., et al.: Classification and Regression Trees. CRC Press, London (1984)
21. Enas, G.G., Chai, S.C.: Choice of the smoothing parameter and efficiency of the k-nearest neighbor classification. Comput. Math. Appl. **2**, 235–244 (1986)
22. Liao, S.H., Chu, P.H., Hsiao, P.Y.: Data mining techniques and applications - a decade review from 2000 to 2011. Expert. Syst. Appl. **39**, 11303–11311 (2012)
23. Quinlan, J.R.: Induction of decision trees. Mach. Learn. **1**, 81–106 (1986)
24. Venables, W.N., Ripley, B.D.: Modern Applied Statistics with S-PLUS. Springer, Berlin (1998)
25. Breiman, L.: Random forests. Mach. Learn. **45**, 5–32 (2001)

Inertial Motion Capture System with an Adaptive Control Algorithm

Michał Pielka(✉), Paweł Janik, Małgorzata Aneta Janik, and Zygmunt Wróbel

Faculty of Computer Science and Material Science,
University of Silesia in Katowice, Sosnowiec, Poland
{michal.pielka,pawel.janik,malgorzata.janik,zygmunt.wrobel}@us.edu.pl

Abstract. Motion capture (MoCap) systems are increasingly used for rehabilitation and diagnostic purposes. However, dissemination of this type of solutions requires further development and reduction of prices, especially for high-precision systems. The paper presents a tested MoCap system architecture, which is based on independent, integrated sensor modules. Some of the cheapest Wi-Fi 2.4 GHz transceivers with a SoC chip were used to build individual modules. This concept made it possible to obtain one of the smallest MoCap modules with a Wi-Fi interface. In turn, the implemented embedded software with an adaptive radio transmission control algorithm allows for optimization of the sensor module operation. The algorithm presented in the paper reduces both the transmission in the sensor network (by about 26%) as well as the router processor load (by about 80%).

Keywords: Data transmissions · IMU motion capture · Sensor network · Sensor power efficiency

1 Introduction

Currently, motion capture systems (MoCap) are quite common. Depending on the requirements and applications, they offer different measurement precision and different data analysis possibilities. MoCap systems can be divided into three main groups: optoelectronic, mechanical-electronic and inertial, using MEMS sensors. Optoelectronic systems are both high-precision solutions, using many fast cameras, e.g. Smart DX from BTS Bioengineering [1], as well as simple, cheap solutions, based on the infrastructure developed for games, e.g. Kinect - Microsoft [2]. This group of motion capture systems concerns stationary solutions, due to the need to deploy cameras and define the scene they monitor. In this context, mechanical-electronic and inertial systems based on Inertial Measurement Unit (IMU) sensors can be treated as an alternative. Solutions that use radio transmission allow for motion capture in a natural environment, limited only by the range of transceivers.

Inertial systems play an important role among wearable solutions. Movement mapping in these systems is carried out by rotating the virtual bones of a computer model. For this purpose, sensors are placed on the human body, allowing

E. Tkacz et al. (Eds.): IBE 2018, AISC 925, pp. 67–74, 2019.
https://doi.org/10.1007/978-3-030-15472-1_8

for measurements of rotation and additionally acceleration or intensity of the magnetic field. The mapping of the entire human body movement is usually carried out using a dozen or so sensors [3]. The inertial MoCap group is dominated by two solutions: sensor costumes and independent modules.

The effectiveness of inertial systems is mainly related to their energy demand for the transmission of large amounts of data. Therefore, it is important to constantly improve the algorithmic methods and new structures of information systems that will allow at least a partial reduction of earlier technological limitations.

Movement monitoring using IMU sensors is used in many aspects. Sensor systems enable to reproduce the movement of the entire human body [4] as well as smaller biomechanical systems such as the hand [5]. Processing of data from sensors requires the use of a sufficiently efficient hardware platform. In this context, an architecture treated as a reference is widely used, in which a dedicated, high-performance microcontroller for data processing such as XSens [6] is applied. Inertial MoCap systems are used in rehabilitation [7], diagnosis and measurement [8] as well as sport [9].

2 System Setup

Low-cost and widely available components were used to build the sensor network. The simplification of the module construction is one of the conditions for reducing production costs. For this reason, a system was designed and tested in which the ESP07 module and the LSM9DS1 sensor were used. The ESP07 module is based on a single ESP8266 chip, which simultaneously supports the radio as well as provides resources of a 32 bit microcontroller μC. To enable communication of the microcontroller with the sensor, the SPI (Serial Peripheral Interface) was used. This structure eliminates the need to use an additional microcontroller for processing data from the sensor.

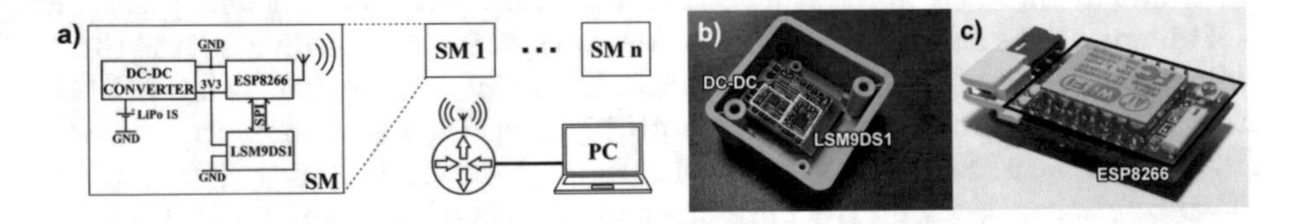

Fig. 1. (a) Block diagram of the sensor module and motion capture system, (b) sensor module looking from the MEMS system side, (c) sensor module looking from the Wi-Fi module side.

The functional structure of a single sensor module (SM) and the entire motion capture system are shown in Fig. 1a. A set of sensor modules transmits data via the ASUS AC1200 Dual Band router to the server (Personal Computer - PC), which in turn visualizes these data, e.g. in a virtual or extended environment.

SM electronics was made in the form of a two-sided PCB circuit sized approx. 26 mm × 16 mm (Fig. 1b and c). SM was powered by a 3.7 V Li-polymer battery with a capacity of 175 mAh. As a result, the presented module (SM) is currently one of the smallest independent sensor network nodes for motion capture using the Wi-Fi technology.

2.1 Adaptive Transmission Algorithm

During the operation of the sensor module program from the LSM9DS1 system, data are continuously collected at the sampling frequency of the gyroscope and accelerometer of 238 Hz. Data are collected from all three sensors, i.e. accelerometer, gyroscope and magnetometer, and represented by the vectors a, ω and m, respectively, described by (1).

$$a = [a_x, a_y, a_z] \quad \omega = [\omega_x, \omega_y, \omega_z] \quad m = [m_x, m_y, m_z] \tag{1}$$

The coefficients a_x, a_y and a_z of the vector a correspond to the acceleration with respect to the x, y and z axes of the sensor. Similarly, ω_x, ω_y, ω_z correspond to the rotational speed around the x, y and z axes, whereas m_x, m_y, m_z correspond to the magnetic field induction values. These coefficients are double-byte integers with a sign. All the data are stored in the FIFO queue (FIFO Frame Buffer) (Fig. 2).

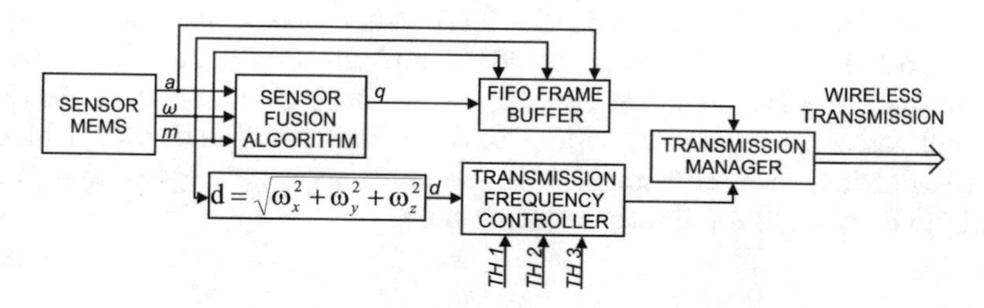

Fig. 2. Logical diagram of the sensor module operation.

In addition, a, ω and m are also input vectors of the sensor fusion algorithm. The Madgwick algorithm was used in the tested system [10] because the author published the original implementation, which facilitates the creation of sensor node software. The output vector of the algorithm is the quaternion q described by (2). The coefficients w, x, y, z of the quaternion q are single-precision floating point numbers (4 bytes) that are placed in the FIFO queue of the frame buffer.

$$q = [w, x, y, z] \tag{2}$$

All the coefficients of the vectors a, ω, m and the quaternion q associated with the same incident of data collection from the LSM9DS1 sensor are an inseparable block, 34 bytes long. The data vector ω collected from the sensor is used to control the transmission frequency. On its basis, the orientation of the object in space is determined. The coefficient d is calculated based on the vector ω according to (3).

$$d = \sqrt{\omega_x^2 + \omega_y^2 + \omega_z^2} \quad (3)$$

By using the coefficient d to determine the transmission frequency, it is possible to take into account the rotational speed of the sensor around all three axes. The adaptive transmission algorithm enables to control the frequency of frame transmission depending on the rotational speed of the monitored object. Transmission control is carried out in such a way that the visualization of rotation in real time is smooth (e.g. in a virtual environment). With a slower orientation change, the frame transmission frequency can be reduced. At the same time, along with the reduction of the transmission frequency, the length of the teletransmission frame, into which the data blocks placed in the FIFO queue are entered, is extended. If the amount of data in the FIFO queue is larger than the capacity of one frame, then two or more frames are sent immediately one after another. After sending the frame, the FIFO buffer is emptied. Four exemplary transmission frequencies were determined in the algorithm for testing purposes: 4 Hz (base frequency), 10 Hz, 30 Hz and 60 Hz. 4 Hz is the lowest adopted transmission frequency, and the increase in frequency is related to the increase in the value of the coefficient d. In addition, three threshold values TH1, TH2, TH3 (positive integers) are also the input data of the algorithm.

Once the coefficient d exceeds the TH1 value, the transmission frequency increases to 10 Hz, the TH2 value - to 30 Hz, and the TH3 value - to 60 Hz. With a sensor sampling frequency of 238 Hz, frames sent at 4 Hz, 10 Hz, 30 Hz and 60 Hz will contain successively twice 30, once 24, once 8 and once 4 blocks of data. In addition, each transmission frame contains a four-byte frame number and a byte defining the number of data blocks.

2.2 Measurements Procedure

The change in the frequency of frames transmitted by the sensor module is related to the rate of its rotation. In order to present the adaptive transmission control, a rotary platform driven by a stepper motor was used. The sensor module was placed on the platform, which allowed for the registration of the change in transmission frequency of the transmission frames depending on the rotational speed. In turn, the time between the reception of subsequent frames was measured using software written for the server.

Figure 3 presents the results of concurrent measurements - angular velocity changes ω_P of the rotating platform, on which the frame transmission frequency f_T depends. As the angular velocity rises, the transmission frequency of variable length frames increases. The algorithm assumes that the frame transmission frequency is proportional to the angular velocity. The method of changing this velocity (linear or non-linear) is declared by the user.

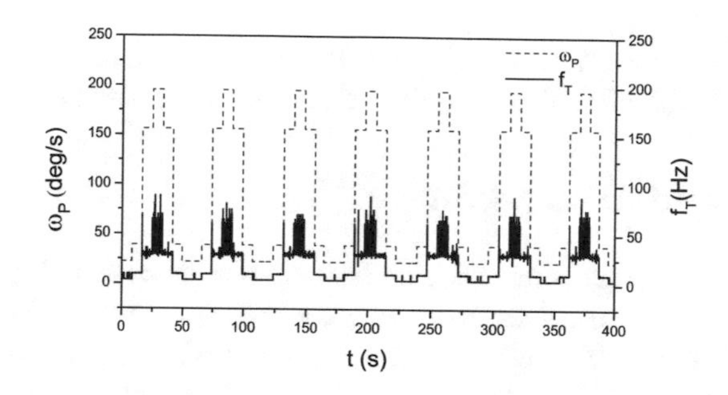

Fig. 3. Dependence of frame transmission frequency on the rotational speed of the sensor module.

During transmission, data are sent using the connectionless UDP protocol, whose header is 8 bytes long. Then a 20-byte-long IPv4 header is added. In the next step, depending on the transmission medium, an Ethernet header or a Wi-Fi header is added to the data. For further analysis, it has been assumed that a 14-byte-long Ethernet header is added to the IPv4 packet. At this stage, the number of lost packets was not investigated and retransmission mechanisms were not introduced. Table 1 presents a summary of the parameters of full transmission frames for the transmission frequencies selected in the adaptive algorithm.

Table 1. Data and frame sizes depending on the transmission frequency.

Number of frames	Frequency [Hz]	Data size [byte]	Frame size [byte]
2	4	1025	1067
1	10	821	863
1	30	277	319
1	60	141	183

Assuming that there is one sensor and a computer (server) in the network and no other traffic except for data transmission by the sensor, the data transfer rates at different frame transmission frequencies are as follows: 4 Hz - 8.34 kB/s, 10 Hz - 8.43 kB/s, 30 Hz - 9.35 kB/s, 60 Hz - 10.72 kB/s. It can be observed that as the transmission frequency increases, the number of transmitted bytes rises. This increase is related to the fact that each individual transmission frame has its header with a total length of 42 bytes, a four-byte number and a byte specifying the number of data blocks in the frame.

The advantage of the proposed algorithm is the fact that despite the change in transmission frequency as well as the number of transmitted bytes, the number of data blocks that reaches the server in 1 s remains constant regardless of the transmission frequency. As a result, despite the reduction of the transmission frequency, there is no reduction in the measurement precision. Decreasing the frame transmission rate affects only the refreshing of real-time information about the movement in the virtual environment on the computer receiving the data.

3 Results and Discussion

An experiment was carried out in which the sensor network load was investigated. For this purpose, 10 transceivers were used, which transmitted frames at two frequencies - low 4 Hz and high 60 Hz. During the experiment, the proportion of the transceivers working at these frequencies changed. A similar situation occurs in practice, because the adaptive algorithm tends to reduce the frame transmission frequency in individual sensor modules. When a given sensor monitors quick changes, the data concerning its rotation are refreshed more often. In the case of slow changes, the refreshing (frame transmission frequency) is proportionally reduced. This procedure enables to limit the system resources required by the sensor network.

Figure 4 presents the characteristics of the router CPU load depending on the ratio between the numbers of transceivers sending data at 4 Hz and 60 Hz, starting from a 10:0 ratio with a step of 1. The ratio was changed every 120 s. Figure 4a shows a steady increase in the router CPU load up to the maximum value, which is reached when data are transmitted in the sensor network only at 60 Hz (0:10 ratio). When data are transmitted in the sensor network only at 4 Hz, the router CPU load is around 3.6%, whereas at 60 Hz it rises to approx. 19.9%, so it is 5.5 times larger.

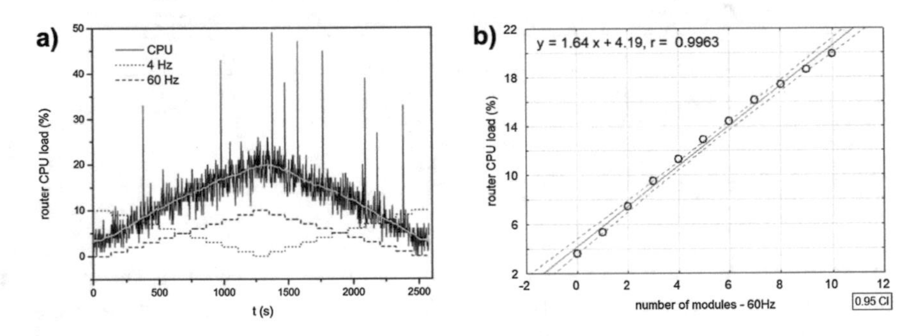

Fig. 4. (a) Characteristics of the router CPU load depending on the ratio between the numbers of transceivers transmitting data at 4 Hz and 60 Hz, (b) results of regression and correlation analysis of data obtained on the basis of Fig. 4a.

Figure 4b presents the results of regression and correlation analysis. The load increase is straight-line with a directional coefficient of 1.64. The increase still remains fivefold, because with 18 modules working at 4 Hz, the load will be about 6.5% (with 10 transceivers the minimum load was about 3.6%, so it was increased by each subsequent module by an average of about 0.36%). Assuming that this tendency will continue with the addition of subsequent modules, the

CPU consumption will be approx. 36% in the case of 18 modules working at 60 Hz (with the increase in the number of modules, the intersection point will increase from approx. 4 to approx. 7). In the reverse situation, i.e. when all 10 transceivers work at the highest refreshing frequency declared in the experiment (60 Hz), and the monitored motion enables to lower the frequency to 4 Hz (for all 10 modules), the router CPU load can be reduced by more than 80%.

In turn, Fig. 5 presents the characteristics of data transfer in the sensor network depending on the previously described ratio between transceivers working at 4 Hz and 60 Hz. The conducted research has shown an increase in the transfer in the network from approx. 83 kB/s (for all transceivers transmitting at 4 Hz) to approx. 112 kB/s (for all sensors transmitting at 60 Hz), which is an increase of almost 35%. In Fig. 5a, it can also be observed that in practice data transmission is carried out with a slightly higher transfer value than it is indicated by theoretical considerations, i.e. when taking into account only the frame length. However, the correlation between theoretical and experimental data is very high ($r = 0.9993$). The directional coefficient of the obtained regression line is 1.14, so the theoretical increase in the transfer of 1 kB/s causes, in practice, an average increase of 1.14 kB/s.

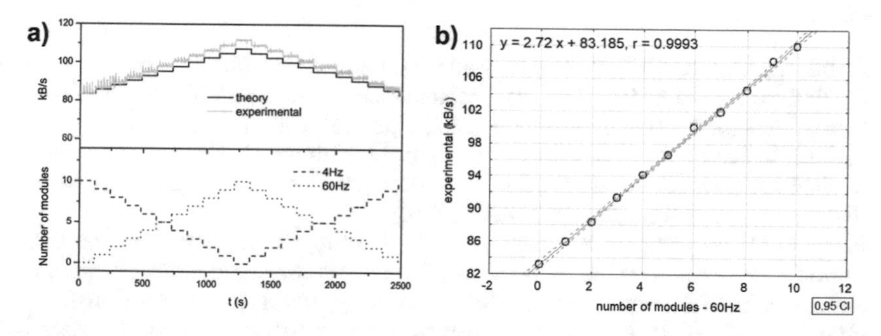

Fig. 5. (a) Characteristics of data transfer in the sensor network depending on the ratio between the numbers of transceivers working at 4 Hz and 60 Hz, (b) results of regression and correlation analysis of data obtained on the basis of Fig. 5a.

In the case of 10 tested transceivers, the algorithm presented in the paper enables to limit network traffic by several dozen percent, whereas in practice, MoCap systems may use, for example, 18 independent sensor modules. The dependence of the transfer on the ratio of low to high-frequency transceivers presented in Fig. 5b is straightforward. Based on the obtained regression equation, it can be concluded that changing the transmission frequency from 4 Hz to 60 Hz in each subsequent module results in a transfer increase of 2.72 kB/s on average. With 10 modules set at 4 Hz, the total transfer was approximately 83.2 kB/s, which gives 8.32 KB/s per 1 module. Thus, with 18 modules working in this way, this transfer will be about 150 kB/s. Assuming that a linear tendency will be maintained when increasing the number of modules above 10, the transfer will reach the value of approx. 199 kB/s at 18 sensors working at 60 Hz (the intersection point will increase from around 83 to about 150). This means an increase of up to approx. 30% in relation to the network load generated by the same sensor modules, all transmitting at 4 Hz. In the reverse situation, i.e. when all 10 transceivers work at the highest refreshing rate (60 Hz) declared in the experiment, and the monitored motion enables to lower the frequency to 4 Hz (for all 10 modules), the data transfer can be reduced by 26%.

4 Conclusions

The paper presents the impact of transmission frequency on the sensor network load. Using the adaptive transmission algorithm, it has been found that the router CPU load depends on the changing ratio between the numbers of transceivers transmitting data at two extreme frequencies of 4 Hz and 60 Hz. The smallest load obtained was approx. 3.6%, with all 10 used transceivers operating at 4 Hz, and increased about 6 times when all transceivers were set at 60 Hz. The described change in the ratio also affects the data transfer, which changes from approx. 83 kB/s to approx. 112 kB/s at 4 Hz and 60 Hz, respectively.

References

1. Rafal, K., Mateusz, K.: Preliminary study on accuracy of step length measurement for CIE exoskeleton. In: IEEE International Conference on Multisensor Fusion and Integration for Intelligent Systems (MFI), pp. 577–581. IEEE (2016)
2. Galan, B., Barry, G., Jackson, D., Mhiripiri, D., Olivier, P., Rochester, L.: Accuracy of the Microsoft Kinect sensor for measuring movement in people with Parkinson's disease. Gait Posture **39**, 1062–1068 (2014)
3. Kuo, M.C., Chiang, P.Y., Kuo, C.C.J.: Coding of motion capture data via temporal-domain sampling and spatial-domain vector quantization techniques. In: Advances in Multimedia Information Processing, PCM, pp. 84–99 (2010)
4. Guo, L., Xiong, S.: Accuracy of base of support using an inertial sensor based motion capture system. Sensors **17**(9), 2091 (2017)
5. Mańkowski, T., Tomczyński, J., Kaczmarek, P.: CIE-DataGlove, a multi-IMU system for hand posture tracking. In: International Conference Automation ICA, Advances in Intelligent Systems and Computing, AISC, vol. 550, pp. 268–276 (2017)
6. Pons-Moll, G., Baak, A., Helten, T., Muller, M., Seidel, H., Rosenhahn, B.: Multisensor-fusion for 3D full-body human motion capture. In: IEEE Conference on Computer Vision and Pattern Recognition (CVPR), pp. 663–670. IEEE (2010)
7. Eke, C., Cain, S., Stirling, L.: Strategy quantification using body worn inertial sensors in a reactive agility task. J. Biomech. **64**, 219–225 (2017)
8. Karatsidis, A., Bellusci, G., Schepers, H., de Zee, M., Andersen, M., Veltink, P.: Estimation of ground reaction forces and moments during gait using only inertial motion capture. Sensors **17**, 75 (2017)
9. Gleadhill, S., Lee, J., James, D.: The development and validation of using inertial sensors to monitor postural change in resistance exercise. J. Biomech. **49**, 1259–1263 (2016)
10. Madgwick, S.: An efficient orientation filter for inertial and inertial/magnetic sensor arrays. Technical report; Report x-io, University of Bristol (2010)

Signal Analysis

Repeatability Investigations of a Handheld Electronic Spirometer

Dawid Kucharski(✉) and Marta Michalska

Faculty of Mechanical Engineering and Management,
Institute of Mechanical Technology, Division of Metrology and Measurement Systems,
Poznan University of Technology, ul. Piotrowo 3, 60-965 Poznan, Poland
dawid.kucharski@put.poznan.pl
http://wbmiz.put.poznan.pl

Abstract. We present a study of the repeatability of an electronic handheld spirometer in the mechanical properties studies of the respiratory system. We use a novel mechanical system for the human respiratory system simulations. The measurement method is described and a statistical data analysis is given. The uncertainty budget of the measurements and conclusions about repeatability of the tested handheld electronic spirometer are provided.

Keywords: Human respiratory system · Handheld spirometer · Measurements · Repeatability

1 Introduction

Spirometry is one of the common method to test the human respiratory system. In many cases used for lung function tests and diseases detections [1–4]. It is based on measuring the volume and speed of the air that can be inhaled and exhaled. At present mainly electronic handheld spirometers are used. Spirometry is non-invasive and painless method. They can easily be use by the patients at home. From a metrological point of view, spirometers are measurement devices with limited accuracy and repeatability [5–8]. Many error sources can influence the spirometer accuracy, such as temperature, pressure, humidity of the air and also the calibration procedure. Every metrological tool should be certified conform to reference calibration values. There are many methods to evaluate the accuracy and repeatability of spirometers [4,9–11]. The producers and users of the spirometers also use the special calibration pumps. These methods not taking into the count the influence of the air flow method, humidity and temperature on the accuracy of the calibration. Without even to control them. Here we propose to use the constructed special, novel, mechatronic device for the human respiratory system simulations, to test the repeatability of the handheld electronic spirometer. The novelty of the presented calibration setup is to simulate as close as possible, the real conditions during the measurements as it is done during the

E. Tkacz et al. (Eds.): IBE 2018, AISC 925, pp. 77–84, 2019.
https://doi.org/10.1007/978-3-030-15472-1_9

patient testing. We propose to precisely control the air flow method, humidity, temperature, volume and air speed. Not only the volume and speed of the air as it is done using the calibration pumps.

2 Setup

2.1 Lung Simulator

The setup consists of three elements: a lung simulator, an electronic spirometer MIR Spirobank II and a PC with free software. All parts are located on a desk with a touch screen monitor for manual operation of all equipment. In Fig. 1 the whole system is shown.

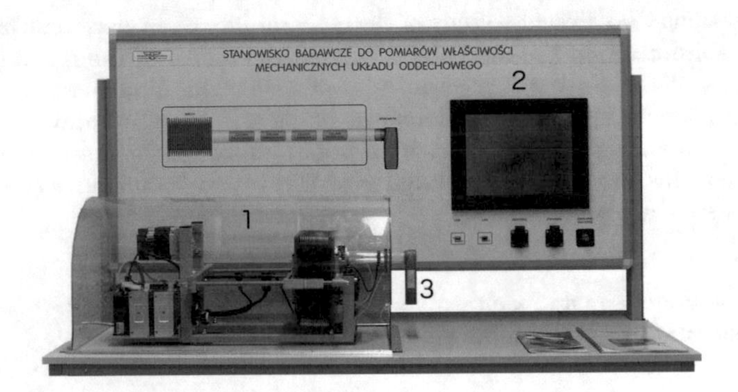

Fig. 1. The experimental setup: 1 - lung simulator, 2 - touch screen, 3 - MIR spirobank II spirometer

The simulator consists of bellows for air flowing, sensors to measure the relative humidity [%], temperature [°C], pressure [bar] and level of the air flow [l/s]. The air flow in the setup is measured by a turbine flow sensor. The human lung simulator (1 in Fig. 1) consists of a bellow, a stepper motor, an electronic controller, a heating system (inside the bellows) for the air and sensors (see Fig. 2).

All parts are protected by a plexiglas shield to eliminate external error sources. The simulator will automatically stop if any part of the system behave strangely, too loud or beyond the mechanical or temperature limits.

Software to control and store the data was written in C. This computer application called "Spirometr", offers such options as: changing the amplitude and the period of the bellow movement and waveform changes. The software offers six variants of the waveform: three basic (sinusoidal, triangular, rectangular) and three corresponding to the human respiratory system ("Spiro 1, 2, 3").

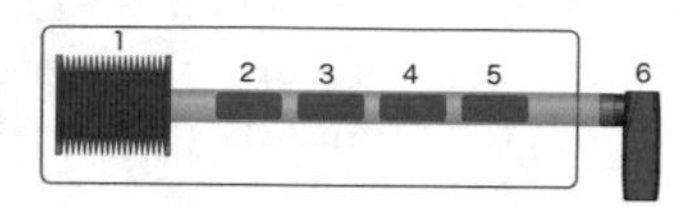

Fig. 2. Scheme of the human lung simulator: 1 - bellows with heating system inside, 2 - humidity sensor, 3 - temperature sensor, 4 - pressure sensor, 5 - flow sensor, 6 - spirometer

Fig. 3. The handheld electronic spirometer MIR Spirobank II

2.2 Handheld Spirometer

An electronic spirometer (see e.g. Fig. 3) is widely used for static and dynamic measurements of the respiratory system. A bidirectional electronic turbine with an infrared (IR) beam chopper sensor is used for the volume and air flow measurement [6,12,13]. We used the WinspiroPRO software freely available on the internet, for fast data acquisition. In this way we increased the speed of the measurements relative to using the software built into the spirometer.

3 Measurement Methods

The repeatability of the handheld spirometer was investigated in the following steps:

- repeatability measurements for the lung simulator in two test modes "Spiro 1" and "Spiro 2";
- repeatability measurements of the MIR Spirobank II spirometer.

All spirometer results are compared with values, derived from standard anthropometric values of a fictive patient (see Tables 1 and 2).

To evaluate the repeatability of the measurements the 3σ statistical rule was used, well known in metrology for the measurements uncertainty evaluations. The three-sigma rule is an empirical rule stating that, for many reasonably symmetric unimodal distributions, almost all of the population lies within three

Table 1. Anthropometric parameters

Age	26
Sex	Male
Height	183 cm
Weight	87 kg
Additional information	Non-smoker

Table 2. The predicted values generated by the WinspiroPRO software for the selected spirometric parameters

Spirometric parameter	Value
Forced vital capacity (FVC)	5.89
Forced expiratory volume in one second (FEV_1)	4.90
Tiffeneau-pinelli index (FEV_1/FVC)	83.6

standard deviations of the mean. For the normal distribution about 99.7% of the population lies within three standard deviations of the mean. This rule is also highly recommended for the unknown (unchecked) distribution of the results as we presented.

3.1 Repeatability of the Lung Simulator

We determined the repeatability of the lung simulator by repeating the measurements 32 times with the same parameters. Two operating modes of the simulator were analysed: "Spiro 1" and "Spiro 2", which are significantly different from each other. The movement sequence for the first one is much more simple. This corresponds to a constant, measured inhale and exhale air, depending on the preset amplitude and frequency. The "Spiro 2" mode corresponds to a much more complicated movement sequence. The movement for inhalation is steady, but during the first second of exhalation the speed is increasing, thereafter slowing down and ending with a very low speed. The "Spiro 2" corresponds to the breathing mode during FVC spirometric tests, it means inhaling with force and calm maximum exhaling.

To test the repeatability of the lung simulator the measurements were repeated 32 times for both the "Spiro 1" and "Spiro 2" modes. The software generated flow and volume versus time graphs (see Figs. 4, 5, 6 and 7). The graphs show also a high repeatability of the air flow character in the human lung simulation behaviour, which is a big advantage compared with the calibration pumps. This, together with specified environmental parameters such as air temperature and humidity gives better conditions for the spirometry calibration than the standard techniques.

3.2 Repeatability of the Handheld Spirometer

The repeatability of the spirometer was tested using fictive patient parameters (Table 1) with the spirometer mounted to the end of the lung simulator.

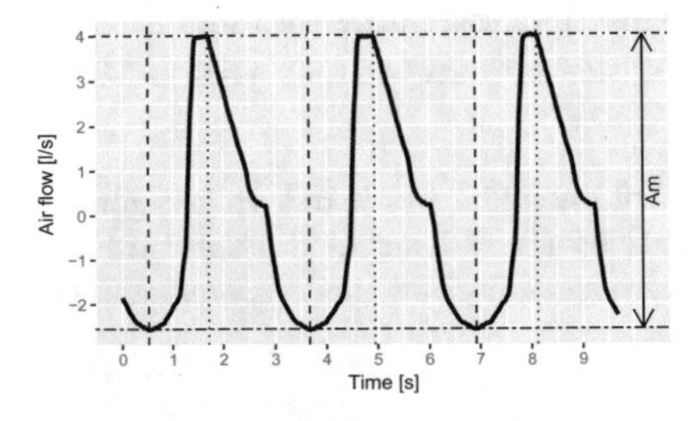

Fig. 4. Air flow vs. time in the lung simulator registered in "Spiro 1" mode. **Am** – amplitude

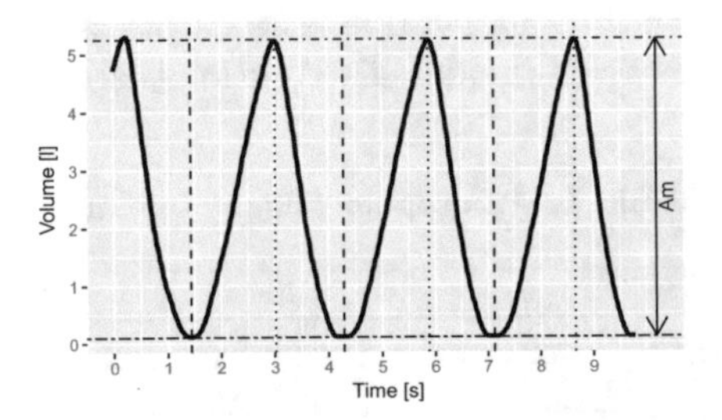

Fig. 5. Air volume vs. time in the lung simulator registered in "Spiro 1" mode. **Am** – amplitude

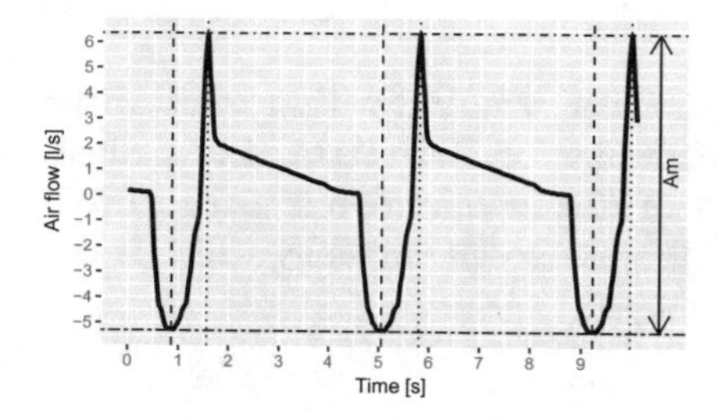

Fig. 6. Air flow vs. time in the lung simulator registered in "Spiro 2" mode. **Am** – amplitude

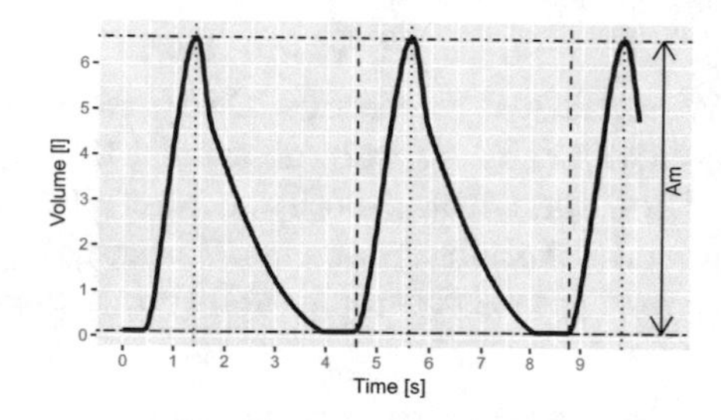

Fig. 7. Air volume vs. time in the lung simulator registered in "Spiro 2" mode. **Am** – amplitude

The simulator was working in the "Spiro 1" and "Spiro 2" modes. Amplitude and a period of the lung movement were the same as during the repeatability tests of the lung simulator. A measurement started simultaneously with the bellow movement. A sound signal from the spirometer signalled to stop the measurement. The results of the spirometer were shown on screen and also stored by the WinspiroPRO software on a separate PC. To test the repeatability 32 measurements were made for both "Spiro 1" and "Spiro 2" modes.

4 Results

4.1 Repeatability of the Lung Simulator

Flow and volume were measured every 0.01 s. Data to be used for the statistical analysis were taken from the "Spirometr" software.

Table 3. Air flow repeatability comparison of the lung simulator for the two test modes

Air flow	"Spiro 1"	"Spiro 2"
Amplitude (Am) [l/s]	6.56 ± 0.05	11.56 ± 0.20
Δt between maxima [s]	3.41 ± 0.78	4.22 ± 0.81
Δt between minima [s]	3.37 ± 0.66	4.20 ± 0.78

The error given is 3σ

Table 4. Volume repeatability comparison of the lung simulator for the two test modes

Air volume	"Spiro 1"	"Spiro 2"
Amplitude (Am) [l]	5.15 ± 0.08	6.61 ± 0.23
Δt between maxima [s]	3.37 ± 0.72	4.23 ± 0.72
Δt between minima [s]	3.38 ± 0.75	3.99 ± 0.96

The error given is 3σ

4.1.1 "Spiro 1" Mode Test

We measured two series. Every measurement was restarted with a reset of the controller. There were no special time synchronization. We started the analysis of the data from 106(8)[1] s after the start of the bellow movement, to avoid initial instability problems. The starting time was chosen manually which is not important for the repeatability investigations.

4.1.2 "Spiro 2" Mode Test

A repeatability test of the lung simulator in the "Spiro 2" mode was done in the same way. Every measurement was restarted with a reset of the controller. We started the analysis of the data from 126(12) s after the start of the bellow movement, to avoid initial instability problems.

4.1.3 A Repeatability Comparison of the Lung Simulator in the Two Test Modes

We compared the results for the "Spiro 1" and "Spiro 2" (see Tables 3 and 4).

The standard deviations are smaller for almost every case analysed in the "Spiro 1" mode. This mode is much more repeatable than the "Spiro 2" mode.

This repeatability tests also proved that the lung simulator is good enough to test the handheld electronic spirometer.

Table 5. Comparison of the average spirometric parameters ($\overline{x} \pm 3\sigma$) tested in the two modes of the lung simulator

Spirometric parameter	"Spiro 1"	"Spiro 2"
FVC [l]	5.84 ± 0.03	7.53 ± 0.08
FEV_1 [l]	4.23 ± 0.03	3.65 ± 0.07
FEV_1/FVC [%]	72.49 ± 0.54	48.50 ± 1.30
$\frac{FVC}{FVC_{predicted}}$ [%]	99.15 ± 0.42	127.82 ± 1.20
$\frac{FEV_1}{FEV_{1(predicted)}}$ [%]	86.41 ± 0.45	74.50 ± 1.30
$\frac{FEV_1/FVC}{FEV_{1(predicted)}/FVC_{predicted}}$ [%]	86.75 ± 0.63	58.01 ± 1.50

The error given is 3σ

4.2 Repeatability of the Handheld Spirometer

The lung simulator was now used to test the MIR Spirobank II spirometer. We tested the repeatability of the following spirometric parameters: FVC, FEV_1, FEV_1/FVC. We measured two series in "Spiro 1 and 2" modes.

[1] 106(8) - notation common in use as: 106 (mean value) $\pm$ 8 (standard deviation σ).

4.2.1 Comparison of the Spirometer Repeatability in the Two Test Modes

To determine the repeatability of the handheld spirometer, the "Spiro 1" and "Spiro 2" modes are compared in Table 5.

As the comparison shows the errors (3σ) are much smaller in the "Spiro 1" mode for all tested spirometric parameters. For the "Spiro 2" test mode the errors (3σ) in the ratios of the parameters are all larger than 1%. These results show a smaller repeatability of the measurements for the spirometer in the "Spiro 2" mode and also a smaller repeatability of the lung simulator itself in this mode.

The spirometric parameters determined in the "Spiro 1" mode are much closer to the predicted values, than in the "Spiro 2", where the parameter FEV_1/FVC differs from the predicted value by 42% (see Tables 2 and 5).

5 Summary

This study confirmed that the MIR Spirobank II handheld electronic spirometer is good enough for respiration testing for a patient at home. It can be used mainly for everyday respiration testing and to give a warning if additional testing should be done a hospital for potential disease detection. A new test setup is presented for spirometer repeatability evaluation. The uncertainty of the measurements were calculated based on average values, standard deviations and error bars of 3σ.

The novel mechanical test setup we used as a good tool for precise and simple spirometer testing, also allowing calibration. With precise automated controlling of the measurement parameters: air volume, air flow, temperature, humidity and whole dynamics, system as presented is more suitable for the spirometers calibration compared to the standard techniques.

Acknowledgments. The authors wish to thank the Mechatronika Wyposażenie Dydaktyczne Company for the construction and programming of the test setup.

Funding. This work was supported by grant 02/22/DSPB/1432.

References

1. Evans, S.E., Scanlon, P.D.: Current practice in pulmonary function testing. Mayo Clin. Proc. **78**(6), 758–763 (2003)
2. Pedersen, O.F., Quanjer, P.H., Tammeling, G.J., Cotes, J.E., Peslin, R., Yernault, J.C.: Lung volumes and forced ventilatory flows. Eur. Respir. J. **6**(Suppl 16), 5–40 (1993)
3. Hutchinson, J.: On the capacity of the lungs, and on the respiratory functions, with a view of establishing a precise and easy method of detecting disease by the spirometer. Med. Chir. Trans. **29**, 137–252 (1846)
4. Ferguson, G.T., Enright, P.L., Buist, A.S., Higgins, M.W.: Office spirometry for lung health assessment in adults: a consensus statement from the National Lung Health Education Program. Chest **117**(4), 1146–1161 (2000)

5. Schermer, T.R.J., Verweij, E.H.A., Cretier, R., Pellegrino, A., Crockett, A.J., Poels, P.J.P.: Accuracy and precision of desktop spirometers in general practices. Respiration **83**(4), 344–352 (2012)
6. Malmberg, L.P., Hedman, J., Sovijärvi, A.R.A.: Accuracy and repeatability of a pocket turbine spirometer: comparison with a rolling seal flow-volume spirometer. Clin. Physiol. **13**(1), 89–98 (1993)
7. Jensen, R.L., Teeter, J.G., England, R.D., White, H.J., Pickering, E.H., Crapo, R.O.: Instrument accuracy and reproducibility in measurements of pulmonary function. Chest **132**(2), 388–395 (2007)
8. Malmberg, P.: Measurement of spirometric indices with a turbine spirometer - repeatability and comparison with a flow volume spirometer. In: Physiology, Clinical, pp. 372–373. Helsinki University Central Hospital, Helsinki, January 1992
9. Miller, M.R.: Standardisation of spirometry. Eur. Respir. J. **26**(2), 319–338 (2005)
10. Enright, P.L., Beck, K.C., Sherrill, D.L.: Repeatability of spirometry in 18,000 adult patients. Am. J. Respir. Crit. Care Med. **169**(2), 235–238 (2004)
11. Standardization of spirometry, 1994 update. American Thoracic Society, September 1995
12. Schneider, A., Gindner, L., Tilemann, L., Schermer, T., Dinant, G.J., Meyer, F.J., Szecsenyi, J.: Diagnostic accuracy of spirometry in primary care. BMC Pulm. Med. **9**, 31 (2009)
13. Gunawardena, K.A., Houston, K., Smith, A.P.: Evaluation of the turbine pocket spirometer. Thorax **42**(9), 689–693 (1987)

Necessity of Telemonitoring in Patients Treated by Means of Cardiac Assist Systems on the Example of Polish Rotary Blood Pump ReligaHeart ROT

Maciej Gawlikowski[1(✉)], Roman Kustosz[1], Malgorzata Gonsior[1], Miroslaw Czak[1], Przemyslaw Kurtyka[1], Jerzy Pacholewicz[2], Michal Zakliczynski[2], Boguslaw Kapelak[3], Karol Wierzbicki[3], and Piotr Siondalski[4]

[1] Foundation of Cardiac Surgery Development, Zabrze, Poland
mgawlik@frk.pl
[2] Department of Cardiac, Vascular and Endovascular Surgery and Transplantology, Medical University of Silesia, Silesian Centre for Heart Diseases, Zabrze, Poland
[3] Jagiellonian University, Medical College, John Paul II Hospital, Krakow, Poland
[4] Department of Cardiovascular Surgery, Medical University of Gdansk, Gdańsk, Poland
http://www.frk.pl

Abstract. Mechanical circulatory support (MCS) became next after heart transplantation clinically recognized way of treatment for patients suffered from end-stage heart failure. Nowadays, long term MCS is realized mostly by means of implantable rotary blood pumps (RBP). Clinically utilized MCS systems have limited capabilities of pump operation supervision and patient's monitoring. A few of them were equipped with remote data transmission to external data base. In this paper the telemonitoring system intended for Polish rotary cardiac assist system ReligaHeart ROT has been presented. The necessity of remote monitoring of supported patient as well as pump operation has been justified by referring to clinical cases of commercial MCS system HVAD (Medtronic) application.

Keywords: Rotary blood pump · Mechanical circulatory support · Heart failure · Telemonitoring

1 Introduction

Nowadays the mechanical circulatory support (MCS) has become clinically recognized method for treatment of patients suffered from end-stage heart failure. The most widely applied systems are based on centrifugal, implantable rotary blood pumps (IRBP) with passively suspended impeller (HVAD/Medtronic - more than 10.000 applications in 2016 [1]) or with active impeller levitation

E. Tkacz et al. (Eds.): IBE 2018, AISC 925, pp. 85–95, 2019.
https://doi.org/10.1007/978-3-030-15472-1_10

(HeartMate III/Abbott - approx. 1000 applications in clinical study [2]). In most cases these low weight, small, efficient and durable continuous-flow RBPs have replaced previously utilized pulsatile-flow ventricular assist devices.

In spite of technical and medical progress the MCS may be risky for patient. The most common problems of MCS are: gastrointestinal bleedings, hemorrhage or ischemic strokes, thrombosis (that could lead even to pump malfunction [3]), wound infections and technical malfunctions of system [4]. In most cases mentioned adverse events are the complications that appear after long-term use. MCS systems are equipped with relatively simple measurement solutions. They can directly measure pump speed and motor power. The most crucial parameters like blood flow, flow pulsation, minimum flow are estimated with accuracy up to 20% and the arterial pressure is not assessed at all. The HVAD system can build 30-days long log-file based on data collected every 15 min, the HeartMate III system routinely register data every 24 h with the ability to shorten time interval between saving data, but for reduced global time covered by log-file. Both systems are able to register simple threshold alarms (like a low flow or high power) and technical events. Neither of mentioned systems can remotely transmit data to the supervising medical staff although the history of supporting contains valuable information about patient status, course of treatment and symptoms of serious adverse events.

2 Clinical Data Analysis

Advanced analysis of signals measured by MCS system allows the early detection of side-effects and avoids serious complications led to deterioration of health or even patient's death.

In our study we based on available medical data only. No additional medical examinations were carried out. In 2017 we analyzed a total of 163 log-files collected from patients supported by HVAD (Medtronic) in 3 hospitals in Poland: Silesian Centre of Heart Diseases in Zabrze, John Paul II Specialized Hospital in Cracow and Medical University of Gdansk. Technical data (pump speed, motor power, mean flow, minimum flow, flow pulsation amplitude and acoustic signal generated by the IRBP) were correlated to patient's health status and selected medical data (ECG, echocardiography, the color of urine, lactate-dehydrogenase LDH concentration). Based on this we selected a set of signals features which indicated the most probable medical complications. Selected cases have been presented and commented below.

2.1 Pump Embolization and Fibrinolytic Treatment Evaluation

The IRBP embolism is associated with the activation of the coagulation system, causing clots that can form: inside the blood pump, inside of outflow graft, and on the region of pump inflow cannula - separately in each place or in few of it in parallel. This danger side-effect may resulted in ischemic stroke. Thromboembolism may be pharmacological treated by administration of heparin or tissue plasminogen activator (tPA). The pump parameters (flow, speed and power) in course of the embolization and thrombolytic treatment as well as spectrum of acoustic signal generated by the pump were presented in Fig. 1.

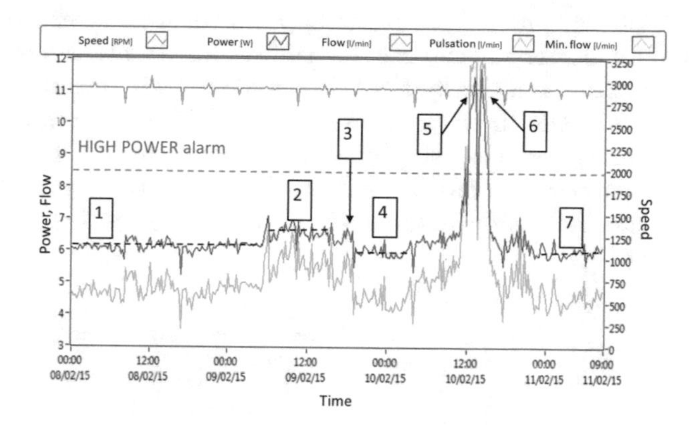

Fig. 1. Course of the pump parameters during embolization and thrombolytic treatment (labels were commented in text)

Before embolization the baseline pump power **(1)** was 6.1 W for 2500 RPM. At the point **(2)** the power rise to 6.5 W (0.3 W above maximum acceptable level for set pump speed) corresponding to brown color of urine was observed. The LDH concentration was 1206 U/l. Dark urine color and elevated LDH are medical symptoms of blood hemolysis. In case of IRBP the listed symptoms are mostly caused by high shear stress to blood cells due to clot presence in the pump. First attempt to fibrinolytic treatment **(3)** by heparin administration before patient's transport to the hospital appeared insufficient: although the power fall to 6.0 W (0.2 W under maximum acceptable level) was observed **(4)** but LDH rise to 1640 U/l then in next 6 hours the sudden power rise **(5)** to 11.2 W appeared. At point **(5)** the fibrinolytic treatment by tPA administration was begun which caused temporary power fall. At point **(6)** second dosage of tPA was administer due to ineffectiveness of previous treatment. It led to power fall **(7)** to 6.0 W (0.1 W under baseline and 0.2 W under maximum acceptable level). In next few days power was stabilized at 6.1 W and LDH concentration decreased to 460 U/l. In relation to correct results of other medical examinations thrombolytic treatment was recognized as sufficient. In commented case, according to Medtronic recommendations, the high power alarm was set to 2.5 W above mean pump power consumption. It allowed to recognize serious embolization only. It should be noted that first event of power increasing was unrecognized.

Together with examinations mentioned above the spectrum of acoustic signal generated by IRBP was assessed. We registered pump noises during chest auscultation by means of Littmann 3200 electronic stethoscope (sampling rate: 4000 Hz, band-pass filter 20–2000 Hz, emphasizes frequencies between 50 and

500 Hz). The comparison of spectrograms during IRBP embolization and after tPA administration were presented in Fig. 2.

The 1st harmonic corresponds to the impeller rotational speed. Impeller consists of 4 canals which causes 4th harmonic in spectrum. The reason of 2nd harmonic is unknown. The 3rd harmonic appears when impeller mass is unbalanced e.g. by thromboembolic material [5], which could be utilized to assess RBP embolization. Because amplitude of 4th harmonic is high and well-measurable, we used it as a reference in order to assess the relative value of 3rd harmonic amplitude. In commented case the relative amplitude of 3rd harmonic during confirmed RBP embolization **(8)** was 32% and 24 h after tPA administration **(9)** was immeasurable. Reference value of 4th harmonic on spectrograms in Fig. 2 were marked as **(10)**.

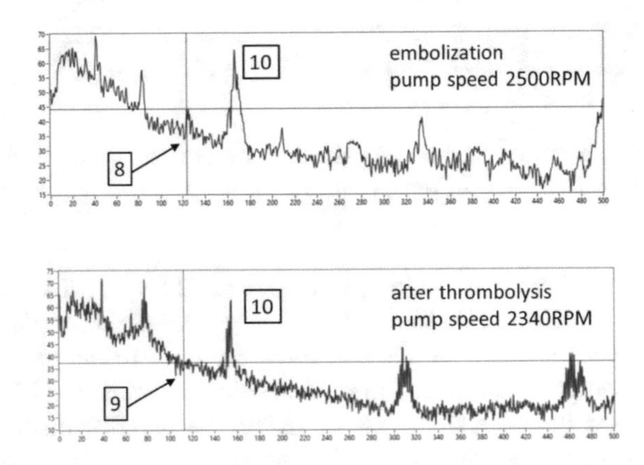

Fig. 2. Spectrograms of acoustic field generated by the pump during embolization and after efficient thrombolytic treatment (labels were commented in text)

Advanced analysis of power trends, focused on detection of slow or sudden power variation, allows to early recognition of IRBP embolization. Sudden power fall may be caused by inflow cannula occlusion by clot located in left ventricle. Slow and gradual power growth may be caused by outflow graft embolization. Sudden power increasing is most frequently related to the embolus located in hydrodynamic bearing. In all listed cases the simple threshold alarm is insufficient to recognize slight power changes, which is first symptom of emerging embolization. Advanced power analysis together with pump acoustic spectrum assessment and medical examinations leads to early detection of thrombus growing inside the pump. It allows to administer thrombolytic treatment at the early stage of embolus growth with prevents ischemic stroke and make therapy sufficient. We used presented methodology 4 times in order to confirm embolization of HVAD and effectiveness of thrombolysis. In all cases the treatment used was successful.

2.2 Flow Related Events

Clinically utilized IRBP have no advanced technical solutions enabling automatic speed adaptation to various hemodynamic conditions. Therefore in some cases (e.g. during physical exercises) heart supporting may be insufficient. Moreover, it is possible that left ventricle comes to be empty (e.g. due to low preload or too high blood draining) and there is lack of blood to be pumped by IRBP. In this case the suction event (intermittent or permanent) occurs which increases risk of thromboembolism and leads to hemodynamic complications. To avoid this effect some pumps can temporary reduce the impeller speed which allows better filling of left ventricle. In other case the pump afterload (arterial pressure) may be too high with respect to pump speed and in consequence the retrograde flow occurs. This kind of flow increases risk of pump embolization and makes supporting insufficient.

The pump flows (mean, minimum and flow pulsation amplitude) in case of suction event and retrograde flow were presented in Fig. 3(a) and (b), respectively. We recognized that typical symptoms of suction event are: sudden fall of minimum flow **(2)** (from 3.0 l/min to 0.1 l/min) corresponding to deterioration of mean flow **(3)** (from 4.3 l/min to 3.2 lmin) and significant rise of flow pulsation amplitude **(1)** (from 4.2 l/min to 6.8 l/min). In contrast to the suction event the retrograde flow **(4)** is characterized by minimum flow below zero. At the same time the character of mean flow and flow pulsation is similar to the case of suction event.

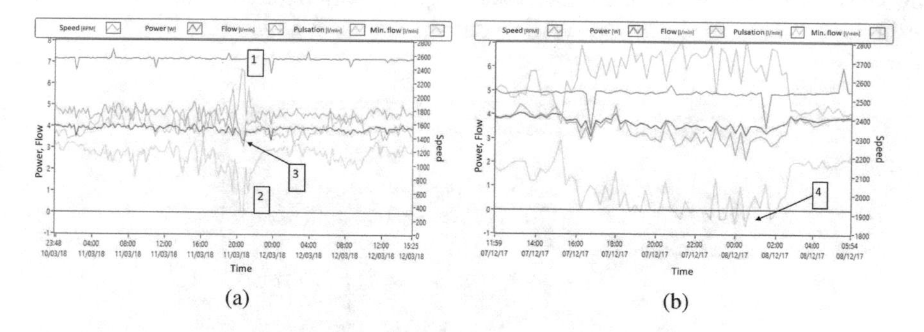

Fig. 3. Course of the intermittent suction event (a) and the retrograde flow (b) (labels were commented in text)

Differences between flow signals in case of suction and retrograde flow are small. Suction may be confirmed by means of echocardiography and assessment of dimensions of left ventricle, however this kind of examination cannot be automated and make by patient in home. Suction and retrograde flow may be detected based on advanced flow signal analysis only. We have been using described methodology during every follow-up visit of patients supported by

HVAD (Medtronic). In 8 cases our analysis carried out in 2017 were an indication to optimization pump speed or to introduce pharmacological treatment. In all cases the negative hemodynamic effects resolved.

2.3 Cardiac Arrhythmia Influence on IRBP Operation

About 70% of patients on MCS therapy have implantable cardioverter-defibrillator (ICD) in order to avoid arrhythmia or ventricular fibrillation (VF). The IRBP and insufficient heart consist an interconnected hemodynamic system so in case of arrhythmia the pump preload decreases, due left ventricle filling decrease, causing pump output reduction. In parallel poor left ventricle filling could lead to suction events.

Course of the pump operation during VF has been presented in Fig. 4. Before VF **(1)** patient's hemodynamic parameters were stable: baseline flow was 4.1 l/min, minimum flow was about 2.1 l/min and flow pulsation was about 3.8 l/min. The region labeled by **(2)** represents persistent VF (lasted 72 hours). After admission to the hospital VF was confirmed in ECG. Echocardiography revealed thrombus adhered to the non-coronary cusp of aortic valve therefore due to the risk of ischemic stroke patient could not been defibrillated before clot lysis. Significant influence of cardiac contraction on pump operation is apparent at a part of the chart labeled as **(3)**: during VF the mean flow fell from 4.1 l/min to 2.3 l/min, minimum flow equaled to mean flow and flow pulsation disappeared. This patient has tolerated persistent VF because he was supported by IRBP. His hemodynamic parameters significantly deteriorated but they were sufficient to assure minimum cardiac output required to survive.

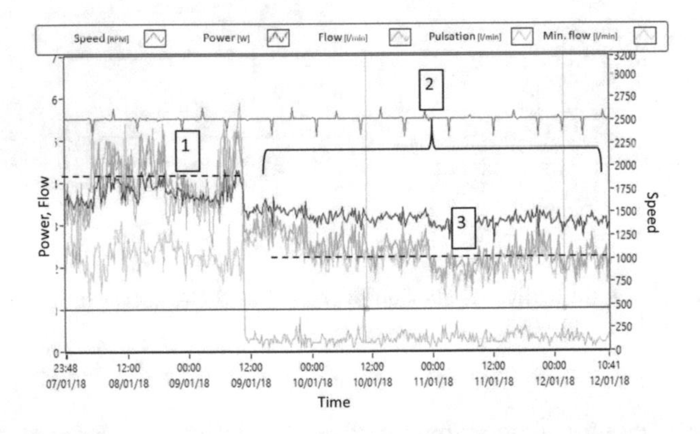

Fig. 4. Course of IRBP operation before and during ventricular fibrillation (labels were commented in text)

Cardiac arrhythmia is not as hazardous as VF, however it causes deterioration of patient's hemodynamics (especially filling of left ventricle), which affects the IRBP operation. In many cases it may cause blood flow deterioration or even intermittent suction. Cardiac arrhythmia is treated pharmacologically or by means of ICD usage. Incorrect settings of ICD may affects the cardiac hemodynamic efficiency what has been presented in Fig. 5. Within the region labeled as **(1)** ICD was set sub-optimally, what caused frequent mean flow drops below of low flow alarm level. The pump speed variations **(2)** had not improved patient's hemodynamic therefore the ICD settings were optimized **(3)**. After this the hemodynamic status has improved which resulted with the low flow alarms disappearance.

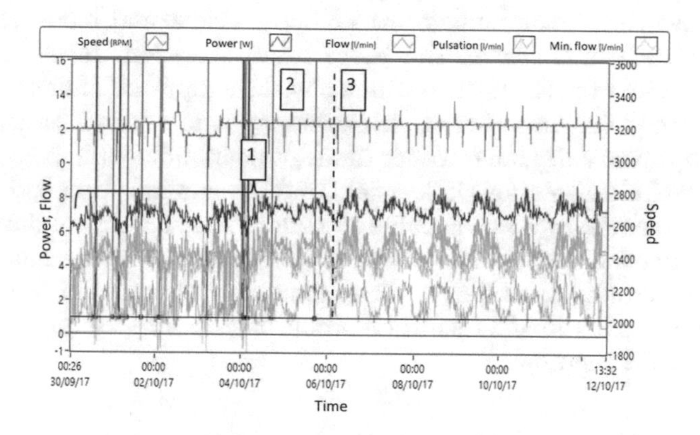

Fig. 5. Influence of ICD settings on IRBP operation (labels were commented in text)

We found correlation between course of flow signals and cardiac arrhythmia. We applied presented methodology in order to patient assessment during every follow up visit. Based on this analysis in 3 cases the ICD settings were optimized and in 2 cases the course of VF treatment were monitored.

3 Necessity of Telemonitoring in MCS Systems

Selected case studies presented above have clearly shown, that advanced analysis of signals measured and collected by MCS system allows for early detection some of side effects. It should be noticed that currently in leading clinically used MCS systems this kind of analysis can be carried out only with hindsight, due to lack of remote access to data stored in IRBP controllers.

The monitoring of patient's biological parameters is as important as pump signals. It was medically proven that the surviving probability of patients on MCS depends on maintaining the mean arterial pressure in the range of 70–80 mmHg [6]. Another crucial issue is pharmacological weakening of coagulation system. Effectiveness of this treatment is routinely assessed by INR ratio measurement from small sample of capillary blood. This kind of examination is made by patient in home but at result must be discussed with physician in order to determine the anticoagulant dose. Another important parameters are patient's weight and body temperature. Gradually increasing of patient's weight may be a symptom of intracellular water retention. Increased body temperature may be caused by wound infection or other inflammatory reaction which influencing coagulation system and sometimes could lead to serious septic problem. Actually, in clinically used MCS systems neither of parameters listed above are not supervised automatically. Routinely a nurse of physicians call to the patient in order to collect his or her actual medical history.

Next important issue is patient's physiotherapy. It has been proved [7] that regular physical activity keeps patient in good health, improves physical endurance, facilitates maintaining proper body weight and blood pressure and prepares patient's organism to the heart transplantation. However, after discharge home, a lot of patients discontinue physiotherapy and limit physical activity. Actually the telemonitoring of physiotherapy is focused on patients suffered from coronary artery disease or other kinds of heart failures.

Modern technical solutions in the field of telecommunication and informatics allow to make complex system capable to remote transmitting of data measured by IRBP controller and external devices (like blood pressure monitor, weigh scale, INR-meter etc.) as well as on-line monitoring of course of the physiotherapeutic session. This kind of solution would significantly improve safety and efficiency of MCS therapy.

4 Telemonitoring System Intended for Rotary Blood Pump ReligaHeart ROT

Polish rotary blood pump ReligaHeart ROT (refer to Fig. 6) has been developed in Foundation of Cardiac Surgery Development [8]. This partially implantable, low-weight and energy saving centrifugal IRBP contains bearing-less suspended impeller and paracorporeal controller, capable to remote data transmission. This solution was a base to develop the idea of complex telemonitoring system, which has been depicted in Fig. 7.

Telemonitoring system allows to wireless transmission of data measured by pump controller and external devices (blood pressure monitor, INR meter, thermometer, weigh scale, pulse-oximeter and ECG) to the central supervising system containing data base. The system creates possibility to transmit statistical data in form of log-file or power and flows waveform (on demand, in case of advanced supporting assessment). In order to link miscellaneous transmission protocols of each of external devices as well as pump controller we used hub device in form of smartwatch. Additional functions of hub are text message displaying and survey handling. Data transmission between pump controller, external devices and hub is held via Bluetooth 2.0, 2.1, 4.0. Hub is connected to the central supervising system via GPRS or WiFi and REST protocol. The controller of ReligaHeart ROT pump is directly connected to the hub via Bluetooth. Physician and nurses can use external terminals in order to communicate with central supervising system.

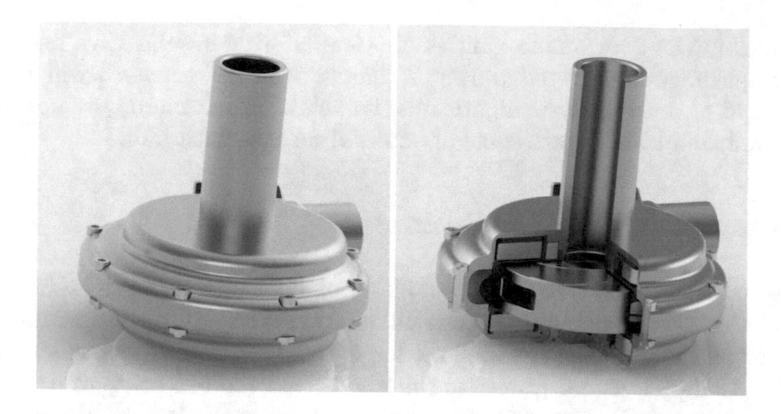

Fig. 6. Polish centrifugal blood pump ReligaHeart ROT

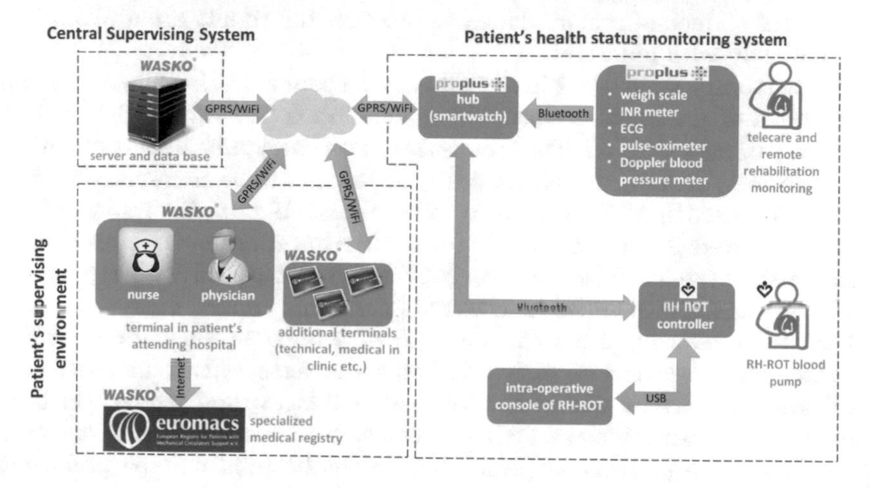

Fig. 7. Idea of telemonitoring system intended for ReligaHeart ROT pump

Unique function of discussed telemonitoring system is remote supervising of physiotherapy session. Supervising itself may be automatic (under control of central monitoring system) or performed by physician. In both cases on-line transmission of ECG (up to 6 leads) is held.

Except functions listed previously the central monitoring system makes advanced processing of collected data (flow and power signals) in order to early detection of symptoms of MCS therapy side-effects (like embolization, suction events, retrograde flow, hemodynamic parameters deterioration due to cardiac arrhythmia) as well as patient's biological data analysis focused on assessment of course of MCS treatment and detection of any health problems. Finally, the central monitoring system stores all available data of supported patients in data base in order to further statistical analysis and cooperation with medical registry.

Currently the core functionalities of telemonioring system have been developed, implement and tested during 2 animal trials. Next work will focus on deployment of data analysis algorithms, launch of physiotherapy remote supervising module and integrate some of external devices with hub.

5 Discussion and Conclusions

In spite of poor measurement capabilities of clinically utilized MCS systems the available data registered by controller allows for non-invasive assessment of course of heart assistance as well as for early detection of side-effects symptoms [9]. Complex signal processing methods and limited computing power of microprocessors make impossible to implement appropriate algorithms in pump controller. Advanced analysis may be carried out by means of external computers, but due to lack of remote data transmission the results are available with delay and only with hindsight.

Contemporary state of art in the field of telecommunication allows to remote transmission of statistical data as well as waveforms (including ECG) independently on patient's location. The legal issues (data encryption, assure of patient's anonymity etc.) are easy to fulfill and they should not to be an formal obstacle to introduce telecare in MCS system. In spite of this, till now, the leaders of MCS have just started implementing the telemonitoring solutions to manufactured cardiac assist systems. The only one MCS system equipped with capability of remote data transmission [10] is marginal.

Remote monitoring of patients on MCS may improve the efficiency of their home-treatment because of early reaction to adverse events. In consequence it will lead to decrease the number of readmission to the hospital which may decrease the treatment costs for the medical care system as well as for the patient. For some patients frequent visits in the hospital may be problematic due to far distance, health reasons on travel logistic. Telecare system would make them live more convenient.

Significant is that the patient's safety on telecare system is continuously supervised which influence the treatment effectiveness. It has been proven that for patients on MCS therapy even routinely telephone interview may improve 6 months patients survival from 90% to 100% and decrease of readmission probability from 67% to 57% [11]. Availability of flow and power waveform data will result in implement advanced methods focused on sophisticated monitoring of patient's hemodynamic status. Even today the methods of flow signal processing in order to detection of aortic valve opening are available [12] as well as patients physical activity assessment [13]. Some investigators try to assess the cardiac contraction and relaxation based on flow waveforms [14].

We hope that as in another areas of medicine (e.g. cardiology, treatment of diabetes) telecare solutions will be introduced into MCS. It will be consecutive milestone on a bumpy road to make long-term MCS outcomes as efficient as heart transplantation.

Acknowledgements. This work has been financially supported by National Centre of Research and Development (grant no. RH-ROT/266798/STRATEGMED–II).

Following commercial companies were involved in telemonitoring project: WASKO S.A., Gliwice, Poland; Pro-PLUS S.A., Warsaw, Poland.

References

1. Warrell, D., Cox, T., Firth, J., Dwight, J.: Oxford Textbook of Medicine: Cardiovascular Disorders, pp. 215–218. Oxford University Press, Oxford (2016)
2. Schroder, J.N., Milano, C.A.: A tale of two centrifugal left ventricular assist devices. J. Thorac. Cardiovasc. Surg. **154**, 850–852 (2017)
3. Antończyk, R., Trejnowska, E., Pacholewicz, J., et al.: Emergency heartware ventricular assist device (HVAD) exchange due to pump thrombosis using minimally invasive technique. Kardiochir. Torakochi. **14**(1), 76–78 (2017)
4. Kirklin, J.K., Pagani, F.D., Kormos, R.L., et al.: Eighth annual INTERMACS report: special focus on framing the impact of adverse events. J. Heart Lunt Transplant. **36**(10), 1080–1086 (2017)
5. Kaufman, F., Hörmandinger, C., et al.: Acoustic spectral analysis for determining pump thrombosis in rotary blood pumps. ASAIO J. **60**(5), 502–507 (2014)
6. Bennett, M.K., Adatya, S.: Blood pressure management in mechanical circulatory support. J. Thorac. Dis. **7**(12), 2125–2128 (2015)
7. Marko, C., Danzinger, G., Käferbäck, M., et al.: Safety and efficacy of cardiac rehabilitation for patients with continuous flow left ventricular assist devices. Eur. J. Prev. Cardiol. **22**(11), 1378–1384 (2015)
8. Major, R., Kustosz, R., Trembecka-Wójciga, K., et al.: Development of surface modification methods for ReligaHeart cardiac support system. Arch. Metall. Mater. **61**(3), 1053–1058 (2016)
9. Gawlikowski, M., Kustosz, R., Glowacki, M., et al.: Non-invasive assessment of thromboembolism in rotary blood pumps: case study. In: Proceedings of SPIE, 12th Conference on Integrated Optics: Sensors, Sensing Structures, and Methods, vol. 10455, p. 104550L (2017)
10. Csepe, T., Kilic, A.: Advancements in mechanical circulatory support for patients in acute and chronic heart failure. J. Thorac. Dis. **9**(10), 4070–4083 (2017)
11. Schlöglhofer, T., Horvat, J., Hartner, Z., et al.: Standardized telephone intervention algorithm for improved ventricular assist device outpatient management. J. Heart Lunt Transplant. **35**(4), S388 (2014)
12. Granegger, M., Masetti, M., Laohasurayodhin, R., et al.: Continuous monitoring of aortic valve opening in rotary blood pump patients. IEEE Trans. Bio-Med. Eng. **63**(6), 1201–1207 (2016)
13. Granegger, M., Schlöglhofer, T., Ober, H., et al.: Daily life activity in patients with left ventricular assist devices. Int. J. Artif. Organs **39**(1), 22–27 (2016)
14. Rich, J.D., Burkhoff, D.: HVAD flow waveform morphologies: theoretical foundation and implications for clinical practice. ASAIO J. **63**(5), 526–535 (2017)

Estimation of Apnea-Hypopnea Index in Sleep Breathing Disorders with the Use of Artificial Neural Networks

Tomasz Walczak[1(✉)], Renata Ferduła[1], Martyna Michałowska[1], Jakub Krzysztof Grabski[1], and Szczepan Cofta[2]

[1] Institute of Applied Mechanics, Faculty of Mechanical Engineering and Management, Poznań University of Technology, Jana Pawła II 24, 60-965 Poznań, Poland
tomasz.walczak@put.poznan.pl
[2] Department of Pulmonology, Allergology and Respiratory Oncology, Poznań University of Medical Sciences, Szamarzewskiego 84, 60-569 Poznań, Poland

Abstract. Sleep Apnea Syndrome (SAS) becomes an important medical and social problem of contemporary societies. It is burdensome, it can be dangerous to health and even cause of death. The most efficient way to detect this syndrome is polysomnography. It gives good results but it is expensive and not commonly available. Main aim of this study is to present another, easier and cheaper way to detect SAS. Proposed method is based on prediction of sleep state using only oximetry and heart rate. The Artificial Neural Network (ANN) algorithm to predict time series was introduced. These networks were used to detect apneas and hypopneas to support diagnose of SAS and to detect whether patient sleeps or not. All data needed to train and test ANN were collected in sleep laboratory for a group of five considered patients with diagnosed SAS. The presented in this work results show that it is possible to predict apneas during sleep with high rate of accuracy, just with use of information about heart rate and blood oxygen saturation. It means that presented method could be effective to diagnose this disease using only simple device with implemented ANN.

Keywords: Sleep apnea · Artificial neural network · Polysomnography

1 Introduction

About one-third of human life is devoted for sleeping [3]. It is time essential for body regeneration and it is extremely important to provide as best as possible quality of sleep, in order to assure proper organism functioning. Permanent sleep breathing disturbance result in significant psychological and physiological problems like: serious cognitive and behavioral changes, problems with concentrations (e.g. while reading), lapses of memory, hallucinations, depression as well as heart arrhythmia, arterial hypertension, cardiac failure or diabetes [1,2].

E. Tkacz et al. (Eds.): IBE 2018, AISC 925, pp. 96–106, 2019.
https://doi.org/10.1007/978-3-030-15472-1_11

The most efficient way to detect sleep disorders and to assess sleep characteristics and quality is polysomnographic (PSG) examination. It is the complex medical assessment of patient vital functions while sleeping (at least 6 h at night), consisted of EMG (electromyography, commonly of chin muscle), EEG (electroencephalography), EOG (electrooculography), ECG (electrocardiography), air flow through upper airway, thoracic and abdominal respiratory movements, heart rate and oximetry signal measurement and analysis [4]. Fragment of results received during polysomnography with marked apnea and desaturation is shown in the Fig. 1. The aim of PSG is usually detection of obstructive or limited respiratory events (hypopnea, arousals) at night what indicates sleep apnea where frequency and duration of these events describe severity of the disease. Generally, sleep apnea is defined as repetitive episodes of inability to breath during sleep [2], but in detail it is calculated in the form of AHI (apnea-hypopnea index) measured by number of apnea or hypopnea episodes per hour of sleeping with regard to level of blood oxygen desaturation during night and level of daily symptoms. However, even if AHI is low, severe apnea syndrome can be diagnosed due to significant desaturation (more than 10% of sleep with less than 90% of oxygen saturation) or due to small amount of deep sleep (less than 5%) [1].

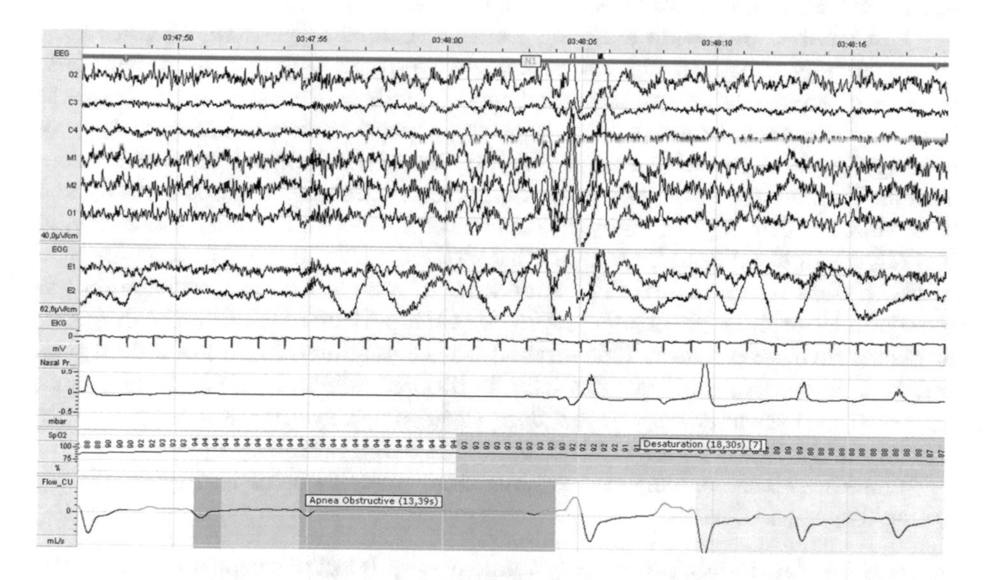

Fig. 1. Fragment of results received during polysomnography with marked obstructive apnea and desaturation (from EMBLA 4000 system). From the top, signal from: six EEG electrodes (O2, C3, C4, M1, M2, O1), two EOG electrodes, EKG, nasal pressure, oximetry and flow

Sleep apnea is caused by excessive relaxation of upper airway muscles what causes narrowing or completely closing an airway and fall in oxygen saturation. As the oxygen saturation decreases, respiratory center in brain sends signal to respiratory muscles to activate them, but it is not enough to overcome the obstruction in upper airway and the heart rate as well as blood pressure begin to rise. When level of oxygen is too low, the person wakes up to breath. After a few seconds of breathing, the person can fall asleep again without remembering awakening [2]. In literature sleep apnea is divided into two main categories [1]:

- central sleep apnea which is caused by a failure of the brain's mechanism that controls breathing,
- obstructive sleep apnea, when the upper airways are obstructed (more common type of sleep apnea),
- mixed sleep apnea (initially central, in continuation obstructive).

In clinical practice of respiratory physicians more than 80% of cases of sleep disturbances is related to obstructive pattern.

The most popular symptoms of people suffering from sleep apnea are morning headaches, feeling of exhaustion and sleeping during the day despite sleeping for seven or more hours at night what highly affects daily activities. These symptoms building up with time can lead even to death during night or early morning hours. Sleep apnea syndrome (SAS) affects 1–2% of children, 2–15% of middle-aged adult and more than 20% of older people [1].

The number of people suffering from sleep apnea syndrome is constantly rising. Among others, it is because knowledge about this disease is rising and obesity of society is rising. More and more people go to Sleep Laboratories to identify their sleep disorders. Then they come again to check whether surgery or oral appliance helped. Other come regularly to check whether their CPAP (Continuous Positive Airway Pressure) works correctly. Demand exceeds capacity and wait time are growing. In many locations around the word patients having clinical signals of sleep apnea syndrome have to wait for months (up to 60 months in the United Kingdom) [7]. The American Academy of Sleep Medicine recommends in-laboratory polysomnography for each patient to confirm the diagnosis of obstructive sleep apnea. This method was checked and has a low failure rate, but it is expensive and patients have limited access to it. So it is clear that an easier and cheaper way to diagnose sleep apnea syndrome is needed. Such a group of devices is called home sleep apnea testing (HSAT) [9].

The American Academy of Sleep Medicine (AASM) name four types of sleep apnea diagnostic devices:

- type 1 – full in-laboratory polysomnography (electroencephalography, electrooculography, chin and leg electromyography, airflow, effort of breathing and oximetry),
- type 2 – comprehesive portable polysomnography with at least seven channels (have to allow to recognize sleep phases and airflow),
- type 3 – portable testing limited to sleep apnea (have to include at least four channels, among others effort, airflow and oximetry).
- type 4 – continuous recording of one or two signals (oximetry and another signal).

All of them are accepted by AASM for prescription CPAP device [5].

Home sleep apnea testing devices collect only some signals from polysomnography, but enough to recognize sleep apnea syndrome. It is possible with class II – IV devices. They can solve the problem of waiting for months (or even years) for examination. But they have some drawbacks. Patient does not have such a knowledge and experience as technician and incorrect usage of device can generate many errors. Also, such devices do not include electroencephalography so technician can't mark whether patient sleeps or not. But estimate time of sleep is important in counting AHI. The American Academy of Sleep Medicine recommended using in such situations REI – respiratory event index (the number of respiratory events per hour of recording time from a HSAT). HSAT can decrease the result of the severity of disorders [6]. AASM published recommendations for using home sleep apnea testing and they much restrict usage of HSAT [8–10].

Considering the fact, that polysomnographic examination requires a lot of technical equipment, that is expensive and hard to available, the authors of this study decided to create and evaluate neural network model in order to propose easier way to detect apnea and hypopnea episodes. The aim was to check whether it is necessary to use so many signals and which of them are really decisive. From many signals from full polysomnography pulse and oxygenation of blood were chosen because they also provide good results and are simple to measure. This type of sleep quality assessment might be useful in home-made monitoring of treatment effectiveness in patients suffering from SAS. The aim of the study was to show how useful can artificial neural networks be in predicting and counting apneas and hypopneas in examination of people with Sleep Apnea Syndrome. The authors of this study hope that this will help in discussions about using simpler devices in Sleep Apnea Syndrome diagnosis.

2 Experimental Procedures

Experimental procedures were carried out in Sleep Laboratory where patients every night have polysomnographic examination. There were 7 EEG electrodes: two central (C3 and C4), two occipital (O1 and O2), two parietal (P3 and P4) and one reference frontal electrode (Fz). There were also two EOG electrodes 1 cm below and above each eye and one on the forehead. On this basis technician marked sleep stages. EKG, nasal pressure, thoracic and abdominal respiratory movements, pulse and oxygenation of blood were also recorded. On this basis technician marked desaturation, apneas and hypopneas.

Five patients took part in the study. There were chosen patients that symptoms suggested severe obstructive sleep apnea syndrome: loud snoring, excessive daytime sleepiness, breathing cessations during sleep, waking with a dry mouth, morning headaches, attention problems, falling asleep during ritual activities. Patients were examined using polysomnography for one night (each of them was examined for at least 6 h). A technician analyzed results of the examination and marked apnea and hypopnea events, desaturation and sleep fazes. All patients had AHI higher than 25.

3 Artificial Neural Network

Artificial neural networks (ANN) are an advanced computational tool enabling solving complex problems, including technical, economic or related to medicine [11]. The main advantage of the algorithms based on ANN is the ability to process a huge amount of data and find certain regularities even when one do not know the mathematical model describing the relations between data in the considered set. They have been used successfully in many issues of classification, regression and prediction also in the health sciences. The ANN application in health areas may concern classification of various symptoms, searching for disease patterns, supporting the diagnostic process, analysis of medical images, as well as predicting the effects of treatments, rehabilitation and pathological changes [12–14]. In this work, the neural network for time series prediction is used to predict the sleep processes and SAS, based on the signals that were collected during sleep examinations of considered patients. Obtained data (pulse, oxygenation of blood, marked sleep fazes, apnea and hypopnea) was saved as a text file with the aim of further data processing in MATLAB. The sampling frequency of this data was 20 Hz. Such a high frequency wasn't needed so every fifth sample was taken. Based on this data artificial neural network's inputs were prepared. There were two parts of research:

- training neural network on the basis of one third of data for one patient and testing it on the basis of two thirds of those data,
- training neural network on the basis of data from four patients and testing it on the basis of data from the fifth.

Each part has two fazes of processing:

- checking whether patient sleeps or not,
- checking when apnea and hypopnea occur.

In both cases, the input was pulse and oxygenation of blood.

The MATLAB Neural Network Toolbox was used to conduct all simulations. Nonlinear autoregressive neural network with external input was used. It was predicting output using current input (pulse, oxygenation of blood – $x(t)$), 10 past values of input data (pulse, oxygenation of blood – $x(t-1), ..., x(t-d)$) and 10 last values of output data (sleep and awakening – $y(t-1), ..., y(t-d)$), where d – number of past values of input and output ($d = 10$) (Fig. 2). Such architecture was chosen because in such signals each sample depends on earlier ones.

The network was created and trained in open loop form. From data intended for training the network 70% of input data were used to train network, 15% were used to measure network generalization (validation) and 15% were used for testing during and after training. There were 10 hidden neurons as shown in (Fig. 3). The network was trained using Levenberg-Marquard algorithm.

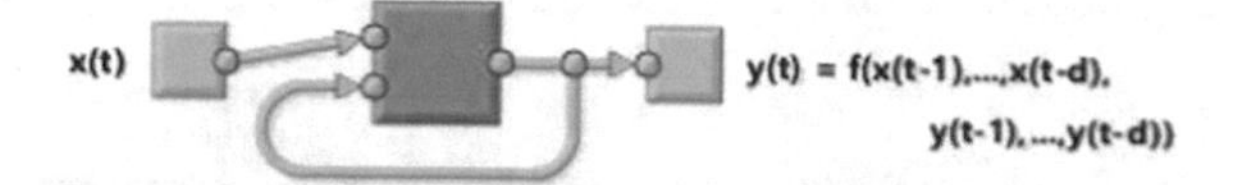

Fig. 2. The scheme of considered artificial neural network

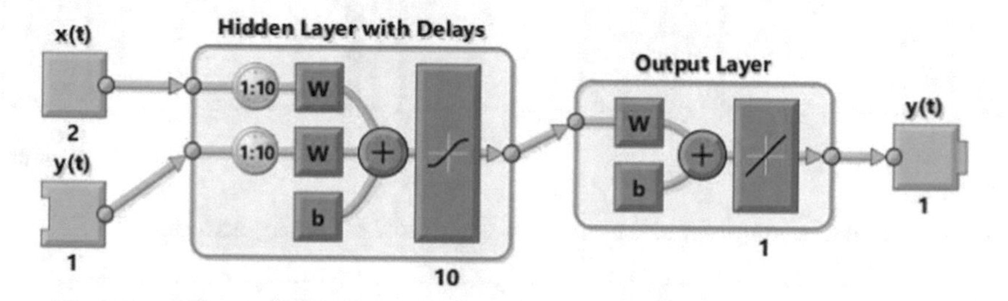

Fig. 3. The architecture of considered artificial neural network

4 Results

All fazes of sleep was changed into binary: 0 means not sleeping and 1 means sleeping. The same was made for apneas and hypopneas – 1 and normal breathing – 0.

Firstly each of five patients were examined separately. First one third of sleep data was taken to train neural network and then this network was used to recognize sleep and apneas, and hypopneas in the next two thirds of those data. The results are shown in the Table 1. First part of the table deals with examination of sleeping or not and the second part of it deals with apnea and hypopnea events. There were calculated mean squared error (MSE), numbers of wrong marked samples and relative error. As it is shown in the Figs. 4 and 5 errors occur only at the beginning and the end of sleeping and hypopnea or apnea. So the number of errors depends on number of awakenings and number of apneas and hypopneas. Only patient 3 has some additional errors and time of sleeping marked by neural network was a bit longer than time of sleeping marked by technician (episode of such errors are shown in the Fig. 6). But the difference was so small that it is negligible in AHI.

Because results of artificial neural network trained on fragment of data was so good, there was decided to tray to train network on data from some patients and check, whether it can recognize sleeping and apnea, and hypopnea events in data from another patient in second part of examination. There were five patients checked. Each time, data from one patient from group of five patients was taken to test network and another four were linked together as an input. Results are shown in the Table 2.

Also in second part of examination number of errors depend on number of wakings, apneas and hypopneas. They always occur at the beginning and at the end of sleeping period. Times of sleeping based on data marked by neural

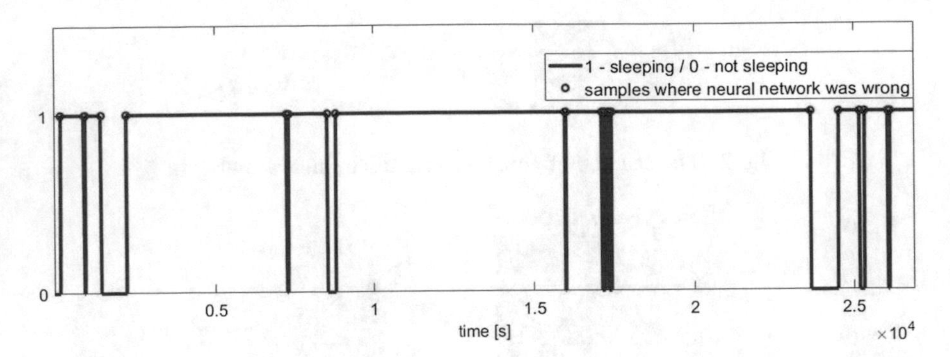

Fig. 4. An example of sleep diagram with marked wrong answers of ANN

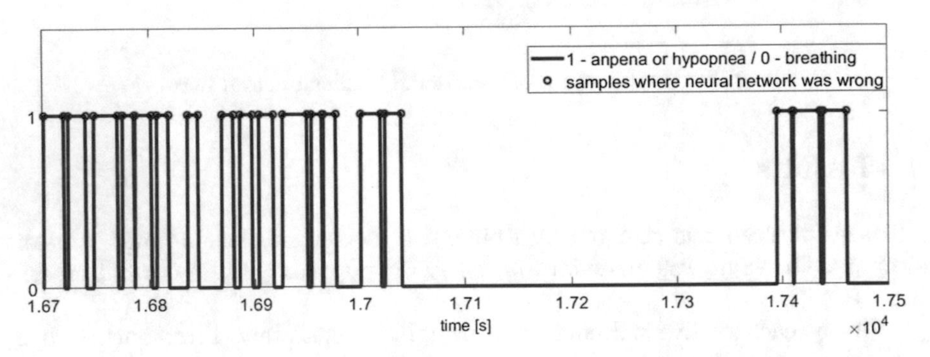

Fig. 5. An example of apnea and hypopnea diagram with marked wrong answers of ANN

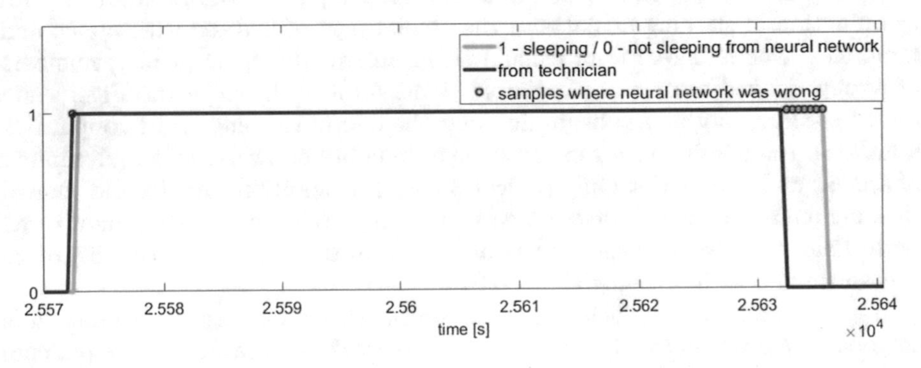

Fig. 6. Fragment of apnea and hypopnea graph for patient 3

network were the same as based on data marked by technician. Also AHI was the same.

As it was shown only device needed to designate number of apnea and hypopnea episodes per hour (AHI) is oximeter. Artificial neural network can easily and fast recognize whether patient sleep or not and whether he is breathing normally

Table 1. Results of artificial neural network trained with one third of patients data and tested with two thirds of them (a–h – apnea or hypopnea)

	Patient 1	Patient 2	Patient 3	Patient 4	Patient 5
Number of training samples	12584	12712	11967	12968	12538
Number of validation samples	2697	2724	2565	2779	2687
Training MSE [10^{-4}]	4.72	3.07	0.83	5.04	5.50
Validation MSE [10^{-5}]	0.48	0.61	0.12	37.18	39.41
Number of testing samples	53934	54481	51291	55578	53736
Testing MSE [10^{-4}]	5.61	6.24	11.86	4.56	4.05
Number of errors	30	26	83	24	21
Relative error [%]	0.056	0.048	0.162	0.043	0.039
Time of sleeping (by network) [h]	6,387	6,040	6,073	6,531	6,730
Time of sleeping (by technician) [h]	6,387	6,040	6,065	6,531	6,730
Number of training samples	12584	12712	11967	12968	12538
Number of validation samples	2697	2724	2565	2779	2687
Training MSE [10^{-3}]	11.74	3.51	0.56	3.71	19.87
Validation MSE [10^{-3}]	7.02	3.24	0.73	3.99	23.64
Number of testing samples	53934	54481	51291	55578	53736
Testing MSE [10^{-2}]	1.47	0.95	0.54	0.71	2.53
Number of errors	804	527	266	388	1408
Relative error [%]	1.49	0.97	0.52	0.70	2.62
Number of a–h (by network)	402	261	131	194	700
AHI (by network)	62,95	43,21	21,57	29,71	104,02
Number of a–h (by technician)	401	261	132	194	700
AHI (by technician)	62,95	43,21	21,76	29,71	104,02

or has apnea or hypopnea. Although there are some errors, but they have no impact on length of sleeping, apneas and hypopneas so the final result of recognition is correct.

Table 2. Results of artificial neural network trained with four patients and tested with fifth (a–h – apnea or hypopnea)

	Patient 1	Patient 2	Patient 3	Patient 4	Patient 5
Number of training samples	150560	150177	152411	149410	150698
Number of validation samples	32263	32181	32659	32016	32293
Training MSE [10^{-4}]	4.57	4.52	4.85	4.27	4.64
Validation MSE [10^{-4}]	3.41	4.35	3.98	5.57	6.56
Number of testing samples	53934	54481	51291	55578	53736
Testing MSE [10^{-4}]	5.56	4.77	21.23	4.32	3.97
Number of errors	30	26	25	24	21
Relative error [%]	0.056	0.048	0.049	0.043	0.039
Time of sleeping (by network) [h]	6,387	6,040	6,065	6,531	6,730
Time of sleeping (by technician) [h]	6,387	6,040	6,065	6,531	6,730
Number of training samples	150560	152411	152411	149410	150698
Number of validation samples	32263	32659	32659	32016	32293
Training MSE [10^{-2}]	1.15	1.28	1.38	1.34	0.89
Validation MSE [10^{-2}]	1.23	1.34	1.44	1.29	0.90
Number of testing samples	53934	51291	51291	55578	53736
Testing MSE [10^{-2}]	1.46	0.94	0.86	0.70	2.73
Number of errors	804	522	272	388	1420
Relative error [%]	1.49	0.96	0.53	0.70	2.64
Number of a–h (by network)	402	261	134	194	702
AHI (by network)	62,95	43,21	22,09	29,71	104,31
Number of a–h (by technician)	401	261	132	194	700
AHI (by technician)	62,95	43,21	21,76	29,71	104,02

5 Summary

The artificial neural networks can be used successfully to support the diagnosis of Sleep Apnea Syndrome. Authors of this study suggest that oximeter is enough to recognize sleep apnea syndrome. This device is easy to use so anyone without much knowledge can operate it. Also problem of recognizing of sleeping was solved using neural network so recognizing of AHI (and not REI) is possible. Such a solution allows patients to be examined at home, even for a few nights and sleep comfortable in domestic environment. Such a device is much cheaper than full polysomnography device so hospitals can afford to buy more oximeters and decrease time of waiting for examination. Also using artificial neural network technician's time spend on analyzing data from examination is decreased.

Acknowledgements. The presented research results were funded with the grant 02/21/DSPB/3513 allocated by the Ministry of Science and Higher Education in Poland.

References

1. American Psychiatric Association: Sleep wake disorders DSM-5 Selections. American Psychiatric Association (2015)
2. Caldwell, J.P.: Sleep: The Complete Guide to Sleep Disorders and a Better Night's Sleep. Firefly Books (1997)
3. Hamet, P., Treablay, J.: Genetics of the sleep-wake cycle and its disorders. Metab. Clin. Exp. **55**(Suppl 2), S7–S12 (2006)
4. Fairbanks, D.N.F., Fujita, S., Ikematsu, T.M.D., Simmons, F.B.: Snoring and Obstructive Aleep Apnea. Raven Press, New York (2016)
5. Department of Health and Human Services, Center for Medicare and Medicaid Services: Decision Memo for Continuous Positive Airway Pressure (CPAP) Therapy for Obstructive Sleep Apnea (OSA). CAG-0093R, 13 March 2008
6. Bianchi, M.T., Goparaju, B.: Potential underestimation of sleep apnea severity by at-home kits: rescoring in-laboratory polysomnography without sleep stagings. J. Clin. Sleep Med. **13**, 551–555 (2017)
7. Flemons, W.W., Douglas, N.J., Kuna, S.T., et al.: Access to diagnosis and treatment of patients with suspected sleep apnea. Am. J. Respirat. Crit. Care Med. **169**, 668–672 (2004)
8. Kapur, V.K., Auckley, D.H., Chowdhuri, S., et al.: Clinical practice guideline for diagnostic testing for adult obstructive sleep apnea: an american academy of sleep medicine clinical practice guideline. J. Clin. Sleep Med. **13**, 479–504 (2017)
9. Collop, N.A., McDowell Anderson, W., Boehlecke, B., et al.: Clinical guidelines for the use of unattended portable monitors in the diagnosis of obstructive sleep apnea in adult patients. J. Clin. Sleep Med. **3**, 737–747 (2007)
10. Epstein, L.J., Kristo, D., Strollo, P.J., et al.: Clinical guideline for the evaluation, management and long-term care of obstructive sleep apnea in adults. J. Clin. Sleep Med. **5**, 263–276 (2009)
11. Tadeusiewicz, R., Korbicz, J., Rutkowski, L., Duch, W.: Sieci neuronowe w inzynierii biomedycznej, Tom 9. (ang. Neural Networks in Biomedical Engineering, vol. 9). Akademicka Oficyna Wydawnicza EXIT, Warszawa (2013)

12. Michałowska, M., Walczak, T., Grabski, J. K., Grygorowicz, M.: Artificial neural networks in knee injury risk evaluation among professional football players. In: AIP Conference Proceedings, Lublin, p. 70002 (2018)
13. Walczak, T., Grabski, J., Grajewska, M., Michałowska, M.: Application of artificial neural networks in man's gait recognition. In: Advances in Mechanics: Theoretical, Computational and Interdisciplinary Issues, pp. 591–594. CRC Press (2016)
14. Grabski, J.K., Walczak, T., Michałowska, M., Cieslak, M.: Gender recognition using artificial neural networks and data coming from force plates. In: Gzik, M., et al. (eds.) Innovations in Biomedical Engineering, IBE 2017. Advances in Intelligent Systems and Computing, vol. 623, pp. 53–60. Springer, Cham (2018)

Voice Fatigue Evaluation: A Comparison of Singing and Speech

Aleksandra Zyśk(✉), Marcin Bugdol, and Paweł Badura

Faculty of Biomedical Engineering, Silesian University of Technology,
Roosevelta 40, 41-800 Zabrze, Poland
olazysk94@gmail.com

Abstract. A study on voice fatigue evaluation is presented in this paper. It concerns the problem of vocal apparatus disorders with a general goal in the computer-assisted proper voice emission teaching support and possible disorder diagnosis. The singing and speech samples provided by 20 singers were recorded before and after a two-hour exhaustive performance. Nine features were extracted from each recording. Each feature was subjected to a statistical analysis using principal component analysis and Wilcoxon test producing feature importance ranking and individual p-values. Finally, a support vector classifier was employed for voice fatigue detection in separate singing and speech experiments, yielding assessment accuracies at 62.9% and 70.9%, respectively. The speech signal proved to reflect the voice fatigue more reliably than singing.

Keywords: Voice analysis · Speech and singing assessment · Voice fatigue · Computer-aided diagnosis

1 Introduction

The vocal apparatus is an indispensable tool for human. By speech production it enables communication, expression of emotions, and artistic expression, in which singing plays a crucial role. For people working with vocal apparatus professionally, the diagnosis of its dysfunctions may affect the course of their careers. Thus, it is necessary to diagnose the disease early and enable rehabilitation.

Singers generally use their voice consciously, since they are taught the correct emission techniques. Therefore, it is difficult to diagnose their vocal disorders. This requires a lot of phoniatric experience. The singers' diseases are caused by e.g.: inadequate vocal skills, maladjustment of vocal parts and singing capabilities, or incorrect vocal classification [1]. The vocal folds nodules belong to the group of disorders caused by improper voice emission and mostly affect people who use vocal skills (singers, teachers, dance and sports instructors) [2]. The groups of singers most susceptible for this disorder are tenors and sopranos. Sometimes, not dangerous soft nodules can appear directly after a demanding performance; they mostly disappear after a short period of voice rest. On the other hand, hard nodules can be caused by excessive voice effort and the emission

E. Tkacz et al. (Eds.): IBE 2018, AISC 925, pp. 107–114, 2019.
https://doi.org/10.1007/978-3-030-15472-1_12

of high frequencies exceeding the vocalist capabilities [3]. The voice burdened by a hard nodule can be described as puffy, trembling, the intonation is unclean, the phonation shortened, and the vibrato loses its flexibility [4]. Similar phenomenon occur in vocal fatigue. Some authors considered it as one of the symptoms of voice disorders [5].

Apart from common diagnostic procedures like the interview, laryngoscopy, and auscultation, complementary methods for voice examination include stroboscopy, photoelectric methods, subglottic pressure measurements, electromyography, glottography, etc. [6]. An approach that brings many facilities and improves patient comfort compared to other methods is acoustic voice analysis. In addition to noninvasiveness, ease to record signal, and small hardware requirements, the method is also promising because of its diagnostic potential [7]. It enables detection of subtle irregularities in the phonation function of the larynx. In [8], the authors also point out that examinations such as auscultation, an interview, and the use of acoustic measures evaluate the voice from diverse perspectives. It was proven that a number of spectral and cepstral parameters determined for normative and pathological voices significantly differ statistically [9]. In [8] a correlation was studied between the commonly used GRBAS3 voice rating system and the Multi-Dimensional Voice Program (MDVP), which is increasingly used by phoniatric clinics [10].

A valuable work broadly covering the issue of acoustic analysis of the voice pathology is Świdziński's habilitation dissertation [10]. The author's main goal was to unify the techniques of voice examination by means of acoustic analysis, divided into various types of voice and speech disorders, and to define the set of acoustic parameters in the assessment of normal and disturbed voices, as well as in detecting organic and functional changes. In the group of 50 patients with disorders, 98% of them were classified correctly.

It should be noted that the speech signal (e.g. a voiced sound) differs from the singing signal for the same person [11]. In [12] a comparison of speech and singing was performed, where mel-frequency cepstral coefficients (MFCC) and the basic frequency curve were used as features. The classification system achieved the accuracy of distinguishing these two signals above 90% for signals with a duration of 2 s. However, to the best of our knowledge no attempt to diagnose the voice pathology based on the singing signal has yet been reported in the literature. It may prove to be a valuable diagnostic tool by providing additional information about the physiological foundations of pathology.

The goal of this study was to compare the diagnostic value of acoustic analysis of speech and singing signal in trained singers. The research material consisted of speech and singing recordings of people with correct emission after (1) initial rehearsal and then (2) after a two-hour voice load, simulating the voice pathology. The recordings were prepared for further analysis. The set of distinctive voice features were defined, subjected to the analysis and selection for each of the two voice fatigue classification tasks: the speech or singing assessment. In either cases the support vector machine (SVM) [13] was used with experimentally adjusted settings.

2 Materials and Methods

The data acquisition and processing framework is presented in Fig. 1.

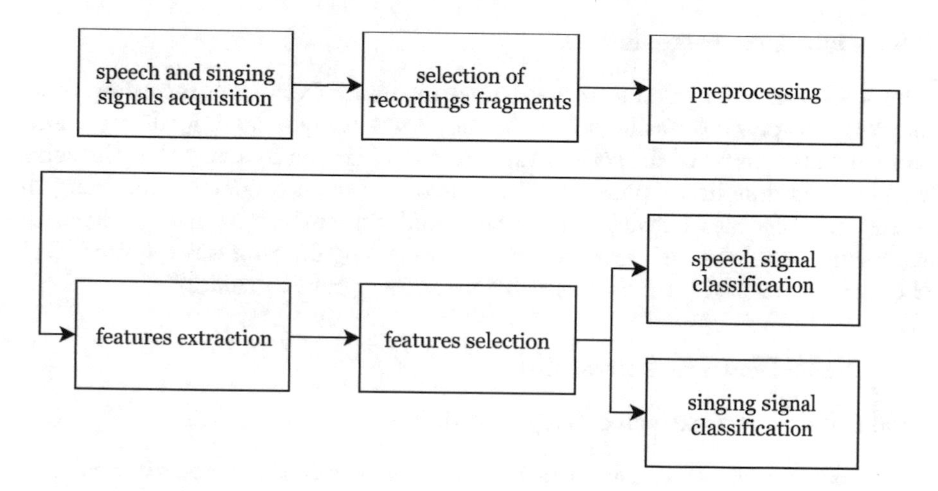

Fig. 1. Workflow of data acquisition and processing.

2.1 Participants

The survey group consisted of 20 chorists: 14 women (10 sopranos, 4 altos) and 6 men (4 tenors and 2 basses). The mean age of participants was 29.0 ± 9.0 (27.2 ± 7.1 and 32.7 ± 11.7 for women and men, respectively).

All the respondents were not professional singers, however, they held exercises on voice emission and participated in choir rehearsals and exhaustive concerts for several years at least once a week for 2 h.

2.2 Experimental Procedure

Prior to the recording, all the respondents took part in 30-min vocal exercises conducted by an experienced conductor. They included warming up the voice apparatus, working on the breath, intonation, vowel equalization, articulation, and involving whole body parts during singing.

The voice acquisition was carried out in a recording studio using an Olympus WS-812 voice recorder with a sampling frequency at 44100 Hz. The voice recorder was ca. 1 m from the subject. Each singer was asked to sing a scale – in case of men the C major scale starting with c^1 sound (261 Hz), and in case of women the D major scale (starting with d^1 sound, 293 Hz). They sang to the "la" syllable. Then, one should say the sentence in Polish: "Ten dzielny żołnierz był z nim razem" (Eng. "This brave soldier was with him together") [10]). For each of the 20 respondents, such recording was made twice: after performing the above-mentioned initial exercises of voice emission and after two hours of exhaustive performance (the Messiah by Georg Friedrich Händel).

2.3 Data Preprocessing

The appropriate fragments were selected manually from each recording of the singing and speech recordings. For singing, it was a signal lasting 105 ms, chosen from the first sound of the scale. Fragments were chosen in a way that the sound was in its middle decay phase. In the case of speech, a fragment containing the sound "a" was also extracted from the word "razem" ("together"). Since the recordings were of a high quality, no noise removing filtering was required. Only the pre-emphasized filter was used before spectrogram calculation.

2.4 Voice Features Extraction

Nine features were extracted every recording:

1. *Shimmer* indicating the variation of tension and the mass of vocal folds described as:

$$shimmer = \frac{\frac{1}{M-1}\sum_{i=1}^{M-1}|A_i - A_{i+1}|}{\frac{1}{M}\sum_{i=1}^{M}A_i}, \tag{1}$$

   where A_i denotes amplitudes of consecutive periods and M is the number of periods.
2. *Jitter* indicating lack of control over vocal folds vibration:

$$jitter = \frac{\frac{1}{M-1}\sum_{i=1}^{M-1}|T_i - T_{i+1}|}{\frac{1}{M}\sum_{i=1}^{M}T_i}, \tag{2}$$

   where T_i stands for consecutive periods durations.
3. Peak amplitude variation vA revealing amplitude fluctuations within the whole signal:

$$vA = \frac{\sqrt{\frac{1}{M-1}\sum_{i=1}^{M}|A_i - \overline{A}|^2}}{\overline{A}}, \tag{3}$$

   where $\overline{A}$ denotes the mean amplitude of beginnings of consecutive periods.
4. Fundamental frequency variation of fundamental frequencies F_{0_i}:

$$vF_0 = \frac{\sqrt{\frac{1}{M-1}\sum_{i=1}^{M}|F_{0_i} - \overline{F_0}|^2}}{\overline{F_0}}, \tag{4}$$

   where $\overline{F_0}$ denotes the mean fundamental frequency originating from next periods of time signal.

5. The above feature is determined also in another way, where fundamental frequencies calculation is based on N consecutive spectrogram frames (vF_{0X}):

$$vF_{0X} = \frac{\sqrt{\frac{1}{N-1}\sum_{n=1}^{N}|F_{0n} - \overline{F_{0X}}|^2}}{\overline{F_{0X}}} \tag{5}$$

6. Harmonic to Noise Ratio HNR describing the contribution of harmonic signal components related to the non-harmonic part:

$$HNR = \frac{max(R)}{R(0)}, \tag{6}$$

where R stands for consecutive peaks of the autocorrelation function.

7. Noise to Harmonic Ratio NHR determined as the ratio of non-harmonic energy originating from one frequency bandwidth and energy sum of the subsequent harmonics from another band:

$$NHR = \frac{E_{inh}}{E_h}. \tag{7}$$

Depending on whether the singing or speech signal was analyzed, different bands were adopted (Table 1).

8. Spectrum Centroid SC coming indicating whether the high or low formants prevail in the power spectral density (PSD):

$$SC = \frac{\sum_{n=1}^{N}(\log_2(F(n)/1000) \cdot PSD(n))}{\sum_{n=1}^{N} PSD(n)}, \tag{8}$$

where $F(n)$ denotes successive frequencies in PSD.

9. Spectrum Spread SS allowing to determine how the energy is distributed in relation to the spectrum centroid:

$$SS = \sqrt{\frac{\sum_{n=1}^{N}((\log_2(F(n)/1000) - SC)^2 * PSD(n))}{\sum_{n=1}^{N} PSD(n)}}. \tag{9}$$

Table 1. Frequency bands for NHR determination.

	Singing	Speech
Non-harmonic band	0–8500	0–4500
Harmonic band	5000–10000	1500–4500

2.5 Features Selection

Two statistical tools were independently used to investigate and select features able to classify voice as fatigued or not: principal component analysis (PCA) and Wilcoxon signed-rank test. PCA helps to reduce the data dimensionality and to reveal some information about data dependencies. As a result, new features are calculated. In this work, instead of using new features, the Pearson correlation coefficients between the obtained principal components and the input features were analysed in order to find the most important voice parameters. The Wilcoxon test is a non-parametric statistical test which can be used for data that do not follow the normal distribution, which is the case in this research.

2.6 Classification

A binary support vector machine classifier was used to detect the voice fatigue [13]. Data standardization was used and two different kernel functions were selected experimentally: the linear kernel was used for speech classification, whereas singing classification employed the Gaussian kernel.

3 Results and Discussion

Table 2 presents the results of the PCA algorithm with the features ordered by the amount of share in the output vector. Significance threshold was set to 0.70 and 0.95. Table 3 presents the results of the Wilcoxon test. Despite receiving different results in PCA selection, there were features placed in the top three of the most important coefficients in both cases. These were: *jitter* and NHR for singing, and vF_0 and vF_{0X} for speech. Wilcoxon test distinguished three important factors ($p < 0.1$) for each signal with the lowest values reached by NHR for singing ($p = 0.004$) and SC for speech (0.011). The test showed an essential share of fundamental frequency fluctuations in both signals.

Table 2. Results of the PCA algorithm for two significance thresholds: 0.70 (left) and 0.95 (right).

0.70 significance threshold				0.95 significance threshold			
Singing		Speech		Singing		Speech	
No.	Feature	No.	Feature	No.	Feature	No.	Feature
1	*jitter*	1	vF_0	1	*jitter*	1	vA
2.	NHR	2.	vF_{0X}	2	vA	2	vF_0
3.	SC	3.	HNR	3	NHR	3	vF_{0X}
				4	SC	4	HNR
				5	SS	5	SC
						6	SS

Table 3. Wilcoxon test results. Features with $p < 0.1$ are distinguished with a bold font.

Singing			Speech		
No.	Features	p value	No.	Features	p value
1.	**NHR**	**0.004**	**1.**	**SC**	**0.011**
2.	**vF$_0$**	**0.052**	**2.**	**vF$_0$**	**0.044**
3.	**jitter**	**0.073**	**3.**	**SS**	**0.057**
4.	vF_{0X}	0.215	4	*jitter*	0.145
5.	SC	0.314	5	HNR	0.145
6.	SS	0.332	6	*shimmer*	0.218
7.	HNR	0.654	7	vA	0.296
8.	vA	0.709	8	NHR	0.627
9.	*shimmer*	0.941	9	vF_{0X}	0.823

The voice fatigue classification was evaluated by means of accuracy, indicating percentage of correct assignments (both non-fatigue and fatigue) related to all cases. The singing approach relied on two features: *jitter* and SC, whereas the speech employed two other: vF_{0X} and HNR. Two k-fold cross-validation schemes were used: $k = 20$ and $k = 40$ (leave-one-sample-out). In the former case, the experiment was repeated 100 times with independent random division between validation. Table 4 shows the obtained results. The voice fatigue detection is apparently more accurate in case of a speech assessment. The selection result, however, makes it clear that the signals in a different way present the load of the voice. This means that the simultaneous use of both methods may give more valuable information than just the speech itself.

Table 4. Voice fatigue detection accuracy summary. Mean and standard deviation of accuracy in [%].

Singing		Speech	
$k = 20$	$k = 40$	$k = 20$	$k = 40$
62.9 ± 2.3	62.5 ± 0.0	70.9 ± 1.2	70.0 ± 0.0

4 Conclusion

A study on voice fatigue assessment and using signal processing tools was described in this paper. The singing and speech signal acquisition protocol applied to experiments involving 20 singers along with a processing framework enabled statistical analysis based on a set of signal and spectral features. Moreover, an attempt to detect the voice fatigue using a support vector machine tool provided an encouraging classification accuracy. Thus, the research will be conducted on vocal apparatus dysfunctions for computer-aided diagnosis and voice emission teaching support.

Acknowledgements. This research was supported by the Polish Ministry of Science and Silesian University of Technology statutory financial support No. BK-209/RIB1/2018.

References

1. Lancer, J.M., Syder, D., Jones, A.S., Le Boutillier, A.: Vocal cord nodules: a review. Clin. Otolaryngol. **13**(1), 43–51 (1988)
2. Yano, J., Ichimura, K., Hoshino, T., Nozue, M.: Personality factors in pathogenesis of polyps and nodules of vocal cords. Auris Nasus Larynx **9**(2), 105–110 (1982)
3. Brodnitz, F.S., Froeschels, E.: Treatment of nodules of vocal cords by chewing method. AMA Arch. Otolaryngol. **59**(5), 560–565 (1954)
4. Holmberg, E.B.: Efficacy of a behaviorally based voice therapy protocol for vocal nodules. J. Voice **15**(3), 395–412 (2001)
5. Yiu, E.M.L., Wang, G., Lo, A.C.Y., Chan, K.M.K., Ma, E.P.M., Kong, J., Barrett, E.A.M.: Quantitative high-speed laryngoscopic analysis of vocal fold vibration in fatigued voice of young karaoke singers. J. Voice **27**(6), 753–761 (2013)
6. Woo, P., Colton, R., Casper, J., Brewer, D.: Diagnostic value of stroboscopic examination in hoarse patients. J. Voice **5**(3), 231–238 (1991). https://doi.org/10.1016/S0892-1997(05)80191-2
7. Kasuya, H., Ogawa, S., Kikuchi, Y.: An acoustic analysis of pathological voice and its application to the evaluation of laryngeal pathology. Speech Commun. **5**, 171–181 (1985)
8. Bhuta, T., Patrick, L., Garnett, J.D.: Perceptual evaluation of voice quality and its correlation with acoustic measurements. J. Voice **18**(3), 299–304 (2004)
9. Boyanov, B., Hadjitodorov, S.: Acoustic analysis of pathological voices. IEEE Eng. Med. Biol. Mag. **16**(4), 74–82 (1997)
10. Świdziński, P.: Przydatność analizy akustycznej w diagnostyce zaburzeń głosu. Rozprawa habilitacyjna (D.Sc dissertation) z Kliniki Foniatrii i Audiologii Katedry Chorób Ucha, Nosa, Gardła i Krtani Akademii Medycznej im. Karola Marcinkowskiego w Poznaniu (1998). (in Polish)
11. Sundberg, J.: The acoustics of the singing voice. Sci. Am. **236**(3), 82–91 (1977)
12. Ohishi, Y., Goto, M., Itou, K., Takeda, K.: Discrimination between singing and speaking voices. In: Interspeech (2005)
13. Cortes, C., Vapnik, V.: Support-vector networks. Mach. Learn. **20**(3), 273–297 (1995)

Approach for Spectral Analysis in Detection of Selected Pronunciation Pathologies

Michał Kręcichwost[1](✉), Piotr Rasztabiga[1], Andre Woloshuk[2], Paweł Badura[1], and Zuzanna Miodońska[1]

[1] Faculty of Biomedical Engineering, Silesian University of Technology, Roosevelta 40, 41-800 Zabrze, Poland
michal.krecichwost@polsl.pl

[2] Weldon School of Biomedical Engineering, Purdue University, 206 S Martin Jischke Dr, West Lafayette, IN 47907, USA

Abstract. A framework for semi-automated detection of selected types of sigmatism is presented in this paper. A database of speech recordings was collected containing sibilant /s/ surrounded by vowels in different articulation phases. Recordings of three pronunciation modes were included into the database: normal, simulated lateral sigmatism, and simulated interdental sigmatism. The data was collected under the supervision of a speech therapy expert, who also provided labelling and annotation of each database entry. Twenty eight features of four types were extracted from each time frame within the sibilant: the mel-frequency cepstral coefficients, filter bank energies, spectral brightness, and zero-crossing rate. A feature aggregation procedure weighing the time frame location influence was proposed to describe each phoneme by a single feature vector. At the three-class classification stage, two tools were employed and compared: the random forest and support vector machine. The latter provides more accurate and repeatable classification results in each articulation phase with a median sensitivity, specificity, and accuracy exceeding 0.71, 0.85, and 0.80, respectively. The results also show that the assessment is generally more efficient when the phoneme is located at the beginning or ending of the word than when in the middle position.

Keywords: Computer-aided pronunciation evaluation · Sibilants · Sigmatism diagnosis

1 Introduction

Misarticulation of sibilant consonants /s, z, ts, dz/, /ʃ, ʒ, tʃ, dʒ/, /ɕ, ʑ, tɕ, dʑ/ (Polish) is a speech disorder called sigmatism (lisping) [1]. Several types of sigmatism can be distinguished depending on the position of the articulators (tongue, lips, etc.), inter alia, interdental or lateral. The diagnosis and treatment of sigmatism is dependent on the disorder subclassification, which is administered by

E. Tkacz et al. (Eds.): IBE 2018, AISC 925, pp. 115–122, 2019.
https://doi.org/10.1007/978-3-030-15472-1_13

a qualified speech therapist. The diagnosis and treatment relies on a subjective audio and visual assessment of articulation. Proper diagnosis of pathological pronunciation of sibilants plays an important role in modern speech therapy.

The working systems designed to assess and determine mispronunciation errors are mainly used to improve the language learning process (CALL systems - Computer-Aided Language Learning) [2]. From a data acquisition point of view, single microphone approaches are most common [3–6]. However, multichannel audio recording systems have been introduced recently, since the spatiotemporal processing of a signal is able to provide valuable information about non-normative sibilant sounds [7–9]. Additionally, articulation learning support can involve multimodal data, e.g. audio and video signals or ultrasound, to observe the articulators. They offer both the possibility of real-time visualization for instant patient feedback and offline analysis of articulator motion, particularly the tongue [10–12]. In recent years, electromagnetic articulography (EMA) became very popular. For example, the correlation analysis between motion trajectories of EMA sensors and trajectories for video markers glued to the upper and lower lips was described in the literature [13]. The use of multimodal systems is, however, associated with higher hardware costs. In addition, a larger number of devices during registration may increase the patient's discomfort or make normal pronunciation difficult e.g. by setting up certain devices or sensors on the speaker's face or articulation organs.

The aim of the study is to design a framework that is able to support sigmatism diagnosis in adults performed by a speech therapist. The system is based on speech audio recordings acquired by a single microphone. The methodology for mispronunciation detection and classification is based on strategies commonly used in computer-aided speech signal analysis and Automatic Speech Recognition (ASR). The main contribution of the study is a computer-aided speech therapy tool that can perform a general analysis. The framework consists of voice recordings, manual segmentation and annotation of the sibilant recordings, audio signal processing in terms of feature extraction, and multi-class classification of sibilants in three different articulation phases.

2 Materials and Methods

2.1 Data Collection

Due to the lack of publicly available pathological human speech databases, a novel collection was created. The database consists of 157 recordings performed by 14 speakers. The speakers did not have any language dysfunctions, but were instructed by a qualified speech therapist to simulate unwanted articulation features. The misarticulations were simulated by misplacing the tongue position during /s/ phoneme execution. Interdental sigmatism was simulated by shifting the tongue into the interdental gap during articulation, whereas lateral sigmatism was simulated by shorting the tongue with the incisor teeth. The dictionary contains words with the problematic /s/ phoneme surrounded by vowels in different articulation phases: at the beginning or end of the pseudoword ('SAS'), or in the middle of it ('ASA'). Each speaker was asked to pronounce the testing sequences with the correct articulation (N), simulated lateral sigmatism (LS), and simulated interdental sigmatism (IS). A total of 87 'ASA' (29 N, 29 LS, 29 IS) and 70 'SAS' (24 N, 23 LS, 23 IS) sequences were recorded and used in the study. The whole set of recordings was acquired using a directional microphone with a linear bandwidth in the frequency range of human speech. The registered audio signal was sampled with a 44100 Hz frequency. Each recording was manually segmented to indicate the boundaries of segments enclosing the dentalized consonant /s/.

2.2 General Workflow

The general workflow of the proposed methodology is presented in Fig. 1. The boundaries of the analysed phoneme /s/ were indicated by the expert. In the next step, the speech signal was preprocessed with digital filters. Then, the features were extracted and aggregated. Finally, the prepared samples were classified using a random forest (RF) and support vector machine (SVM).

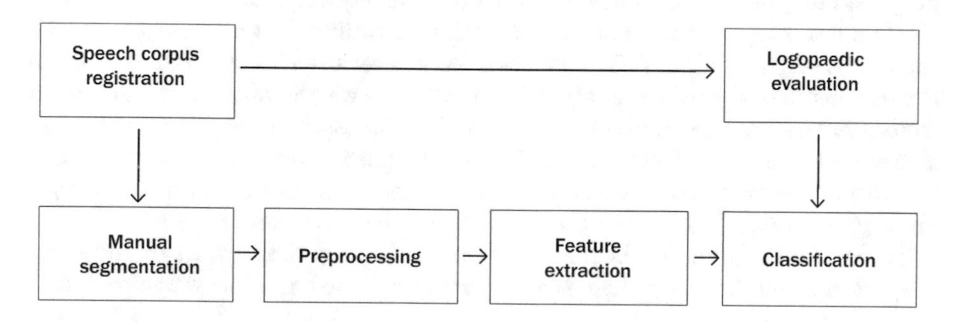

Fig. 1. The general workflow of the methodology

2.2.1 Preprocessing

The first step of the preprocessing was signal rescaling followed by high-pass FIR filtering. A filter with 65 dB stopband attenuation, maximum passband ripple of 0.5 dB and a cut-off frequency of 4 kHz (the lowest threshold of dentalised sounds in the human speech signal [14]) was used. Then, pre-emphasis filtering was applied. The signal was partitioned into 25 ms frames with 10 ms overlap using the Hamming window.

2.2.2 Feature Extraction

Each frame of the analysed signal was subjected to the extraction of the following features: 13 mel-frequency cepstral coefficients (MFCC), 13 filter bank energy coefficients (FBE), spectral brightness coefficient (SpB), zero-crossing rate value

(ZCR). The MFCCs [15,16] and FBEs [17] are a widely and successfully used set of features, especially when considering speech signal analysis. The spectral brightness coefficient carries information about the spectrum energy in the frequency bands above a given cut-off frequency in relation to the whole spectrum energy [18]. In this study, the cut-off frequency was set at 8 kHz. The ZCR coefficient was notably useful since the analysed voiceless phoneme /s/ reaches high values and has a strong correlation to energy spread in relation to frequency [19].

2.2.3 Aggregation of Features

At this stage of processing, a feature vector (FV) was determined for each frame included in the /s/ segment. In order to build one vector of attributes (SFV) to describe each segment, an aggregation method was proposed (Fig. 2). This method also made it possible to modify the values of individual features while taking into account the articulation moment of a given sound [20] and to lower the influence of features from edge frames. Due to the difficulty in defining the boundaries between successive articulation moments of the /s/ sound (normally determined using a time-lapse analysis of the audio-video image) [20,21], it was proposed to modify the values of the features using weights.

The first stage of the aggregation method is to match the Gaussian curve to the analyzed phoneme /s/. The position parameter used by the μ-gaussis is half the number of frames of the analyzed segment. The weight values c for subsequent frames are determined based on the matched Gaussian curve. These values are also scaled to a range from 0 to 1. The middle part of the frames correspond to the top of the voice (the phase where the speech organs are actually arranged for a given sound), therefore the weights for these frames are close to 1. The features coming from the beginning frame and the ending frame of the sound are not omitted, however their effect is reduced by using a low weight value.

2.2.4 Classification

The feature vectors from the recordings obtained in the previous stages were labelled as belonging to one of three groups: normative articulation (N), simulated lateral sigmatism (LS), and simulated interdental sigmatism (IS). The final stage of the presented methodology was the design of two classifiers dedicated to multi-class classification: random forest [22] and support vector machine [23]. The RF is a combination of individual decision tree predictors with each tree created independently based on a random sample dataset and a random split selection. In this study, a forest consisting of 100 trees was used. A one-against-one modification of binary SVM was employed with a linear kernel function. Scaling of input data was omitted.

3 Results and Discussion

In order to determine the ability of the classifier to detect a given type of pathology, sensitivity (TPR), specificity (SPC), and accuracy (ACC) were used. The

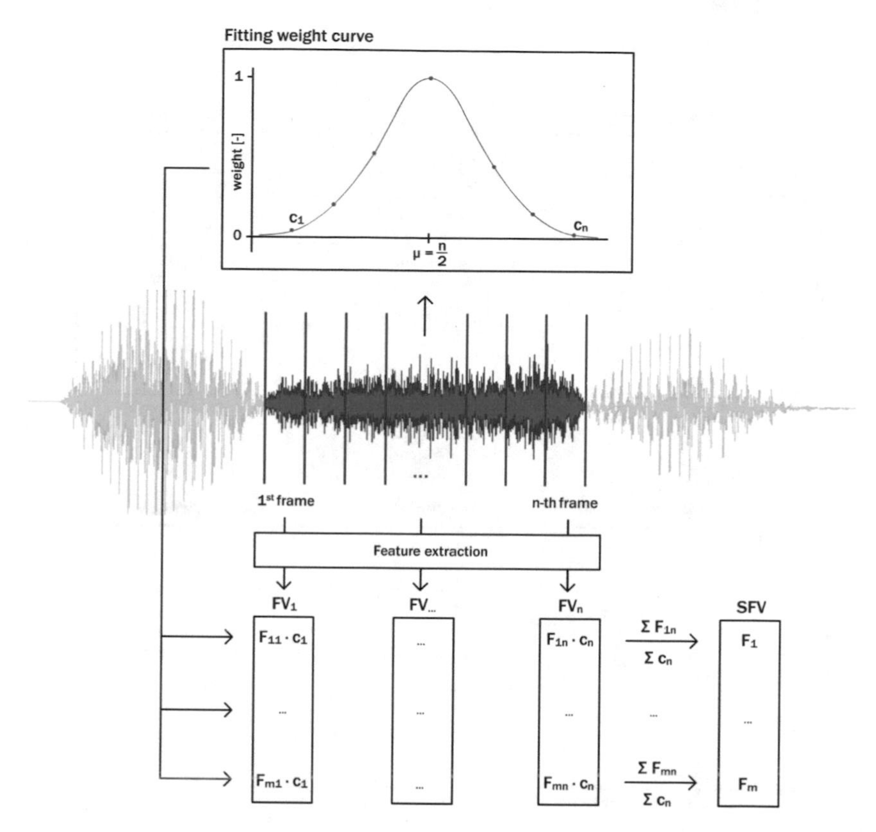

Fig. 2. Illustration of the feature aggregation procedure. n denotes the number of frames within the segment, F, c stand for the feature and weight, respectively

experimentation goal was to assess the classification for each prediction model for the three articulation classes (N, LS, IS). The extracted sibilants were grouped into three articulation phase-related groups: from 'ASA' only (middle), first (beginning), and second (ending) 'S' in 'SAS'. According to Sect. 2, the feature vector for each phoneme consisted of 28 items of four types (MFCC, FBE, SpB, and ZCR). Then, the obtained data was classified using both the SVM and RF algorithms, using 10-fold cross validation. Each experiment was repeated 20 times and the obtained metrics are gathered and presented in Fig. 3.

The proposed method for Gaussian weight-based aggregation of feature vectors within a phoneme yields median $TPR/SPC/ACC$ values exceeding 0.71/0.85/0.80, respectively. By taking into account particular phases of the dentalized phoneme realization it differentiates between the influence of the middle and outer time frames. There are differences in detection accuracy obtained in various articulation phases. The phoneme duration is longer when /s/ is in the middle of the pseudoword ('ASA') than when it is outside ('SAS'). That influences the duration of particular phoneme phases. With /s/ in the first/last

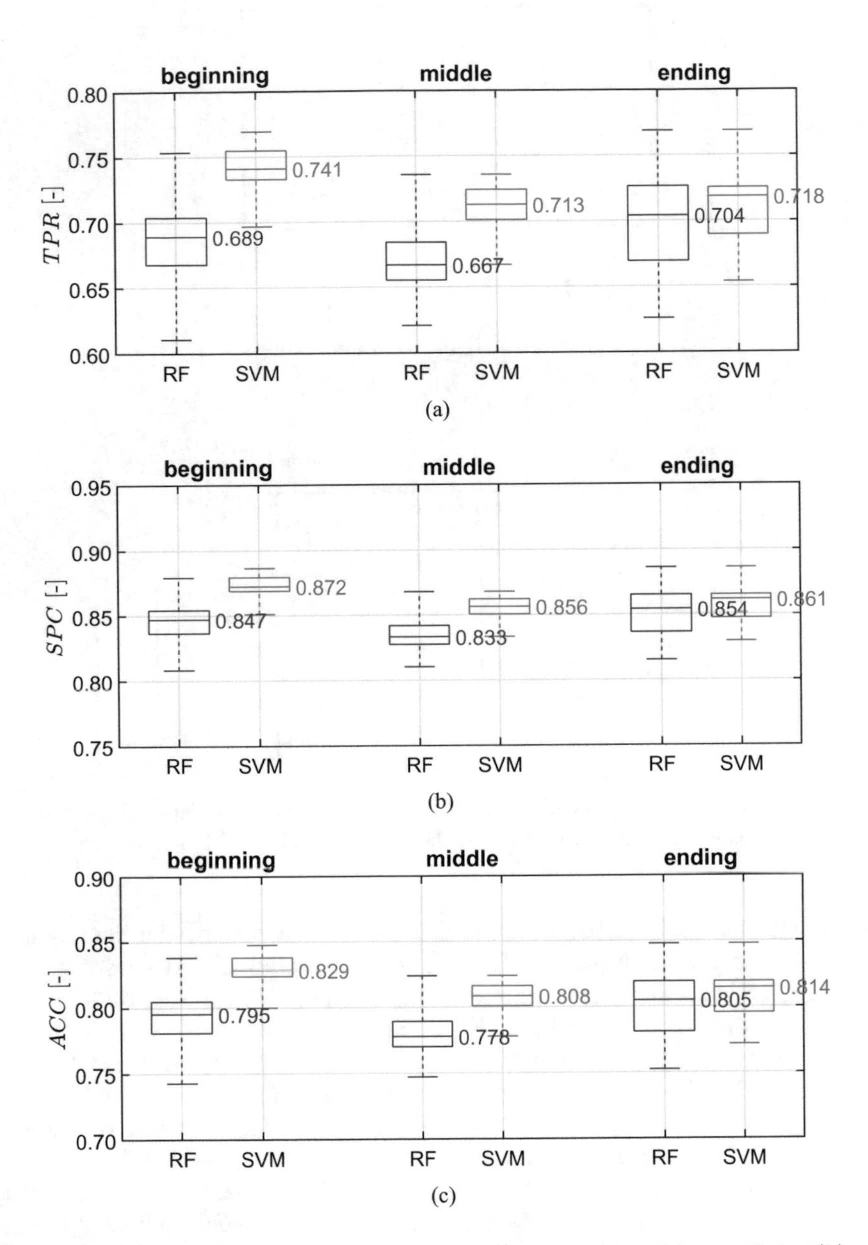

Fig. 3. Boxplots for classification accuracy metrics: sensitivity (a), specificity (b), and accuracy (c). Results for RF (red) and SVM (green) classifiers compared throughout three articulation phases: beginning, middle, and ending. Each box covers a 25th to 75th percentile range with median value given and indicated by a central line and extreme values bordered by whiskers

pseudoword position, the phase with articulatory organs position appropriate for the phoneme is relatively longer. Thus, the feature vector aggregation relies more on this segment. When in the middle position within the pseudoword, the phoneme is affected by the neighboring vowels in terms of feature extraction due to vowel coarticulation [24,25].

According to the obtained results, the assessment of the /s/ phoneme realization was more accurate when using the SVM algorithm, which produced higher median values of each sensitivity, specificity, and accuracy in every experiment. The obtained results gave a promise to prepare the methodology to perform some further improvements in the future, e.g.: (1) collecting a greater number of recordings from real speech disorder patients; (2) preparing new testing sequences, involving another set of sibilants; (3) defining feature vectors independent of noise and speaker's voice pitch.

4 Conclusion

A framework for detection of phoneme /s/ mispronunciation was described in this paper. A set of spectral features extracted and aggregated throughout a phoneme's time frames provided information used by the classifier to efficiently detect lateral or interdental mispronunciation. The SVM classifier yielded detection sensitivity over 0.71, specificity over 0.85, and accuracy over 0.80 in each articulation phase, which stands for a reliable performance for computer aided sigmatism diagnosis.

Acknowledgements. This research was supported by the Polish Ministry of Science and Silesian University of Technology statutory financial support No. BK-209/RIB1/2018.

References

1. Lobacz, P., Dobrzanska, K.: Opis akustyczny glosek sybilantnych w wymowie dzieci przedszkolnych. Audiofonologia **15**, 7–26 (1999). (in Polish)
2. Miodońska, Z., Kręcichwost, M., Szymańska, A.: Computer-aided evaluation of sibilants in preschool children sigmatism diagnosis. In: Information Technologies in Medicine, pp. 367–376. Springer (2016)
3. Wielgat, R., Zielinski, T., Wozniak, T., Grabias, S., Król, D.: Automatic recognition of pathological phoneme production. Folia Phoniatr Logop **60**(6), 323–331 (2008). Spoken Language Technology for Education
4. Valentini-Botinhao, C., Degenkolb-Weyers, S., Maier, A., Nöth, E., Eysholdt, U., Bocklet, T.: Automatic detection of sigmatism in children. In: WOCCI, pp. 1–4 (2012)
5. Seddik, A.F., El Adawy, M., Shahin, A.I.: A computer-aided speech disorders correction system for arabic language, pp. 18–21, September 2013
6. Bodusz, W., Miodońska, Z., Badura, P.: Approach for spectrogram analysis in detection of selected pronunciation pathologies. In: Innovations in Biomedical Engineering, vol. 623, pp. 3–11. Springer (2018)

7. Kostera, K., Więclawek, W., Kręcichwost, M.: Prototype measurement system for spatial analysis of speech signal for speech therapy. In: Innovations in Biomedical Engineering, vol. 623, pp. 79–86. Springer (2018)
8. Kręcichwost, M., Miodońska, Z., Trzaskalik, J., Pyttel, J., Spinczyk, D.: Acoustic mask for air flow distribution analysis in speech therapy. In: Information Technologies in Medicine, pp. 377–387. Springer (2016)
9. Król, D., Lorenc, A.: Acoustic field distribution in speech with the use of the microphone array. Tarnowskie Colloquia Naukowe **3**(4), 9–16 (2017)
10. Sebkhi, N., Desai, D., Islam, M., Lu, J., Wilson, K., Ghovanloo, M.: Multimodal speech capture system for speech rehabilitation and learning. IEEE Trans. Biomed. Eng. **64**(11), 2639–2649 (2017)
11. Aron, M., Berger, M.-O., Kerrien, E., Wrobel-Dautcourt, B., Potard, B., Laprie, Y.: Multimodal acquisition of articulatory data: geometrical and temporal registration. J. Acoust. Soc. Am. **139**(2), 636–648 (2016)
12. Opielinski, K.J., Gudra, T., Migda, J.: Computer ultrasonic imaging of the tongue shape changes in the process of articulation of vowels. In: Computer Recognition Systems 2, pp. 629–636. Springer, Berlin (2007)
13. Wielgat, R., Mik, L., Lorenc, A.: Correlational and regressive analysis of the relationship between tongue and lips motion - an EMA and video study of selected polish speech sounds, pp. 509–514, June 2017
14. Martony, J.: On the synthesis and perception of voiceless fricatives. STL-QPSR **3**(1), 17–22 (1962)
15. Young, S.J., Kershaw, D., Odell, J., Ollason, D., Valtchev, V., Woodland, P.: The HTK Book Version 3.4. Cambridge University Press, Cambridge (2006)
16. Huang, X., Acero, A., Hon, H.-W.: Spoken Language Processing: A Guide to Theory, Algorithm, and System Development, 1st edn. Prentice Hall PTR, Upper Saddle River (2001)
17. Paliwal, K.K.: Decorrelated and liftered filter-bank energies for robust speech recognition. In: EUROSPEECH (1999)
18. Jensen, K., Andersen, T.H.: Real-time beat estimation using feature extraction. In: Computer Music Modeling and Retrieval, pp. 13–22. Springer, Berlin (2004)
19. Bachu, R.G., Kopparthi, S., Adapa, B., Barkana, B.D.: Voiced/unvoiced decision for speech signals based on zero-crossing rate and energy. In: Advanced Techniques in Computing Sciences and Software Engineering, pp. 279–282. Springer, Dordrecht (2010)
20. Reidy, P.F.: Spectral dynamics of sibilant fricatives are contrastive and language specific. J. Acoust. Soc. Am. **140**(4), 2518–2529 (2016)
21. Klesla, J.: Analiza akustyczna polskich spolglosek tracych bezdzwiecznych realizowanych przez dzieci nieslyszace. Audiofonologia Problemy teorii i praktyki **26**, 107–118 (2004). (in Polish)
22. Breiman, L.: Random forests. Mach. Learn. **45**(1), 5–32 (2001)
23. Cortes, C., Vapnik, V.: Support-vector networks. Mach. Learn. **20**, 273–297 (1995)
24. Soli, S.D.: Second formants in fricatives: acoustic consequences of fricative vowel coarticulation. J. Acoust. Soc. Am. **70**(4), 976–984 (1981)
25. Sereno, J.A., Baum, S.R., Marean, G.C., Lieberman, P.: Acoustic analyses and perceptual data on anticipatory labial coarticulation in adults and children. J. Acoust. Soc. Am. **77**(S1), S7–S8 (1985)

Heart Beat Detection from Smartphone SCG Signals: Comparison with Previous Study on HR Estimation

Szymon Sieciński(✉) and Paweł Kostka

Department of Biosensors and Biomedical Signals Processing,
Silesian University of Technology, Zabrze, Poland
szymon.siecinski@polsl.pl

Abstract. Seismocardiography (SCG) is a non-invasive method of analyzing and recording cardiovascular activity as vibrations transmitted to the chest wall. Mobile devices offer the possibility to monitor health parameters thanks to embedded sensors. Various applications have been proposed for SCG, including heart rate calculation. Our aim is to detect heart beat on seismocardiograms using improved algorithm and compare its performance with results obtained in previous study.

Algorithm proposed in this study consists of signal preprocessing, RMS envelope calculation and peak finding. Algorithm performance was measured as sensitivity (Se) and positive predictive value (PPV) of beat detection on 4 signals acquired from 4 subjects.

We achieved average $Se = 0.994$ and $PPV = 0.966$ and in the best case $Se = 0.995$ and $PPV = 0.970$. Results prove major improvement of beat detection from smartphone seismocardiograms since the previous study.

Keywords: Seismocardiography · AO detection · Heart beat detection · Smartphone

1 Introduction

Availability of inexpensive, high quality and sensitive accelerometers in combination with low cost computational power provides the reasons for using analysis of cardiovascular vibrations in clinical practice [24]. MEMS technology has provided small and cheap accelerometers that may be used in medicine, including remote patient activity monitoring, fall detection [10], long term heart monitoring [6] and estimating hemodynamic parameters [8]. Smartphones have the capability to acquire and analyze vital parameters thanks to embedded sensors and powerful processors [3], which was proved by Ramos-Castro et al. [14] in the case of seismocardiography.

Seismocardiography (SCG) is a non-invasive method of recording and analyzing vibrations generated by cardiovascular activity transmitted to the chest wall

E. Tkacz et al. (Eds.): IBE 2018, AISC 925, pp. 123–130, 2019.
https://doi.org/10.1007/978-3-030-15472-1_14

[9]. Recordings are performed on subjects in supine position with the accelerometer placed on the sternum [22,23]. In literature, most studies on SCG consider only analyzing the dorso-ventral component (Z axis). 3D SCG can be achieved by using tri-axial accelerometers [7]. Our study considers dorso-ventral component.

SCG signal consists of the following points: mitral valve closure (MC), aortic valve opening (AO), onset of rapid enjection (RE), aortic valve closure (AC), mitral valve opening (MO), the peak of rapid filling (RF) and atrial systole (AS) [23]. Aortic valve opening has the highest amplitude [20]. Figure 1 presents a SCG cycle with annotated fiducial points.

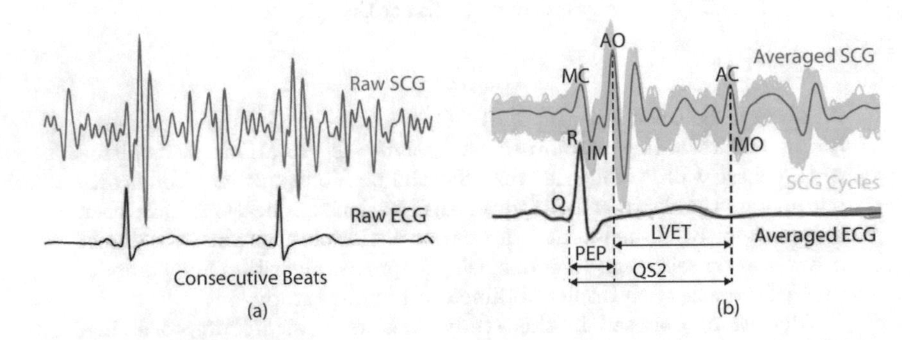

Fig. 1. SCG vs. ECG by Ghufran Shafiq et al. Image retrieved from [20]. License: CC-BY 4.0. Part (a) shows raw SCG signal (above) and ECG signal (below). Part (b) presents annotated SCG and ECG ensemble averages (dark lines) and superimposed SCG and ECG beats (light shades).

Various applications have been proposed for SCG, including heart rate variability (HRV) analysis, detecting heart arrhythmias, myocardial ischemia [7,21,24]. Our aim is to assess the performance of algorithm proposed and developed in papers [18,19] on data analyzed in paper from 2018 [17].

2 Methods

2.1 Experiment Protocol

The authors use data acquired for the study described in the paper published in 2018 [17]: SCG signals acquired from 4 healthy subjects (with no heart disease diagnosed) lying down in supine position at rest. Each measurement lasted 193.09–203.42 s and was performed using LG H340n smartphone, the same device as in paper [17]. In that paper "LG H440n" has appeared by mistake instead of "LG H340n". The authors used Accelerometer Meter mobile application available on Google Play to register precordial accelerations. The application acquired signals using sampling frequency $f_s = 100\,\text{Hz}$ and sensitivity of $0.009\,\text{m/s}^2$. The smartphone was placed loosely on sternum according to Fig. 2.

Technical specifications of accelerometer embedded in mobile device used for signal acquisition are shown in Table 1.

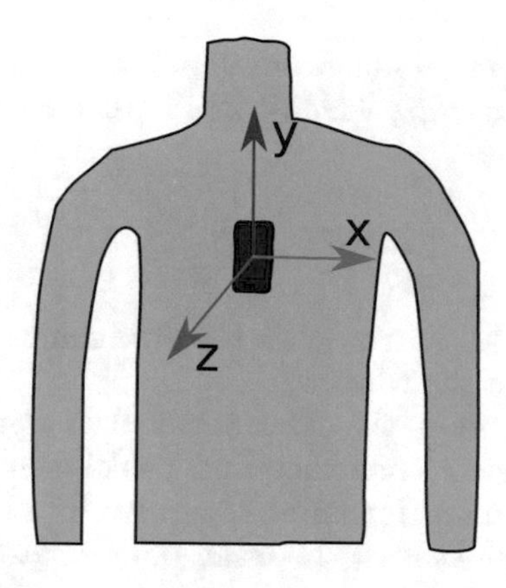

Fig. 2. Smartphone placement and coordinate system.

Table 1. Accelerometer technical specifications acquired by Sensor Multitool mobile application [16] and selected parameters listed in BMA255 Datasheet [1].

Sensor name	LGE Accelerometer Sensor
Vendor	BOSCH
Current consumption	0.13 mA
Resolution	0.00958 m/s^2
Maximum range	156.88 m/s^2
Total supply current in normal mode	130 μA = 0.13 mA
Resolution	0.98 mg $\approx$ 0.0096138 m/s^2 [2]
Sensitivity (g_{FS16g}, $T_A = 25\,°C$)	128 LSB/g
Zero-g offset	±60 mg
Maximum bandwidth	1000 Hz
Maximum range	16 g $\approx$ 156.96 m/s^2

2.2 Signal Processing

The SCG signal registered by a smartphone consists of three orthogonal components X, Y, Z. In this study we consider only Z component. Recordings were exported to text files and analyzed off-line using MATLAB software (The Mathworks, Inc., Natick, MA, USA).

The authors use heart beat detection algorithm proposed in [18] and improved in paper [19]. All the steps of proposed algorithm are shown in Fig. 3. The first step of the algorithm is band-pass filtering using FFT brick-wall filter and the frequency range 5–25 Hz. We chose this range proposed by Landreani et al. [11, 12] to filter out respiratory artifacts and out-of-band noise.

Then, we computed upper RMS envelope with 50 samples ($f_s/2$) window of each component. Segments with values over the standard deviation of upper RMS envelope or the mean RMS envelope times two were zeroed out.

RMS envelope (Root mean square envelope) measures the energy of the signal and is calculated using moving window RMS (root mean square) according to the Eq. (1) [4,5]:

$$RMS = \sqrt{\sum_{i=1}^{T} x_i^2(t)} \tag{1}$$

where T is window length and $x_i(t)$ is the i-th sample of the signal centered around t as seen through the window.

After cancelling the noise, the authors calculate upper RMS envelope with 10-sample ($f_s/10$) window. Then they find peaks using MATLAB `findpeaks` function on signal to find local maxima separated by at least 50 samples ($f_s/2$) and with an amplitude of at least the mean of calculated envelope.

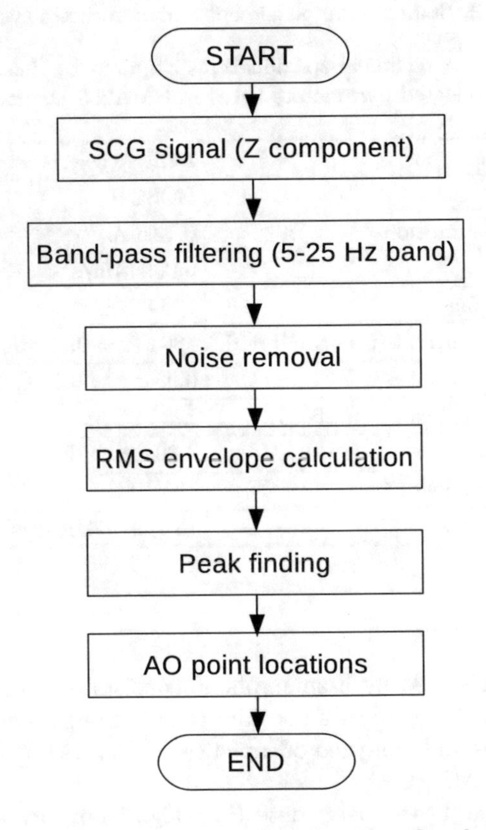

Fig. 3. Heart beat detection algorithm flowchart.

3 Results

In this study we consider 4 recordings of seismocardiograms acquired from 4 subjects described in the paper published in 2018 [17]. To assess the performance of heart beat detection algorithm, we annotated manually the heart beat occurrences based on SCG curve annotation presented in [23] and calculated the true positives (TP), false positives (FP), false negatives (FN).

Table 2 contains subjects description. Table 4 shows number of beats in each analyzed signal, true positives (TP), false positives (FP), false negatives (FN), sensitivity (Se) and positive predictive value (PPV) acquired in this study. Results from earlier analysis on the same subjects are shown in Table 3. True positive is defined as the AO point detected correctly. False negative is considered if proposed algorithm omits the AO point and false positive is determined for misclassified AO points. Figure 4 presents an example of heart beat detection results on subject 3.

Sensitivity (Se) is defined as $Se = \frac{TP}{TP+FN}$ and positive predictive value (PPV) is defined as $PPV = \frac{TP}{TP+FP}$. The number of beats is the sum of TP and FN.

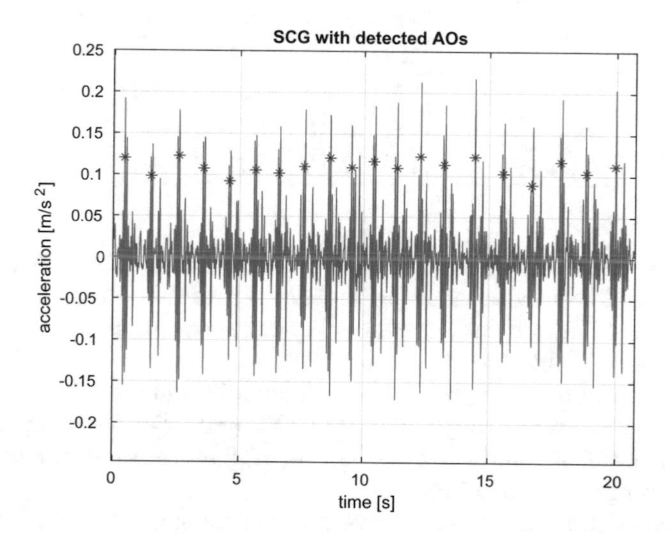

Fig. 4. Example of detection results on 20 s SCG signal fragment acquired from subject 3. Asterisks indicate detected heart beat locations.

Table 2. Subjects description.

Subject number	Age [years]	Gender	Recording length [s]
1	62	Male	193.09
2	63	Female	203.48
3	25	Male	203.42
4	66	Female	203.33

Table 3. Algorithm performance parameters acquired in previous study [17].

Subject number	TP	FP	FN	Beats	Se	PPV
1	164	36	27	191	0.859	0.820
2	186	145	54	240	0.775	0.562
3	184	5	1	185	0.995	0.974
4	183	126	34	217	0.843	0.592
Total	717	312	116	833	0.861	0.697
Average	179	78	29	208	0.868	0.737

Table 4. Number of beats in analyzed signals and algorithm performance parameters.

Subject number	TP	FP	FN	Beats	Se	PPV
1	196	6	1	197	0.995	0.970
2	186	3	4	190	0.980	0.985
3	184	7	0	184	1.000	0.963
4	222	13	0	222	1.000	0.965
Total	795	29	5	800	0.994	0.965
Average	199	7	1	200	0.994	0.966

PPV rates shown in Table 4 are lower than Se rate in all cases except subject number 2. The difference is caused by higher FN rate than FP rate for this signal. Lower PPV rate for subject 3 is caused by higher occurrence of false positives than indicated in Table 3.

Sensitivity and positive predictive value for all annotated 833 beats in previous study (Table 3) was strongly affected by FP and FN rates of female subjects. In this study lower FP and FN rates indicate a significant performance improvement using proposed algorithm.

Observed differences between male and female subject are smaller than in previous study. Lower sensitivity of beat detection on signals from female subject may be caused by smartphone size and position, as indicated by Landreani et al. [11]. Higher number of beats shown in Table 3 is caused by misclassifying some beat occurrences in previous study.

4 Discussion

Our assessment of algorithm proposed in paper [18] proves its high performance of heart beat detection (average Se = 0.994, PPV = 0.966; in the best case Se = 0.995, PPV = 0.970). We observed higher performance in all analyzed signals except for subject 3. Similar results were achieved by Pouymiro et al. [15] and Li et al. [13].

Due to slightly lower algorithm performance on female subjects, more suitable smartphone position may be considered. Performance indexes achieved in this study seem encouraging and show the need of further studies using that algorithm in another experimental set-ups.

References

1. Bosch Sensortec: BMA255, Digital, Triaxial Accelerometer. BMA255 Datasheet, 1 August 2014
2. Bosch Sensortec: BMA255 (n.d.). https://www.bosch-sensortec.com/bst/products/all_products/bma255. Accessed 24 Jan 2018
3. Bruining, N., et al.: Acquisition and analysis of cardiovascular signals on smartphones: potential, pitfalls and perspectives. Eur. J. Prev. Cardiol. **21**(2 Suppl.), 4–13 (2014). By the Task Force of the e-Cardiology Working Group of the European Society of Cardiology
4. Caetano, M.F., Rodet, X.: Improved estimation of the amplitude envelope of time domain signals using true envelope cepstral smoothing. In: IEEE International Conference on Acoustics, Speech and Signal Processing, Czech Republic, May 2011, pp. 11–21 (2011)
5. De Luca, C.J.: Electromyography. In: Webster, J.G. (ed.) Encyclopedia of Medical Devices and Instrumentation, pp. 98–109. Wiley (2006). https://doi.org/10.1002/0471732877.emd097
6. Di Rienzo, M., Vaini, E., Castiglioni, P., Meriggi, P., Rizzo, F.: Beat-to-beat estimation of LVET and QS2 indices of cardiac mechanics from wearable seismocardiography in ambulant subjects. In: 2013 35th Annual International Conference of the IEEE Engineering in Medicine and Biology Society, EMBC, Osaka, pp. 7017–7020 (2013). https://doi.org/10.1109/EMBC.2013.6611173
7. Inan, O.T., Migeotte, P.F., Park, K.S., Etemadi, M., Tavakolian, K., Casanella, R., Zanetti, J., Tank, J., Funtova, I., Prisk, G.K., Di Rienzo, M.: Ballistocardiography and seismocardiography: a review of recent advances. IEEE J. Biomed. Health Inform. **19**(4), 1414–27 (2015)
8. Komorowski, D., Pietraszek, S., Darlak, M.: Pressure and output flow estimation of pneumatically controlled ventricular assist device (VAD) with the help of both acceleration and gyro sensors. In: World Congress on Medical Physics and Biomedical Engineering 2006, IFMBE Proceedings, vol. 14, no. Part 7, pp. 719–722 (2007)
9. Korzeniowska-Kubacka, I.: Sejsmokardiografia—nowa nieinwazyjna metoda oceny czynności lewej komory w chorobie niedokrwiennej serca. Folia Cardiol. **10**(3), 265–268 (2003)
10. Kostka, P., Tkacz, E.: Modern MEMS acceleration sensors in tele-monitoring systems for movement parameters and human fall remote detection. In: Kapczyński, A., Tkacz, E., Rostanski, M. (eds.) Internet - Technical Developments and Applications 2. Advances in Intelligent and Soft Computing, vol. 118, pp. 271–277. Springer, Heidelberg (2012). https://doi.org/10.1007/978-3-642-25355-3_23
11. Landreani, F., Martin-Yebra, A., Casellato, C., Frigo, C., Pavan, E., Migeotte, P.F., Caiani, E.G.: Beat-to-beat heart rate detection by smartphone's accelerometers: validation with ECG. In: 38th Annual International Conference of the IEEE Engineering in Medicine and Biology Society, EMBC, FL, Orlando, pp. 525–528 (2016)
12. Landreani, F., Morri, M., Martin-Yebra, A., Casellato, C., Pavan, E., Frigo, C., Caiani, E.G.: Ultra-short-term heart rate variability analysis on accelerometric signals from mobile phone. In: 2017 E-Health and Bioengineering Conference, EHB, Sinaia 2017, pp. 241–244 (2017). https://doi.org/10.1109/EHB.2017.7995406

13. Li, Y., Tang, X., Xu,Z.: An approach of heartbeat segmentation in seismocardiogram by matched-filtering. In: 2015 7th International Conference on Intelligent Human-Machine Systems and Cybernetics, August 2015, vol. 2, pp. 47–51. https://doi.org/10.1109/IHMSC.2015.157
14. Ramos-Castro, J., Moreno, J., Miranda-Vidal, H., García-González, M., Fernández-Chimeno, M., Rodas, G., Capdevila, L.: Heart rate variability analysis using a seismocardiogram signal. In: 2012 Annual International Conference of the IEEE Engineering in Medicine and Biology Society, EMBC, August 2012, pp. 5642–5645 (2012)
15. Pouymiro, I.R., Cordova, E.V., Perez, F.E.V.: Robust detection of AO and IM points in the seismocardiogram using CWT. IEEE Lat. Am. Trans. **14**(11), 4468–4473 (2016). https://doi.org/10.1109/TLA.2016.7795816
16. Wered Software: Sensor Multitool (Version 1.3.0) (2017). https://play.google.com/store/apps/details?id=com.wered.sensorsmultitool&hl=pl. Accessed 23 Jan 2018
17. Siecinski, S., Kostka, P.: Determining heart rate beat-to-beat from smartphone seismocardiograms: preliminary studies. In: Innovations in Biomedical Engineering. Advances in Intelligent Systems and Computing, vol. 623, pp. 133–140. Springer, Cham (2018). https://doi.org/10.1007/978-3-319-70063-2_15
18. Siecinski, S., Kostka, P.: Influence of gravitational offset removal on heart beat detection performance from Android smartphone seismocardiograms. In: Pietka, E., Badura, P., Kawa, J., Wieclawek, W. (eds.) Information Technology in Biomedicine, ITIB 2018. Advances in Intelligent Systems and Computing, vol. 762, pp. 337–344. Springer, Cham (2019). https://doi.org/10.1007/978-3-319-91211-0_30
19. Siecinski, S., Kostka, P.S., Tkacz, E.J.: Heart rate variability analysis on CEBS database signals. In: 40th Annual International Conference of IEEE EMBS, Honolulu, HI, USA, 17-21 July 2018, pp. 5697–5700. https://doi.org/10.1109/EMBC.2018.8513551
20. Shafiq, G., et al.: Automatic identification of systolic time intervals in seismocardiogram. Sci. Rep. **6**, 37524 (2016). https://doi.org/10.1038/srep37524
21. Tadi, M.J., et al.: A real-time approach for heart rate monitoring using a Hilbert transform in seismocardiograms. Physiol. Meas. **37**, 1885 (2016)
22. Zanetti, J.M., Salerno, D.M.: Seismocardiography: a technique for recording precordial acceleration. In: Proceedings of Fourth Annual IEEE Symposium of Computer-Based Medical Systems, Baltimore, MD, USA, pp. 4–9 (1991)
23. Zanetti, J.M., Salerno, D.M.: Seismocardiography: waveform identification and noise analysis. In: Computers in Cardiology, Venice, Italy, pp. 49–52 (1991)
24. Zanetti, J.M., Tavakolian, K.: Seismocardiography: past, present and future. In: Proceedings of 35th Annual International Conference of the IEEE EMBS, Osaka, Japan, 3–7 July (2013)

Improving the Automated Detection of Silent AF Episodes Based on HR Variability Measures

Janusz Wróbel(✉), Krzysztof Horoba, Janusz Jeżewski, Adam Matonia, and Tomasz Kupka

Institute of Medical Technology and Equipment ITAM, 118 Roosevelt Str., 41-800 Zabrze, Poland
januszw@itam.zabrze.pl

Abstract. Atrial fibrillation (AF) is one of the most common types of heart arrhythmias and it leads to increased risk of stroke and heart attack. We proposed improved method for automated AF detection with additional aggregation of short AF episodes to provide more reliable information on risk for the patient. The aggregation parameters underwent optimization process and evaluation study has been carried out with a use of two databases provided by PhysioNet.org. Final results showed that the new method for automated detection of AF episodes was capable to detect more than 96% of all reference AF provided by the MIT-BIH Atrial Fibrillation Database, and above 93% for the Long-Term AF Database. At the same time less than 7% of the detected AF were false episodes for both databases. The proposed method can be easily implemented in the mobile instrumentation for long-term monitoring where the computational power is usually limited.

Keywords: Atrial fibrillation · Automated AF detection · Heart rate variability measures

1 Introduction

Atrial fibrillation (AF) is a serious heart condition when the atria contracts irregularly at rates of 400 to 600 beats per minute. Symptoms are not always evident but often include palpitations, irregular heartbeat, shortness of breath, chest pains and others. Atrial fibrillation is one of the most common types of heart arrhythmias and it leads to increased risk of stroke and heart attack. Although, the atrial fibrillation is not a very common disease – currently its occurrence among adults is estimated at 1.5 to 2.5%, but it is highly correlated with an age. As reported in the study [1] concerning the European population, AF was found on average in 5.5% of subjects, while in the age group above 85 years it was almost 18%. Recognition of silent (asymptomatic) AF is a big challenge, since the patient has no symptoms, and the diagnosis is usually made on the

E. Tkacz et al. (Eds.): IBE 2018, AISC 925, pp. 131–140, 2019.
https://doi.org/10.1007/978-3-030-15472-1_15

basis of accidentally performed electrocardiography (ECG). The AF episode are recognized from the ECG analysis, basing on a lack or abnormal P-wave, or estimating the changes of interval between two consecutive R-waves (RR) by means of various irregularity measures. The most efficient diagnosis of silent AF are provided by techniques of long-term heart rate monitoring, like Holter monitor, continuous telemetry [2–5], or implementable devices with internal memory [6]. They enable not only detection of silent AF episodes, but also determination of its duration, which in turn relates to so-called AF burden, defined as the percentage of AF in the heart rate recording in a certain time period – usually a day. According to [7], the silent AF was estimated at 10.5% in a group of above 75 years when using 24-h Holter monitoring.

Visual analysis of 24-h recording requires a lot of time and efforts from the cardiologists, thus the methods for automated detection of atrial fibrillation which could improve the objectivity of classification are needed. Because of usually low quality of the long-term ECG signal, involving the RR irregularity by means of various irregularity measures, seems to be more useful for the automated detection, then analysis of low amplitude P-wave. What's more the information of RR intervals (heart rate) may be obtained not only from electrocardiographic signal but also from photopletysmographic wave [8,9]. However the latter technique still requires further research work to ensure a continuous long-term recording of the heart rate (HR) when using comfortable and simple devices [10]. In the field of long term monitoring of the patient's biophysical signals many additional issues have to be considered, like design and control requirements [11,12], extraction of the useful signals in a presence of increased noise level [13–16], as well as preliminary signal interpretation [17,18]. There are algorithms based on the variance of RR intervals or variance of differences between neighbouring intervals [19,20]. Other detection methods process the heart rate signal in time-frequency domain with the wavelet transform [21] and autoregressive modelling [22] showing significant advantages in AF episodes detection. Among statistical approaches the following method have been proposed: the tests to explore whether probability density functions for R-R intervals differ [23], turning points ratio to characterize arrhythmia condition [24], as well as the geometric features determined from the density histogram of both RR intervals and their differences [25]. Another methods use artificial neural network with interval transition matrices as an input to differentiate among two AF states [26].

When conducting the project being aimed at development of the wristband recorder for long-term AF monitoring, the authors created the algorithm for automated detection of AF episode based on the linear classifier [27]. However, when monitoring the patients in a testing group, the short accidental changes between AF and non-AF heart beats quite often occurred that were recognized by the clinical experts involved as incorrect and making interpretation more difficult. In this paper we proposed improved method for automated AF detection, which includes an additional stage leading to aggregation of short AF episodes to provide more reliable information on risk for the patient. The aggregation parameters underwent optimization process and evaluation study has been carried out with a use of two databases provided by PhysioNet.org.

2 Methodology

The method for automated AF detection has been designed to be applied in wearable devices for long term patient's monitoring and thus to cope with limited computational power [28]. Therefore, it has been based exclusively on easily accessible information about heart beat intervals and the features extracted by simple processing. The features set is composed of the heart rate value HR, median value MED, median absolute deviation from median MAD, quantile of the established order QNT, and proportion coefficient PRP. These features has been selected in a series of preliminary investigations carried out among more possible features. Each consecutive heart beat is marked as AF or non-AF by the proposed binary classifier with the feature set led to its input. The first stage of the proposed method is calculation of consecutive heart beat values basing on the intervals between consecutive heart beats:

$$HR_i\,[bpm] = \frac{60000}{RR_i\,[ms]} \tag{1}$$

where: RR_i is the *i-th* interval between two consecutive heart beats expressed in milliseconds. Then, the remain features are determined in symmetrical moving window whose width $l = 2k+1$ has been set at 21 heart rate values:

- $MED_i = \text{median}\{HR_{(i-k)}, \ldots HR_{(i+k)}\}$;
- $MAD_i = \text{median}\{x_{(i-k)}, \ldots x_{(i+k)}\}$, where $x_i = |HR_i - MED_i|$;
- QNT_i – represents the quantile of order 0.7 estimated over 21 values of heart rate;
- PRP_i – is the ratio of number of HR values within the range of 120 to 160 bpm, to their total number.

The control parameter values (window width, quantile order and HR range) were determined in a series of experiments performed earlier, where for a predefined set of values of the given control parameter, the area of the overlapping part of probability density functions for no atrial fibrillation and atrial fibrillation episodes was calculated [27].

The status of consecutive heart beat HB_i is set according to the formula:

$$HR_i = \begin{cases} AF & \text{for } y_i > 0 \\ non\text{-}AF & \text{for } y_i \leq 0 \end{cases} \tag{2}$$

When y_i is the output of the proposed binary classifier obtained for a given feature set:

$$y_i = a_5 \cdot HR_i + a_4 \cdot MED_i + a_3 \cdot MAD_i + a_2 \cdot QNT_i + a_1 \cdot PRP_i + a_0 \tag{3}$$

The linear coefficients a_5 to a_0 were determined using ten-fold cross-validation on the training data [29]. This procedure was repeated ten times and the results were verified on the testing set. The training and testing set was obtained from randomly chosen data (AF annotations) provided by the MIT-BIH Atrial Fibrillation Database [30,31]. The obtained classification accuracy defined as the number of correctly classified heart beats to their total number was 93.32% for the testing data set and 93.28% for the whole data set.

The final stage – aggregation of the classified heart beats has been considered as a crucial step, which should lead to removing accidental changes of heart beat status, and thus to obtain more reliable information on AF episodes. During aggregation process each heart beat status is validated. Its value stays unchanged if the number of the heart beats with the same status exceeds the established threshold. We defined the percentage threshold for AF = 1, as the number of heart beats classified as AF to the number of all beats in the analysed window. The optimization process had to be carried out to find the values of the window width and the percentage threshold ensuring the best AF detection performance. The performance of the automated detection of atrial fibrillation episodes is determined mainly by two components: the ability to detect the true occurrences of AF, which is measured by the sensitivity (Se), and the smallest number of false arrhythmias – resulting from incorrect classification of HR irregularity, which is expressed by the positive predictive value (PPV). The optimization is performed by controlling the distance D between the point defined by Se and PPV obtained for a given set of window width and the percentage threshold to the point of maximum performance (Se = 100%, PPV = 100%):

$$D(w,p) = \sqrt{(100 - Se(w,p))^2 + (100 - PPV(w,p))^2} \tag{4}$$

where: Se - sensitivity in the range 0 to 100%, PPV - positive predictive value in the range 0 to 100, w - the window width expressed in heart beats, p - percentage threshold of AF beats in the analysed window. The algorithm provides the best detection performance for this window width and percentage threshold for which the distance D reaches the minimum value D_{min}. Two databases available through Physionet.org have been used: the MIT-BIH Atrial Fibrillation Database [30,31] and The Long-Term AF Database [31,32]. The first database includes 25 ECG recordings of 10 h in duration, whereas the second one 84 long-term recordings whose duration varies from 24 to 25 h. In both databases all recordings are accompanied by the files with the time markers of detected QRS complexes, and with the AF episode annotations, as their beginning and end. Basing on this information for the task of this study, the HR values in bpm were calculated for each recording, and then using AF annotations they were marked as reference AF (value 1 assigned) and reference non-AF (0). For the optimization process, which may refer to the training phase, 12 recordings from MIT-BIH and 42 from Long-Term databases were randomly selected. While the evaluation of the method performance – the testing phase was accomplished using the whole data set.

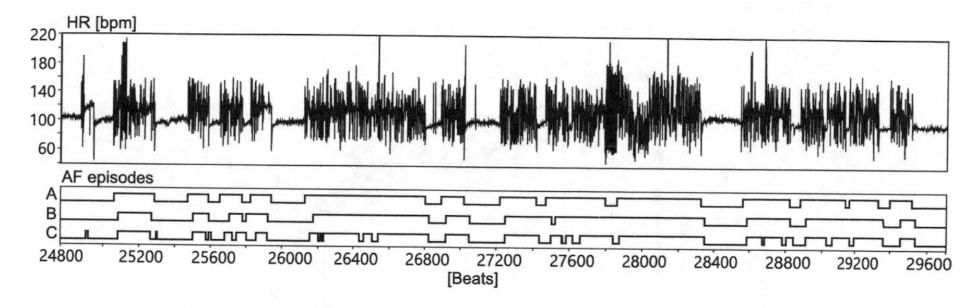

Fig. 1. An example of a record from the database with: determined heart rate values (upper graph), reference AF episodes (A), and AF detected by the proposed method with (B) and without aggregation (C)

3 Results

Randomly selected 54 recordings provided 5 111 305 reference heart beats, among them 2 941 617 (57.55%) were annotated as the AF type. The reference and detected AF episodes are expected to overlap each other, but usually they overlap only partially, thus the true positive, true negative, false positive and false negative cases were recognized to calculate the values of Se and PPV. Figure 1 present the influence of the aggregation stage on the matching the reference and detected AF episodes in the exemplary recording. During optimization process the window width was changed from 10 to 190 beats with step of 10 beats, while the threshold from 5 to 95% with 5% step. Those established ranges and steps being applied to aggregation of annotated heart beats into AF episode led to 361 pairs of Se and PPV, that are represented as the points in Fig. 2. The best performance expressed by the Se equal to 93.14%, and PPV to 92.88%, was obtained for $D_{min} = 9.89$ relating to window width of 150 beats and percentage threshold equal to 45%. These aggregation parameters have been applied for the testing phase, where the whole data set was used.

Table 1. Detection performance obtained without (A) and with (B) the aggregation optimized by the window width 150 beats and percentage threshold 45%, using the MIT-BIH Atrial Fibrillation Database with 1 221 559, and Long-Term AF Database with 8 525 075 reference annotated heart beats

		MIT-BIH		Long-Term	
		A	B	A	B
Sensitivity	[%]	90.34	96.10	88.67	93.20
Specificity	[%]	95.46	95.62	90.09	90.71
PPV	[%]	93.64	94.21	92.32	93.10
NPV	[%]	93.03	97.07	85.53	90.84
Accuracy	[%]	93.28	95.83	89.27	92.14

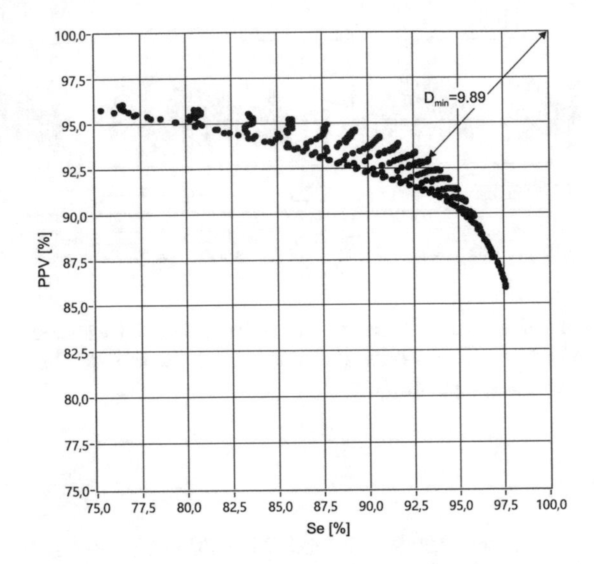

Fig. 2. Influence of the aggregation parameters (window width w and percentage threshold p) on the method performance expressed by the Sensitivity and PPV values. The minimum distance $D_{min} = D$ for $w = 150$ and $p = 45\%$ corresponds to Se = 93.14% and PPV = 92.88%

Results obtained during the testing phase are listed separately for both databases in Table 1. The values of the performance measures: sensitivity, specificity, positive predictive value PPV, negative predictive value NPV, and accuracy (the number of properly classified AF and non-AF episodes to all episodes), have been calculated with and without aggregation stage. Significant increase of the sensitivity after applying the aggregation has been noted for both databases, while the PPV improved slightly. The method performance was quantified basing on all AF episodes detected in all recordings from two databases by relating them to reference AF episodes. However, it could be expected that detection effectiveness of the automated method depends on the recordings themselves, as they may come from different patients. Thus, we tested how the Se and PPV provided by the method changed across the recordings (patients). The proposed method for automated detection of AF episodes showed the Se at least 95% for 64% of the recordings in the MIT-BIH Atrial Fibrillation Database, and for 60% in the Long-Term AF Database (Fig. 3). From the other hand, for 75% of the recordings from the first database the method ensured the Se above 85%, while for the second database above 78%. The same tendency was observed for the PPV (Fig. 4), where the results obtained for the MIT-BIH AF Database were better than for the Long-Term AF Database. The method ensured the PPV at least 95% for 56% of recordings from the first, and for 52% from the second database.

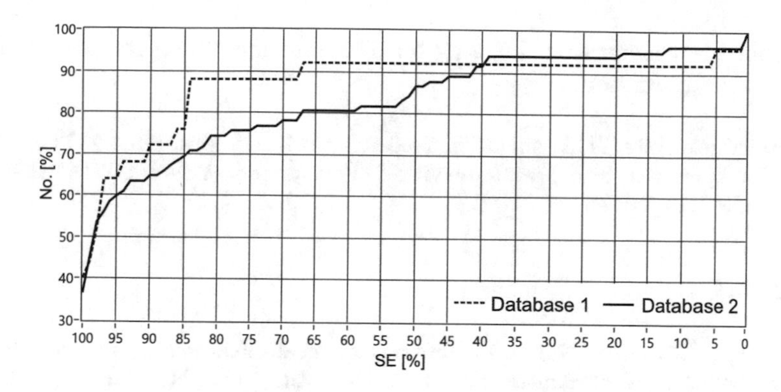

Fig. 3. Graphs showing the number (percentage) of recordings for which the particular algorithms provided the given Se value for MIT-BIH Atrial Fibrillation Database 1 and Long-Term AF Database 2

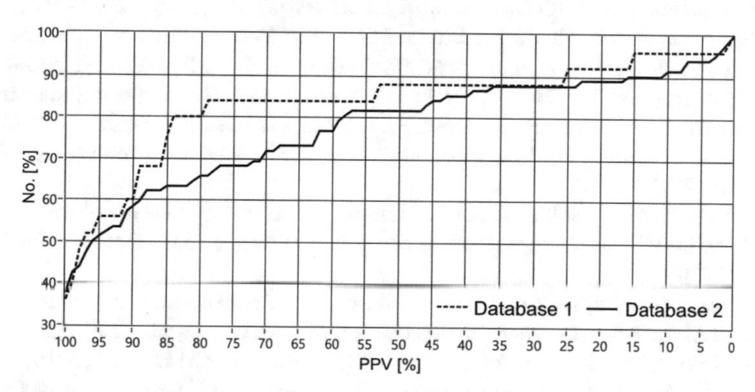

Fig. 4. Graphs showing the number (percentage) of recordings for which the particular algorithms provided the given PPV value for MIT-BIH Atrial Fibrillation Database 1 and Long-Term AF Database 2

4 Conclusion

The aggregation of the AF episodes with parameters optimization led to considerable increase of sensitivity and improvement of the positive predictive value in relation to previously developed method. Final results showed that the new method for automated detection of AF episodes was capable to detect more than 96% of all reference AF provided by the MIT-BIH Atrial Fibrillation Database, and above 93% for the Long-Term AF Database. At the same time less than 7% of the detected AF were false episodes for both databases. When analysing the method performance across particular recordings within the research material, we noted that the new method provided satisfying sensitivity of at least 95% for about 60% of recordings, whereas the PPV exceeding 95% was provided for about 50% of recordings. The proposed method can be easily implemented in the mobile instrumentation for long-term monitoring where the computational power is usually limited.

Acknowledgements. This scientific research work is supported by The National Centre for Research and Development of Poland (grant No. STRATEGMED2/269343/18/NCBR/2016).

References

1. Heeringa, J., van der Kuip, D., Hofman, A., et al.: Prevalence, incidence and lifetime risk of atrial fibrillation: the Rotterdam study. Eur. Heart J. **27**(8), 949–953 (2006)
2. Chan, P.H., Wong, C.K., Poh, Y.C., et al.: Diagnostic performance of a smartphone-based photoplethysmographic application for atrial fibrillation screening in a primary care setting. J. Am. Heart Assoc. **5**(7), e003428 (2016)
3. Desteghe, L., Raymaekers, Z., Lutin, M., et al.: Performance of handheld electrocardiogram devices to detect atrial fibrillation in a cardiology and geriatric ward setting. Europace **19**, 29–39 (2016). https://doi.org/10.1093/europace/euw025
4. Haberman, Z.C., Jahn, R.T., Bose, R., et al.: Wireless smartphone ECG enables large-scale screening in diverse populations. J. Cardiovasc. Electrophysiol. **26**(5), 520–526 (2015)
5. Vaid, J., Poh, M.Z., Saleh, A., et al.: Diagnostic accuracy of a novel mobile application (Cardio Rhythm) for detecting atrial fibrillation. J. Am. Coll. Cardiol. **65**(10), A361 (2015)
6. Hindricks, G., Pokushalov, E., Urban, L.: Performance of a new leadless implantable cardiac monitor in detecting and quantifying atrial fibrillation results of the XPECT trial. Circ. Arrhythm. Electrophysiol. **3**, 141–147 (2010)
7. Camm, J.: Atrial fibrillation and risk. Clin. Cardiol. **35**(Suppl. 1), 1–2 (2012)
8. Lu, S., Zhao, H., Ju, K., et al.: Can photoplethysmography variability serve as an alternative approach to obtain heart rate variability information. J. Clin. Monit. Comput. **22**(1), 23–29 (2008)
9. Liang, K., Sun, Y., Tian, F., Ye, S.: Research on algorithm of extracting PPG signal for detecting atrial fibrillation based on probability density function. Open Biomed. Eng. J. **9**(1), 179–184 (2015)
10. Roj, D., Wrobel, J., Matonia, A., Sobotnicka, E.: Hardware design issues and functional requirements for smart wristband monitor of silent atrial fibrillation. In: Proceedings of 24th International Conference on Mixed Design of Integrated Circuits and Systems MIXDES 2017, Bydgoszcz, 22–24 June, pp. 596–600 (2017)
11. Wrobel, J., Matonia, A., Horoba, K., Jezewski, J., et al.: Pregnancy telemonitoring with smart control of algorithms for signal analysis. J. Med. Imag. Health Inform. **5**(6), 1302–1310 (2015)
12. Wrobel, J., Jezewski, J., Horoba, K., Pawlak, A., Czabanski, R., Jezewski, M., et al.: Medical cyber-physical system for home telecare of high-risk pregnancy – design challenges and requirements. J. Med. Imag. Health Inform. **5**(6), 1295–1301 (2015)
13. Matonia, A., Jezewski, J., Kupka, T., Horoba, K., Wrobel, J., Gacek, A.: The influence of coincidence of fetal and maternal QRS complexes on fetal heart rate reliability. Med. Biol. Eng. Comput. **44**(5), 393–403 (2006)

14. Jezewski, J., Matonia, A., Kupka, T., Roj, D., Czabanski, R.: Determination of the fetal heart rate from abdominal signals: evaluation of beat-to-beat accuracy in relation to the direct fetal electrocardiogram. Biomed. Eng. **57**(5), 383–394 (2012)
15. Jezewski, J., Horoba, K., Roj, D., Wrobel, J., Kupka, T., Matonia, A.: Evaluating the fetal heart rate baseline estimation algorithms by their influence on detection of clinically important patterns. Biocybern. Biomed. Eng. **36**(4), 562–573 (2016)
16. Wrobel, J., Roj, D., Jezewski, J., Horoba, K., Kupka, T., Jezewski, M.: Evaluation of the robustness of fetal heart rate variability measures to low signal quality. J. Med. Imag. Health Inform. **5**(6), 1311–1318 (2015)
17. Wrobel, J., Horoba, K., Pander, T., Jezewski, J., Czabanski, R.: Improving the fetal heart rate signal interpretation by application of myriad filtering. Biocybern. Biomed. Eng. **33**(4), 211–221 (2013)
18. Czabanski, R., Jezewski, J., Horoba, K., Jezewski, M.: Fetal state assessment using fuzzy analysis of the fetal heart rate signals – agreement with the neonatal outcome. Biocybern. Biomed. Eng. **33**(3), 145–155 (2013)
19. Ghodrati, A., Murray, B., Marinello, S.: RR interval analysis for detection of Atrial Fibrillation in ECG monitors. In: 30th Annual International Conference of the IEEE EMBS, pp. 601–604 (2008). http://ieeexplore.ieee.org/stamp/stamp.jsp?arnumber=4649224
20. Logan, B., Healey, J.: Robust detection of atrial fibrillation for a long term telemonitoring system. In: Computers in Cardiology, pp. 619–622. IEEE (2005). http://ieeexplore.ieee.org/stamp/stamp.jsp?arnumber=1588177
21. Duverney, D., Gaspoz, J.M., Pichot, V., Roche, F., et al.: High accuracy of automatic detection of atrial fibrillation using wavelet transform of heart rate intervals. Pacing Clin. Electrophysiol. **25**(4), 457–462 (2002)
22. Cerutti, S., Mainardi, L.T., Porta, A., Bianchi, A.M.: Analysis of the dynamics of RR interval series for the detection of atrial fibrillation episodes. In: Computers in Cardiology, pp. 77–80. IEEE (1997). http://ieeexplore.ieee.org/stamp/stamp.jsp?arnumber=647834
23. Tateno, K., Glass, L.: Automatic detection of atrial fibrillation using the coefficient of variation and density histograms of RR and ΔRR intervals. Med. Biol. Eng. Comput. **39**(6), 664–671 (2001)
24. Dash, S., Chon, K., Lu, S., Raeder, E.: Automatic real time detection of atrial fibrillation. Ann. Biomed. Eng. **37**(9), 1701–1709 (2009)
25. Petrucci, E., Balian, V., Filippini, G., Mainardi, L.T.: Atrial fibrillation detection algorithms for very long term ECG monitoring. In: Computers in Cardiology, pp. 623–626. IEEE (2005). http://ieeexplore.ieee.org/stamp/stamp.jsp?arnumber=1588178
26. Artis, S.G., Mark, R., Moody, G.: Detection of atrial fibrillation using artificial neural networks. In: Computers in Cardiology, pp. 173–176. IEEE (1991)
27. Henzel, N., Wrobel, J., Horoba, K.: Atrial fibrillation episodes detection based on classification of heart rate derived features. In: Proceedings of 24th International Conference on Mixed Design of Integrated Circuits and Systems, MIXDES 2017, Bydgoszcz, 22–24 June, pp. 571–576 (2017)
28. Roj, D., Wrobel, J., Matonia, A., Horoba, K., Henzel, N.: Control and signal processing software embedded in smart wristband monitor of silent atrial fibrillation. In: Proceedings of 24th International Conference on Mixed Design of Integrated Circuits and Systems, MIXDES 2017, Bydgoszcz, 22–24 June, pp. 585–590 (2017)
29. Hastie, T., Tibshirani, R., Friedman, J.: The Elements of Statistical Learning - Data Minining, Inference, and Prediction. Springer Series in Statistics, 2nd edn. Springer, New York (2009)

30. Moody, G.B., Mark, R.G., et al.: A new method for detecting atrial fibrillation using R-R intervals. Comput. Cardiol. **10**, 227–230 (1983)
31. Goldberger, A.L., Amaral, L.A., Glass, L., Hausdorff, J.M., et al.: PhysioBank, PhysioToolkit, and PhysioNet: components of a new research resource for complex physiologic signals. Circulation **101**(23), e215–e220 (2000)
32. Petrutiu, S., Sahakian, A.V., Swiryn, S.: Abrupt changes in fibrillatory wave characteristics at the termination of paroxysmal atrial fibrillation in humans. Europace **9**, 466–470 (2007)

Hot Spot Analysis by Means of Continuous Wavelet Transform and Time-Frequency Filtering

Anna Tamulewicz(✉) and Ewaryst Tkacz

Department of Biosensors and Biomedical Signal Processing, Faculty of Biomedical Engineering, Silesian University of Technology, Roosevelta 40, 41-800 Zabrze, Poland
anna.tamulewicz@polsl.pl

Abstract. In the paper, the analysis of hot spots in proteins with the aid of digital signal processing methods was conducted. The algorithm employs time-frequency filtering and continuous wavelet transform (CWT); its aim is to find amino acid regions where the characteristic frequency is dominant by detecting peaks in energy plot. The research showed that the choice of wavelet function has big impact on the results. The best results were achieved by using CWT with the Morlet wavelet and the sixth order derivative of a Gaussian wavelet.

Keywords: Hot spot · Protein interactions · Continuous wavelet transform

1 Introduction

Most of the biological processes in cells are controlled by protein complexes, which can interact by so-called interfaces. The energy of this interface is not uniformly distributed – some of the amino acid residues have the biggest impact on the binding energy of the protein complex and are called hot spots.

The biological method of hot spots identification is called alanine scanning mutagenesis (ASM) [2]. It enables to determine an energy which each of the amino acid residues contribute to the binding between protein and its target. Hot spots determined in ASM experiment are deposited in alanine scanning energetic database (ASEdb) [18]. The ASM, however, requires specialistic laboratory equipment, is time-consuming and expensive, which induced the need of developing computational tools.

Apart from standard methods of hot spots prediction – which are based on the knowledge of three-dimensional structure of proteins or physico-chemical properties of their residues – digital signal processing methods are getting more and more popular. The biggest advantage of such algorithms comparing to standard methods is an ability of using only primary structure (i.e. amino acid sequence)

E. Tkacz et al. (Eds.): IBE 2018, AISC 925, pp. 141–148, 2019.
https://doi.org/10.1007/978-3-030-15472-1_16

of proteins. Among the digital signal processing methods used for hot spots analysis the following can be distinguish: Fourier transform [3], Continuous Wavelet Transform (CWT) [4,8,11,13], Short-Time Fourier Transform (STFT) [10], S transform [12], digital filtering [9], and others. Although, the CWT was used before, the full potential of this tool has not been exploited, which was the motivation for this research.

1.1 Digital Signal Processing in the Context of Hot Spot Identification

The tools of digital signal processing can be applied to the hot spots analysis on the basis of resonant recognition model (RRM) [3,20], which assumes the correlation between protein function and numerical representation of its amino acid. Such representation can be obtained by assigning each of the amino acids the numerical value relevant for protein biological activity. The numerical values can represent physical, chemical, thermodynamic, structural or statistical properties of amino acids, but it has been proven [7] that electron-ion interaction potential values (EIIP, Table 1) are the most suitable for protein analysis with the aid of digital signal processing methods.

Table 1. EIIP values of amino acids.

Amino acid	Code	EIIP [Ry]	Amino acid	Code	EIIP [Ry]
Alanine	A	0.0373	Cysteine	C	0.0829
Aspartic acid	D	0.1263	Glutamic acid	E	0.0058
Phenylalanine	F	0.0946	Glycine	G	0.0050
Histidine	H	0.0242	Isoleucine	I	0.0000
Lysine	K	0.0371	Leucine	L	0.0000
Methionine	M	0.0823	Asparagine	N	0.0036
Proline	P	0.0198	Glutamine	Q	0.0761
Arginine	R	0.0959	Serine	S	0.0829
Threonine	T	0.0941	Valin	V	0.0057
Tryptophan	W	0.0548	Tyrosine	Y	0.0516

The studies show [4] that proteins with common biological function have common frequency components in the energy distribution, which are called characteristic frequencies of the biological process. Proteins in the protein complex have the same characteristic frequency but an opposite phase. The characteristic frequency can be obtained by finding a set of proteins functionally related with the analysed molecule, and calculating the modulus of multiple cross-spectral function (so-called consensus spectrum) for this set. The frequency of significant peak in the consensus spectrum denote the characteristic frequency for the set of analysed proteins.

2 Methodology

In the work, the hot spots analysis was conducted with the aid of one of the digital signal processing method – continuous wavelet transform.

2.1 Continuous Wavelet Transform

Continuous wavelet transform (CWT) allows for signal analysis in the time-frequency plane. It can be defined as the correlation between the analysed signal $x(t)$ and the the function called wavelet $\psi(t)$, which can be scaled (stretched or compressed) and shifted. The CWT can be described as follows:

$$C(b,a) = \int_{-\infty}^{+\infty} x(t)\psi_{a,b}^{*}(t)dt = \frac{1}{\sqrt{a}}\int_{-\infty}^{+\infty} x(t)\psi^{*}\left(\frac{t-b}{a}\right)dt \qquad (1)$$

where: $\psi(t)^*$ is the complex conjugate of the analysing mother wavelet $\psi(t)$, $C(b,a)$ denotes CWT coefficient, which is the function of the parameters: a – scale parameter, and b – position parameter.

Apart from a and b parameters, CWT coefficients depends on the choice of the wavelet function. Six different wavelet functions were considered, including: Morlet wavelet, second (Mexican hat), fourth, and sixth order derivatives of Gaussian wavelets, Paul and Bump wavelets [6,15,16,19].

2.2 The Algorithm and Data Set Used in the Analysis

2.2.1 Data Set

Primary sequences of the analyzed proteins were obtained from Protein Data Bank (PDB) [1] and UniProt database [5,17]. The data set (Table 2) consisted

Table 2. Proteins in the analysed data set.

PDB id	Chain	Molecule	Characteristic frequency	Number of hot spots
4fgf	A	Basic fibroblast growth factor	0.4520	4
3hhr	A	Human growth hormone	0.1350	3
3hhr	B	Human growth hormone receptor (hGHbp)	0.1350	5
1rcb	A	Interleukin-4	0.2935	2
1brs	A	Barnase	0.1605	6
1brs	D	Barstar	0.1605	3
1bxi	A	Colicin E9 immunity protein	0.0950	6
1ulo	A	Endoglucanase C	0.0465	3
1nt3	A	Neurotrophin-3	0.0345	2

of 9 proteins with known characteristic frequency and was used to present the performance of the algorithm. In a previous study [14], the authors have shown that the choice of related proteins set is important to determine the characteristic frequency, which is critical for the identification of hot spots. Because of the problems in gathering proper sets of related proteins, the proteins with characteristic frequency presented in the literature [9,12,13] were chosen. For each of the protein in the data set, the hot spot residues were determined according to ASEdb, whereas the rest of the amino acid were treated as non-hot spot residues. Thus, the data set was found to be comprised of 1226 residues in which 34 were hot spots and 1192 were non-hot spots.

2.2.2 The Hot Spot Analysis Algorithm

Algorithm used in the analysis is employed to detect hot spots by applying time–frequency filtering and is based on the methods described in [9,12,14]:

1. Create numerical signal by converting the given amino acid sequence by using EIIP values.
2. Compute CWT coefficients (by using different wavelet functions) of the analyzed amino acid sequences.
3. Determine the proper scale corresponding to the characteristic frequency.
4. Select the characteristic frequency from the spectra by using proper filtering matrix.
5. Plot the squared magnitude of the filtered signal and find the localization of hot spots by locating the energy peaks.

3 Results and Discussion

The performance of the algorithm was assessed by comparing obtained results with the ASEdb database. In order to present application of the method, the following statistical measures were calculated (2): accuracy (ACC), sensitivity (Sn) and specificity (Sp).

$$ACC = \frac{TP + TN}{TP + TN + FP + FN}, \tag{2}$$

$$Sn = \frac{TP}{TP + FN}, \tag{3}$$

$$Sp = \frac{TN}{TN + FP} \tag{4}$$

where: TP and TN is the number of correctly identified hot spots and non-hot spots, respectively, whereas FP and FN is the number of non-hot spots and hot spots incorrectly identified, respectively.

The overall obtained results are presented in Table 3, which summarizes accuracy, sensitivity, and specificity acquired by using CWT with different wavelet functions. To demonstrate the performance of the algorithm, the results obtained

Table 3. The statistical measures for the identification of hot spots.

Wavelet function	ACC [%]	Sn [%]	Sp [%]
Morlet	85.9	76.5	86.2
Mexican hat	84.2	64.7	84.7
Fourth order derivative of Gaussian	84.5	64.7	85.1
Sixth order derivative of Gaussian	85.7	76.5	86.0
Paul	84.5	67.6	85.0
Bump	84.8	67.6	85.3

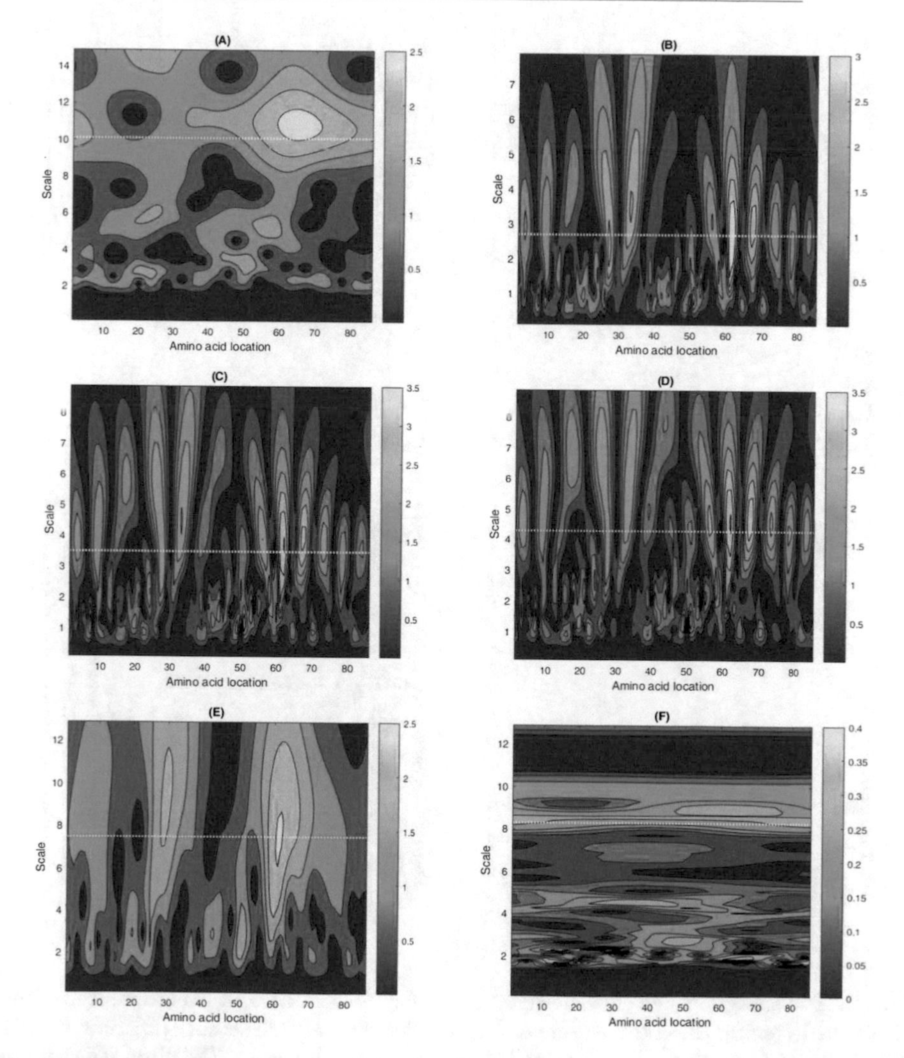

Fig. 1. Amplitude spectra of the Colicin E9 immunity protein obtained with the aid of CWT by using different wavelets: (A) Morlet, (B) Mexican hat, (C) fourth order derivative of Gaussian, (D) sixth order derivative of Gaussian, (E) Paul, (F) Bump. The scales corresponding to the characteristic frequency are marked by white horizontal lines.

for the protein Colicin E9 immunity protein (1*bxi*) are presented in Figs. 1 and 2. Figure 1 shows the amplitude spectra acquired by using CWT with different wavelets. The calculated scales corresponding to the characteristic frequency (0.0950 for 1*bxi*) are marked by lines.

In order to detect hot spot residues, the energy at the characteristic frequency was calculated. The hot spot locations can be determined by finding peaks in the energy plot. To separate significant from insignificant peaks in the energy

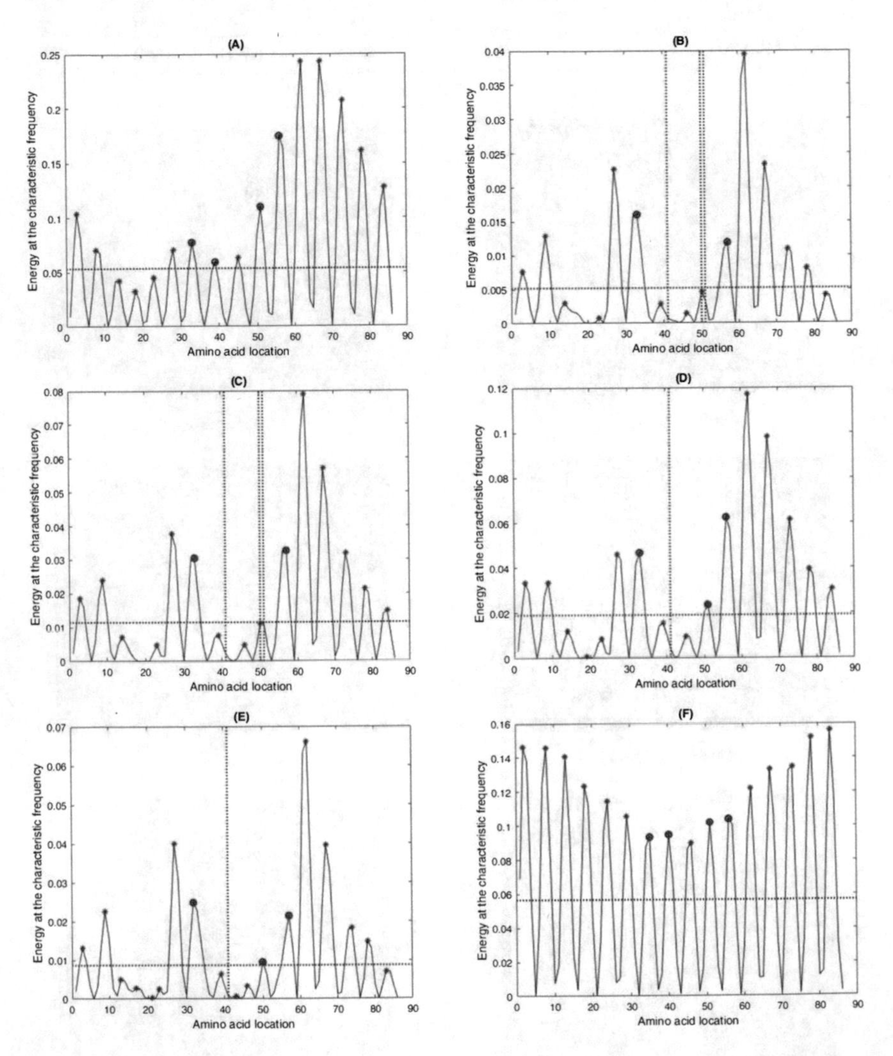

Fig. 2. The energy at the characteristic frequency for the Colicin E9 immunity protein obtained by using different wavelets: (A) Morlet, (B) Mexican hat, (C) fourth order derivative of Gaussian, (D) sixth order derivative of Gaussian, (E) Paul, (F) Bump. The determined thresholds are marked by horizontal lines, whereas correctly identified and unidentified hot spots are marked by circles and vertical lines, respectively.

plot, the concept presented by [9,12] was followed, that is, the peak-to-average ratio as an evaluation criterion was assumed. The threshold t_p for hot spots and non-hot spots separation was calculated as 90% of the average energy values. Each energy peak that is more than t_p is deemed as hot spot. Figure 2 shows the determined energy at the characteristic frequency obtained from CWT spectra with different wavelets for the protein $1bxi$.

The obtained results suggest that the choice of analyzing wavelet is very important. The best performance was acquired with the Morlet wavelet and the sixth order derivative of Gaussian wavelet – ~86% of non-hot spots and ~76% of hot spots reported in ASEdb were correctly identified. The Morlet wavelet function was reported [4] to be the best for use with protein sequences, which is consistent with the obtained results. For the Paul and the Bump wavelets, ~68% of hot spots were correctly identified, whereas the lowest sensitivity (~65%) was achieved with the use of the second- (Mexican hat) and the fourth order derivatives of Gaussian. Because non-hot spots were overrepresented in the analyzed data set, obtained accuracies and specificities are slightly bigger than sensitivities and are ~84%–86% for accuracy and ~85%–86% for specificity.

As mentioned before, the CWT was used for hot spot analysis in works of Cosic [4], Pirogova [8], Rao [11], and Shakya [13]. In the [4,8,11], however, the exact identification of hot spots is missing, which did not allowed for the evaluation of the performance and comparison with the method considered in this paper. Comparing to [13] – which achieved the accuracy of 67%, sensitivity of 70%, and specificity of 65% – the proposed algorithm with the Morlet and the sixth order derivative of Gaussian wavelet functions allowed to achieve better results in all considered statistical measures, whereas ACC and Sp have bigger values for all wavelet functions used.

4 Summary and Conclusion

The considered algorithm uses the time–frequency filtering. It focuses on the finding regions in the protein sequence where the characteristic frequency is dominant. A number of wavelet functions were tested, and the Morlet and the sixth order derivative of Gaussian wavelets were found to be the most appropriate for hot spot analysis. The obtained results show that the considered model improved hot spot detection compared with other wavelet-based methods, either by allowing for more accurate hot spots identification or by achieving better statistical measures values.

References

1. Berman, H.M., Westbrook, J., Feng, Z., Gilliland, G., Bhat, T.N., Weissig, H., Shindyalov, I.N., Bourne, P.E.: The protein data bank. Nucleic Acids Res. **28**(1), 235–242 (2000)
2. Clackson, T., Wells, J.A.: A hot spot of binding energy in a hormone-receptor interface. Science **267**(5196), 383–386 (1995)

3. Cosic, I.: Macromolecular bioactivity: is it resonant interaction between macromolecules?–Theory and applications. IEEE Trans. Biomed. Eng. **41**(12), 1101–1114 (1994)
4. Cosic, I.: Analysis of HIV proteins using DSP techniques. In: 2001 Conference Proceedings of the 23rd Annual International Conference of the IEEE Engineering in Medicine and Biology Society, pp. 2886–2889. IEEE (2001)
5. European Bioinformatics Institute (EMBL-EBI), Swiss Institute of Bioinformatics (SIB), Protein Information Resource (PIR). The Universal Protein Resource (UniProt). http://www.uniprot.org
6. Komorowski, D., Pietraszek, S.: The use of continuous wavelet transform based on the fast fourier transform in the analysis of multi-channel electrogastrography recordings. J. Med. Syst. **40**(1), 10 (2015)
7. Lazovic, J.: Selection of amino acid parameters for Fourier transform-based analysis of proteins. Comput. Appl. Biosci. **12**(6), 553–562 (1996)
8. Pirogova, E., Fang, Q., Akay, M., Cosic, I.: Investigation of the structural and functional relationships of oncogene proteins. Proc. IEEE **90**(12), 1859–1867 (2002)
9. Ramachandran, P., Antoniou, A.: Identification of hot-spot locations in proteins using digital filters. IEEE J. Sel. Top. Signal Process. **2**(3), 378–389 (2008)
10. Ramachandran, P., Antoniou, A., Vaidyanathan, P.: Identification and location of hot spots in proteins using the short-time discrete Fourier transform. In: Conference Record of the Thirty-Eighth Asilomar Conference on Signals, Systems and Computers, vol. 2, pp. 1656–1660. IEEE (2004)
11. Rao, K.D., Swamy, M.N.S.: Analysis of genomics and proteomics using DSP techniques. IEEE Trans. Circuits Syst. I Regul. Pap. **55–I**(1), 370–378 (2008)
12. Sahu, S.S., Panda, G.: Efficient localization of hot spots in proteins using a novel S-transform based filtering approach. IEEE/ACM Trans. Comput. Biol. Bioinform. **8**(5), 1235–1246 (2011)
13. Shakya, D.K., Saxena, R., Sharma, S.: Identification of hot spots in proteins using modified Gabor wavelet transform. Pertanika J. Sci. Technol. **22**(2), 457–470 (2014)
14. Tamulewicz, A., Tkacz, E.: Human fibroblast growth factor 2 hot spot analysis by means of time-frequency transforms. In: Information Technologies in Medicine. Advances in Intelligent Systems and Computing, vol. 472, pp. 147–159. Springer (2016)
15. The MathWorks, Inc. Continuous wavelet transform and scale-based analysis. https://www.mathworks.com/help/wavelet/gs/continuous-wavelet-transform-and-scale-based-analysis.html
16. The MathWorks, Inc. CWTFT. https://www.mathworks.com/help/wavelet/ref/cwtft.html
17. The UniProt Consortium. UniProt: a hub for protein information. Nucleic Acids Res. **43**(Database issue), D204–D212 (2015)
18. Thorn, K.S., Bogan, A.A.: ASEdb: a database of alanine mutations and their effects on the free energy of binding in protein interactions. Bioinformatics **17**(3), 284–285 (2001)
19. Torrence, C., Compo, G.P.: A practical guide to wavelet analysis. Bull. Am. Meteorol. Soc. **79**(1), 61–78 (1998)
20. Veljkovic, V., Cosic, I., Dimitrijevic, B., Lalovic, D.: Is it possible to analyze DNA and protein sequences by the methods of digital signal processing? IEEE Trans. Biomed. Eng. **32**(5), 337–341 (1985)

Analysis of the Influence of Selected Thiosemicarbazides, 1,2,4-Triazoles and 1,3,4-Thiadiazoles' Structure on Their Microbiological Activity Against Candida Albicans ATCC30028 and Candida Albicans Clinical Isolate 26

Anna Filipowska[1(✉)] and Michał Jóźwiak[2]

[1] Department of Biosensors and Processing of Biomedical Signals, Silesian University of Technology, Roosevelta 40, 41-800 Zabrze, Poland
anna.filipowska@polsl.pl

[2] Department of Biochemistry, First Faculty of Medicine, Medical University of Warsaw, 02-097 Warsaw, Poland

Abstract. Considering the increasing incidence of fungal infections, in silico research of selected thiosemicarbazides, 1,2,4-triazoles and 1,3,4-thiadiazoles exhibiting varied biological activity against the Candida albicans fungi has been conducted in relation to this paper. The determined molecular descriptors for all 30 examined compounds and the published microbiological research served as the basis for obtaining multilinear QSAR models describing the relationship between biological activity against the *Candida albicans* ATCC 30028 and *Candida albicans* clinical isolate 26 fungi and the structure of examined compounds. The Leave-One-Out Cross Validation method was used to verify obtained models. The obtained Quantitative Structure-Activity Relationships are characterized by high determination coefficients and good prediction power. Additionally, the results of cluster analysis of compounds on the basis of their physicochemical parameters have also been presented.

Keywords: Thiosemicarbazides · 1,2,4-triazoles · 1,3,4-thiadiazoles · Quantitative Structure-Activity Relationship · Multiple linear regression · Cluster analysis

1 Introduction

In the recent years fungal infections have become an ever greater clinical challenge. Patients suffering from congenial or acquired immunological defects (including people infected with HIV), neoplastic diseases (including blood neoplasms) or diabetes are particularly susceptible to fungal infections. Factors increasing the chance of endogenous fungal infections also include:

E. Tkacz et al. (Eds.): IBE 2018, AISC 925, pp. 149–158, 2019.
https://doi.org/10.1007/978-3-030-15472-1_17

- taking immunosuppressive medications (e.g. in post-transplantation patients),
- radiotherapy,
- chemotherapy [1].

The most common etiological factors are fungi of the *Candida* (about 60–80% of infections) and *Aspergillus* genera [2]. Therefore, research is conducted to find new antifungal therapeutic agents. In order to limit costs, traditional in vitro research is supported by in silico research including, among others, the Quantitative Structure-Activity Relationships (QSAR) analysis, as well as classification and clustering (grouping) of compounds based on similarities.

This paper concerns determination of physicochemical parameters describing properties of 30 compounds from the thiosemicarbazides, triazoles and thiadiazoles categories, whose obtaining method and biological activity have been described in detail in a paper by Jóźwiak [3]. The analyzed compounds show varied biological activity against pathogenic *Candida albicans* ATCC 30028 and *Candida albicans* clinical isolate 26 strain fungal strains. The determined molecular descriptors and biological activity against examined fungal strains were used as basis for chemometric analyses and attempts at creating QSAR models describing relationships between biological activity against analyzed strains and compound structure based on molecular descriptors.

2 Experimental Works and Methods

Physicochemical parameters in the form of molecular descriptors have been determined for all 30 analyzed compounds from the thiosemicarbazides, triazoles and thiadiazoles categories (listed in the Table 1) by means of the HyperChem ver. 7.5 software [4]. Extensive conformational search was performed at molecular mechanics level with Optimized Potentials for Liquid Simulations (OPLS) force field. The most stable structures obtained were subsequently optimized to the closest local minimum at the semiempirical level using PM3 parametrizations. Convergence criteria were set to 0.1 and 0.01 $\frac{\text{kcal}}{\text{mol}\cdot\text{Å}}$ for OPLS and PM3 calculations, respectively. The selected descriptors were then used to develop a QSAR models [3,5,6].

The multiple linear regression (MLR) method was used for modeling of quantitative structure-activity relationships (QSAR). Minimal inhibitory concentrations obtained for analyzed fungal strains, expressed as $\log(1/MIC[\mu\text{M}])$, were used as dependent variables, and selected molecular descriptors—as independent variables. The statistical calculations were conducted at confidence level of 95% ($p < 0.05$). The analyses were conducted by means of multiple backward regression which entailed successively rejecting the least statistically significant (having the highest p value) structural parameters (molecular descriptors). Statistical verification of the model was performed by means of T-test and Fisher-Snedecor F-test. T-test is used to verify significance of each parameter of a model (independent, explanatory variables), and F-test is used for checking statistical significance of the entire model [7]. Conditions for obtaining models by means of the multiple linear regression where estimators were calculated by means of the least squares method (model linearity, coincidence condition, uniform dispersion (homoscedasticity) condition, normal distribution of model residue) were checked for all determined equations.

Statistical parameters for the resulting regression equations: correlation coefficient (R), determination coefficient (R^2), adjusted determination coefficient for calibration (R^2_{adj}), a standard error of estimate (SEE), determination coefficient of LOO validation (Q^2) are presented in Table 3.

The obtained QSAR mathematical models were internally validated by means of the Leave-One-Out Cross Validation. Successively one compound was excluded from the data set (N) used as a basis for a model and it was used to validate the resulting new model containing $N-1$ elements. The Q^2 validation coefficient was calculated from the formula 1 [7,8]:

$$Q^2 = 1 - \frac{\sum_{i=1}^{n}\left(y_{exp,i} - y_{pred,i}\right)^2}{\sum_{i=1}^{n}\left(y_{exp,i} - y_{ave,i}\right)^2} \tag{1}$$

where: y_{exp}—experimental output value for the i-th compound, y_{pred}—predicted output value for the i-th compound, $y_{ave,i}$—average value calculated for $(N-i)$ compounds.

Cluster analysis was conducted in order to determine similarities between compounds described by the full set of variables in the form of physicochemical parameters. The analysis of similarities included standardization of data. As a result, the influence of data with minor deviation has been increased and the influence of data with major standard deviation has been decreased. The complete linkage method and the squared Euclidean distance metric were utilized for cluster analysis.

Furthermore, the standardization procedure eliminates the influence of different units of measurement and renders the data dimensionless [9,10].

Based on the calculated physicochemical parameters (19 molecular descriptors determined for all analyzed compounds) and microbiological research, the analysis of similarities between the analyzed compounds was conducted and QSAR models describing relationship between antifungal activity against the standard *Candida albicans* ATCC 30028 strain and the *Candida albicans* clinical isolate 26 strain were determined. The structure of analyzed compounds is presented in Table 1, and results of microbiological research are presented in the [3] paper.

3 Results

Based on the calculated physicochemical parameters (19 molecular descriptors determined for all analyzed compounds) and microbiological research, the analysis of similarities between the analyzed compounds was conducted and QSAR models describing relationship between antifungal activity against the standard *Candida albicans* ATCC 30028 strain and the *Candida albicans* clinical isolate

Table 1. Structures of derivatives of thiosemicarbazides (compound 1–10), 1,2,4-triazoles (compound 1–10a) and 1,3,4-thiadiazoles (compound 1–10b) used in the study [3].

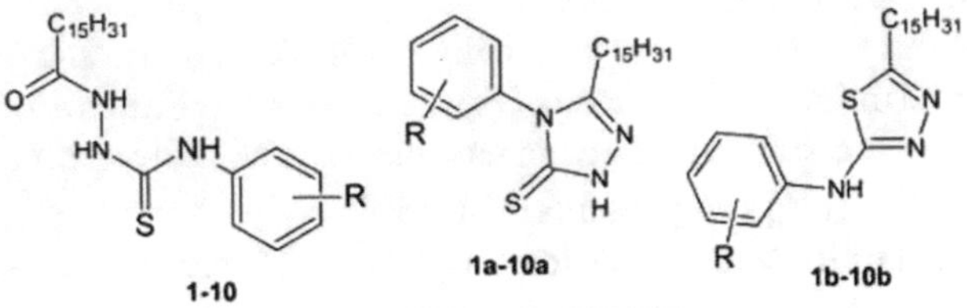

1-10 1a-10a 1b-10b

Compound	R
1	*H*
2	2-Cl
3	3-Cl, 4-F
4	3-Br
5	3-Cl, 4-CH_3
6	3-Cl
7	2-Br
8	3-Cl, 4-Cl
9	4-Br
10	4-Cl

26 strain were determined. The structure of analyzed compounds is presented in Table 1, and results of microbiological research are presented in the [3] paper.

Values of selected molecular descriptors used to create QSAR models are presented in Table 2. Figure 1 presents a diagram of compound grouping based on their physicochemical parameters and antifungal activity in the form of a color feature—compound map. Numbers of analyzed compounds are placed along the Y-axis, and molecular descriptors and antifungal activity are included along the X-axis. The scale describing features (molecular descriptors, antifungal activity) is dimensionless. Upon analyzing the diagram presented in Fig. 1 it is clear that there should be a simple linear correlation between the antifungal activity and electronic descriptors concerning molecular stability E_T—total energy, E_B—binding energy, E_{IA}—isolated atomic energy, E_E—electronic energy, I_{CC}—core-core interaction, and the HF parameter describing molecule's heat of formation. The listed electronic descriptors are correlated with one another which was confirmed by the analysis of similarities (cluster analysis).

Table 3 includes obtained results for linear QSAR models for the 30 analyzed chemical compounds.

The Table 3 presents QSAR models where antifungal activity against the analyzed fungal strains depends on the following physicochemical parameters: $\log P$—partition coefficient octanol/water, SA—surface area - approx., Rf—molar refractivity, E_B—binding energy, I_{CC}—core-core interaction. In all obtained equations the antifungal activity depends on molar refractivity Rf and on one descriptor describing the electronic properties belonging to the group of descriptors described by the $E_T = E_B + E_{IA} = I_{CC} + E_E$ equation. Molar

Table 2. Molecular descriptors of thiosemicarbazides (compound 1–10), 1,2,4-triazoles (compound 1–10a) and 1,3,4-thiadiazoles (compound 1–10b) used in the study [3].

Compound	SA [Å^2]	log P	I_{CC} [kcal/mol]	Rf [Å^{-3}]	E_B [kcal/mol]
1	678.53	7.48	833397	123.41	−6468
2	629.88	8.00	936389	128.21	−6451
3	708.18	8.14	964893	128.43	−6462
4	692.40	8.27	904648	131.03	−6436
5	593.68	8.46	1027373	133.26	−6733
6	622.99	8.00	940545	128.21	−6449
7	705.39	8.27	910269	131.03	−6435
8	740.48	8.51	962474	133.02	−6435
9	676.98	8.27	908276	131.03	−6436
10	618.63	8.00	933938	128.21	−6450
1a	561.80	8.27	817208	120.50	−6255
2a	524.19	8.79	894591	125.30	−6239
3a	642.93	8.93	915872	125.52	−6246
4a	678.34	8.85	817007	127.89	−6217
5a	639.53	9.25	922260	130.34	−6524
6a	623.79	8.79	852920	125.30	−6239
7a	540.92	9.06	896889	128.12	−6220
8a	651.42	9.31	917074	130.11	−6220
9a	599.94	9.06	866838	128.12	−6219
10a	610.86	8.79	869020	125.30	−6237
1b	557.53	8.06	801371	120.27	−6249
2b	601.69	8.57	848427	125.07	−6230
3b	679.88	8.71	874345	125.29	−6242
4b	678.34	8.85	817007	127.89	−6217
5b	712.63	9.04	870447	130.11	−6519
6b	621.52	8.57	835217	125.07	−6234
7b	624.94	8.85	836767	127.89	−6216
8b	695.11	9.09	865275	129.88	−6216
9b	678.60	8.85	821676	127.89	−6218
10b	667.88	8.57	819273	125.07	−6236

refractivity may be interpreted as the measure of molecule electrons' mobility or a measure of steric interactions between compounds [11], i.e. it determines the volume occupied by an atom or a group of atoms. Strict correlation between Rf and biological activity of the analyzed compounds may indicate that the strength of the chemical bond of the compound with the polar surface area is

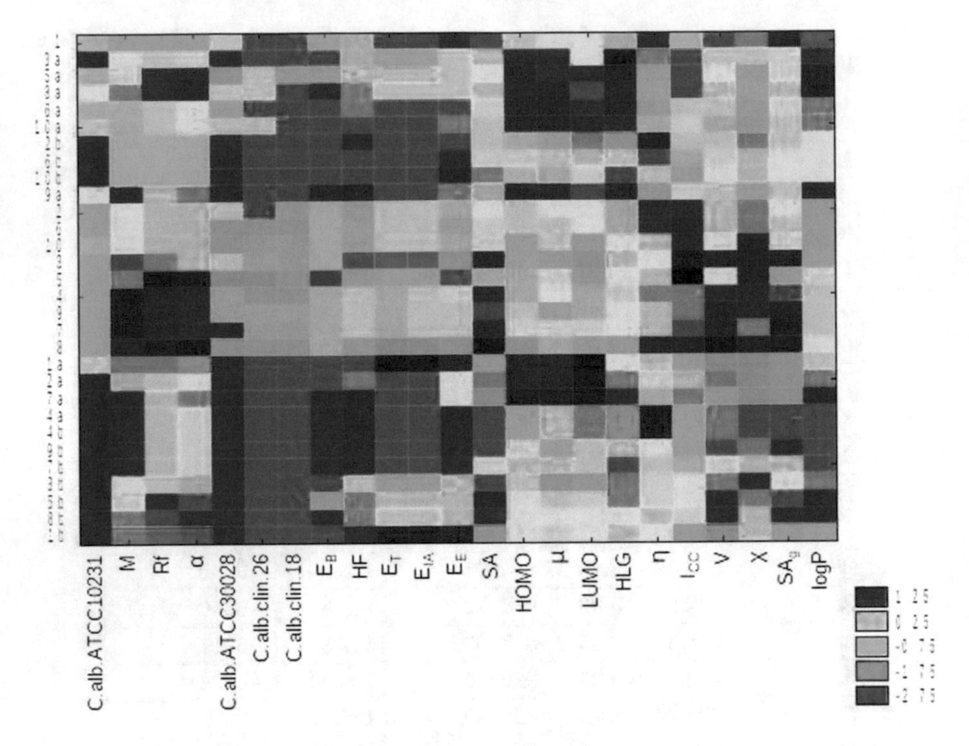

Fig. 1. Color map of the feature-compound representations. X axis—names of molecular descriptors and antifungal activity; Y axis—numbers of analyzed compounds

Table 3. Formulas describing the relationship between activity against *Candida albicans* ATCC 3002, *Candida albicans* clinical isolate 26 and molecular descriptor.

Strain	Equation	R	R^2	$R^2_{adj.}$	F	p	SEE	Q^2
Candidada albicans ATCC30028	$\log(1/MIC_{C.alb.ATCC30028}) = -2.8 \cdot 10^{-12}(\pm 1.0 \cdot 10^{-12})SA^4 + 0.0000012(\pm 0.0000006)Rf^3 - 1.8 \cdot 10^{-11}(\pm 4 \cdot 10^{-12})E_B^3 - 9.05(\pm 0.8)$	0.848	0.720	0.687	42.5	$4.6 \cdot 10^{-7}$	0.31	0.600
Candidada albicans Clinical isolate 26	$\log(1/MIC_{C.alb.clin.26}) = -0.50(\pm 0.2)\log P + 0.0000016(\pm 0.0000006)Rf^3 + 2.6 \cdot 10^{-12}(\pm 9 \cdot 10^{-13})I_{CC}^2 - 3.34(\pm 1.4)$	0.847	0.718	0.685	22.0	$2.6 \cdot 10^{-7}$	0.33	0.621

the factor influencing activity [12,13]. For the *Candida albicans* clinical isolate 26 strain, a model with $\log P$ as one of the independent variables has been obtained. This parameter describing compound lipophilicity is connected with compound's ability to permeate biological membranes, and, as a result, with the speed of distribution and absorption processes [11]. For the obtained QSAR models, positive linear correlation of such descriptors as Rf, ICC^2 and negative linear correlation of such descriptors as $\log P^2$, SA^4, E_B with biological activity against the selected fungal strains can be observed. All equations presented in Table 3 meet the condition of acceptability $R^2 > 0.6$ and the condition of prediction for mod-

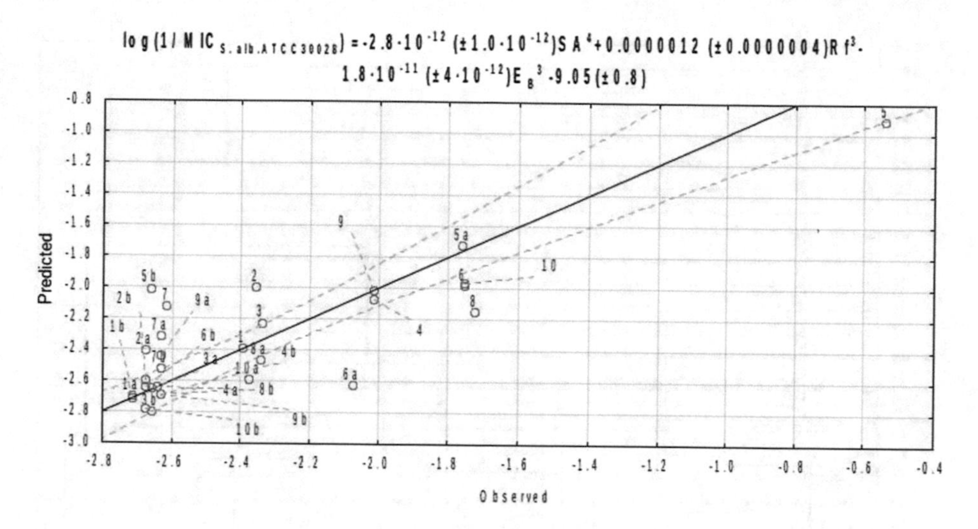

Fig. 2. Correlation of activity against *Candida albicans* ATCC30028 and SA^4, $log_P{}^2$, Rf^3, E_B^3.

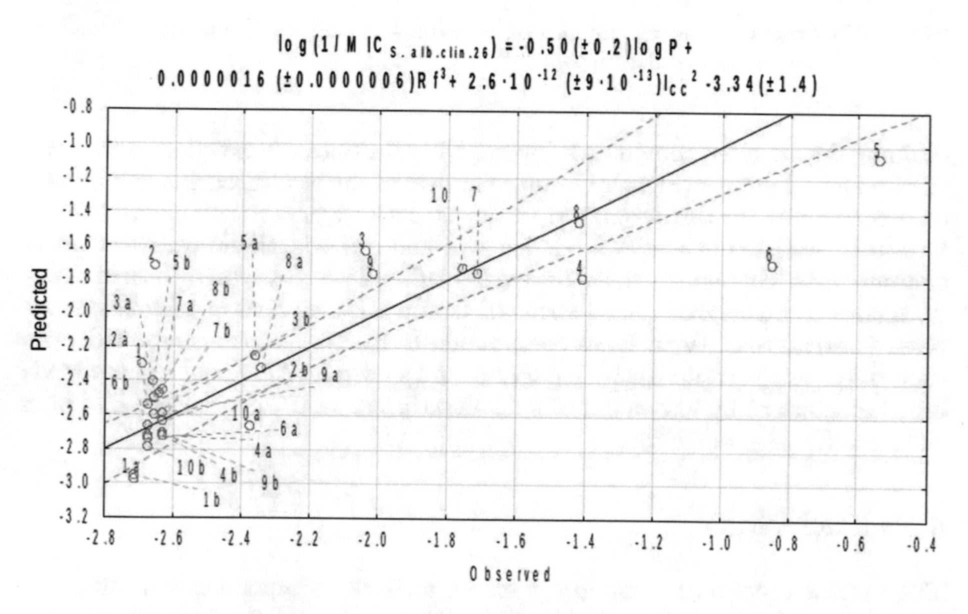

Fig. 3. Correlation of activity against *Candida albicans* Clinical isolate 26 and $\log P$, Rf^3, ICC^2.

els $Q^2 > 0.5$ [14,15]. Figures 2 and 3 show diagrams of predicted and observed values for analyzed fungal strains and obtained QSAR models (Table 3).

The conducted analysis of similarities, entailing categorizing compounds based on $\log P$, Rf, SA, I_{CC}, EE, μ descriptors describing lipophilic, steric and electronic parameters and for which linear QSAR models were obtained, enabled

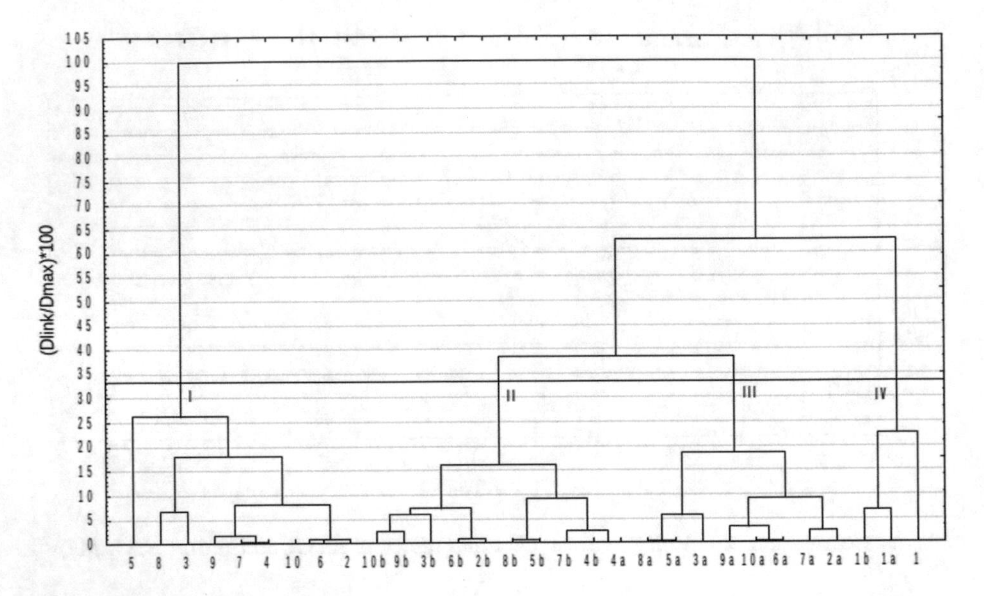

Fig. 4. Dendrogram showing similarities between the analyzed compounds based on $\log P$, SA, Rf, I_{CC}, E_E, μ descriptors.

dividing the analyzed compounds (by applying the Sneath's criterion at the level of 33%) into 4 groups (Fig. 4). Group I consists of thiosemicarbazide derivatives. The compounds in this group are the most active against the analyzed fungi. Group II consists of thiadiazole derivatives and one triazole derivative, namely compound 4a (characterized, similarly to thiadiazole derivatives, by lack of activity against the analyzed fungal strains). Group III comprises of triazole derivatives. Lastly, group IV includes compounds 1, 1a, 1b, i.e. one compound from each of the analyzed derivative categories. As shown in Fig. 1, these compounds are characterized by low values of such descriptors as Rf, $\log P$, I_{CC}, and high values of the E_E descriptor.

4 Conclusions

The obtained molecular descriptors for all analyzed compounds and microbiological research were used as the basis for determining 2 QSAR models describing linear relationship between biological activity of the analyzed compounds against the analyzed fungal strains (*Candida albicans* ATCC 30028 and *Candida albicans* clinical isolate 26) and the following molecular descriptors: Rf, ICC^2, $\log P^2$, SA^4, E_B. The analysis of similarities confirms that there is a relationship between the analyzed compounds' structure and values of electronic descriptors described by Eq. 2:

$$E_T = E_B + E_{IA} = I_{CC} + E_E \quad (2)$$

The obtained models may be used to predict biological activity of new thiosemicarbazide, triazole and thiadiazole derivatives against the analyzed fungal strains.

References

1. Ciszewski, M., Czekaj, T.: Healthcare-associated fungal infections—a rising threat. Nowa Medycyna **2**, 73–76 (2014)
2. Biliński, P., Seferyńska, I., Warzocha, K.: Diagnosis and treatment of fungal infections in oncohematology. Onkol. Prak. Klin. **4**, 15–24 (2008)
3. Jóźwiak, M., Stpięń, K., Wrzosek, M., Olejarz, W., Kubiak-Tomaszewska, G., Filipowska, A., Filipowski, W., Struga, M.: Synthesis, structural studies and biological evaluation of connections of thiosemicarbazide, 1,2,4-triazole and 1,3,4-thiadiazole with palmitic acid. Molecules **23**, 822 (2018)
4. Hyper Chem: Hyper Chem (TM), Professional, Hypercube, Inc., 1115 NW 4th Street, Gainesville, Florida 32601, U.S.A. (2012)
5. Stefanska, J., Nowicka, G., Struga, M., Szulczyk, D., Koziol, A.E., Augustynowicz-Kopec, E., Napiorkowska, A., Bielenica, A., Filipowski, W., Filipowska, A., Drzewiecka, A., Giliberti, G., Madeddu, S., Boi, S., La Colla, P., Sanna, G.: Antimicrobial and anti-biofilm activity of thiourea derivatives incorporating a 2-aminothiazole scaffold. Chem. Pharm. Bull. (Tokyo) **63**, 225–236 (2015). https://doi.org/10.1248/cpb.c14-00837
6. Kusmierz, E., Siwek, A., Kosikowska, U., Malm, A., Stefanska, J., Dzitko, K., Wujec, M.: Antibacterial activity and structure-activity relationship studies of 4- substituted-5-(diphenylmethyl)-2,4-dihydro-3H-1,2,4-triazole-3-thiones. Lett. Drug Des. Discov. **10**, 95–101 (2013). https://doi.org/10.2174/157018013804725198
7. Filipowska, A., Filipowski, W., Tkacz, E.: Study of structure-cytotoxicity relationships of thiourea derivatives containing the 2-aminothiazole moiety. In: Innovations in Biomedical Engineering. Advances in Intelligent Systems and Computing, vol. 526, pp. 276–285. Springer (2017). https://doi.org/10.1007/978-3-319-47154-9_32
8. Khaledian, S., Saaidpour, S.: Quantitative structure-property relationship modelling of distribution coefficients (logD7.4) of diverse drug by sub-structural molecular fragments method. Orient. J. Chem. **31**, 1969–1976 (2015). https://doi.org/10.13005/ojc/310414
9. Astela, A., Biziukb, M., Przyjazny, A., Namieśnik, J.: Chemometrics in monitoring spatial and temporal variations in drinking water quality. Water Res. **40**, 1706–1716 (2006). https://doi.org/10.1016/j.watres.2006.02.018
10. Filipowska, A., Filipowski, W., Tkacz, E., Wujec, M.: Statistical analysis of the impact of molecular descriptors on antimicrobial activity of thiourea derivatives incorporating 3-amino-1,2,4-triazole scaffold. In: Gzik, M., Tkacz, E., Paszenda, Z., Piętka, E. (eds.) Innovations in Biomedical Engineering. Advances in Intelligent Systems and Computing, vol. 623, 1st edn, pp. 276–285. Springer, Cham (2018). https://doi.org/10.1007/978-3-319-70063-2_19
11. Patrick, G.L.: An Introduction to Medicinal Chemistry, 5th edn, pp. 383–406. Oxford University Press, Oxford (2013). ISBN 9780199697397
12. Todeschini, R., Consonni, V.: Molecular Descriptors for Chemoinformatics. Methods and Principles in Medicinal Chemistry Ed. by Mannhold, R., Kuloinyi, H., Folkers, G., vol. 41. Wiley (2009)

13. Kubinyi, H.: QSAR; Hansch Analysis and Related Approaches. Methods and Principles in Medicinal Chemistry Ed. by Mannhold, R., Krogsgaard-Larsen, P., Timmerman, H., vol. 1. Wiley (1993)
14. Golbraikh, A., Tropsha, A.: Predictive QSAR modeling based on diversity sampling of experimental datasets for the training and test set selection. J. Comput. Aided Mol. Des. **16**, 357–369 (2002). PMID: 12489684
15. Filipowska, A., Filipowski, W., Tkacz, E., Nowicka, G., Struga, M.: Statistical analysis of the impact of molecular descriptors on cytotoxicity of thiourea derivatives incorporating 2-aminothiazole scaffold. Chem. Pharm. Bull. **64**, 1196–1202 (2016). https://doi.org/10.1248/cpb.c16-00317

Modelling and Simulations in Biomechanics

Repeatability of Selected Kinematic Parameters During Gait on Treadmill in Virtual Reality

Piotr Wodarski(✉), Mateusz Stasiewicz, Andrzej Bieniek, Jacek Jurkojć, Robert Michnik, Miłosz Chrzan, and Marek Gzik

Faculty of Biomedical Engineering, Department of Biomechatronics, Silesian University of Technology, Gliwice, Poland
{piotr.wodarski,andrzej.bieniek,jacek.jurkojc, robert.michnik,milosz.chrzan,marek.gzik}@polsl.pl, mjfstasiewicz@gmail.com

Abstract. Gait is the most important form of human locomotion and can be decribed by kinematics quantities and stabilographics paramaters. It is possible to determie those parameters from the distribution of forces on the ground. The purpose of this study is determining of repeatability of gait on treadmill in virtual reality. Thirty-two healthy adults have been exmained. In this study each person has been tested 6 times (natural gait on the threadmill and 5 times gait on the treadmil in virtual reality). Between each part of the study was 240 s break, during that person was stimulated by various stimuli, like increseed or decresed speed of treadmill or virtual scenery. All this to disperse the examined person and to maximize the destabilization of the posture of this person. In this study was concluded that gait on the treadmill in virtual reality is repetitive in reference to measured values of selected kinematic parameters determined by distribution of forces on the ground. Results of this study were confronted with literature review. The study are first attempt describing of impact of VR environments on human gait.

Keywords: Human gait · Gait kinematic parameters · Virtual reality

1 Introduction

Gait is a characteristic form of locomotion of human beings. Disfunctions of gait are associated with reduced comfort of life, in mental and physical spaces. In conclusion, life quality of people with gait impairment is low. Virtual reality (VR) creates new chances for the people [1,2]. Development of special methods of gait analysis allows to create new methods of diagnostic and treatment gait abnormalities with systems using virtual reality. In those studies, gait can be described by kinematics quantities and stabilographics parameters [1].

In [1] authors show the positive effect of using VR in rehabilitations process for post-stroke people. Effectiveness of this type of training is greater then normal therapy. This analysis show significant improvements in mobility, balance

E. Tkacz et al. (Eds.): IBE 2018, AISC 925, pp. 161–168, 2019.
https://doi.org/10.1007/978-3-030-15472-1_18

and walking speed. But effectiveness of this method is different for people with different level of motor impairments. Research review suggests that VR rehabilitation is motivating and more involving than normal rehabilitation [3,8,9].

The authors [3] compare of gait parameters between normal gait and gait in immersive environment. This study shows that gait parameters are influenced by immersive environment. People in VR conditions walk slower than normal condition. This effect is a combinations of decreased stride length and different frequency of stepping. Comparison between gait in VR and gait in perturbed scenery VR shows that people walking in perturbed VR conditions walking faster than people walking in unperturbed VR. This study show that walk in virtual reality is able to influence on gait and stabilography parameters. Conclusions from this study allow to think that VR could be useful tool for a functional evaluation of selected gait parameters.

Observation of relative movements of objects in the field of human vision *(optic flow)* provides him with a lot of information about his position in space, the direction of his body movements and speed [7]. Manipulation of these parameters in the world of virtual reality through changes of the displayed scene may affect the human gait and stabilometric parameters [4–6].

It is slightly different in the study of gait on the treadmill [6] in unperturbed VR scenery conditions and perturbed VR scenery conditions. This scenery moves faster or slower than it should. Significant differences were identified for the ankle joint ROM (range of motion), hip joint ROM and the stride interval. In the case of knee joint ROM, no significant differences were found. Observed changes occurred despite the fact that scientists in this study used a real-time feedback mechanism that allowed the control of selected parameters by the subject.

Literature review presents many new facts about the differences between natural gait and gait in virtual reality. However, it should be checked whether the impact of VR on humans's gait on treadmill does not change over time. There is a need to check the repeatability of gait in VR, or there are any differences between next attempt in the VR world. The literature review indicates that these differences can be detected, among other, by analyzing the way of balancing the body, eg by measuring the distribution of pressure forces on the ground. The study are first attempt describing of impact of VR environments on human gait.

Approval

This study was previously approved by the Ethics in Research Committee of the Academy of Physical Education in Katowice (protocol number 11/2015).

2 Methodology

Thirty-two healthy adults (6 men, 26 women) without any lower limb dysfunction and motor dysfunction have been examined ($age, 22.5 \pm 1.78 lat; height, 169.91 \pm 8.7; bodyweight, 62.56 \pm 15.21\,\mathrm{kg}, BMI, 21.47 \pm 3.42\,\mathrm{kg/m^2}$). The virtual reality environment used in the study was prepared using the Unity3D software and

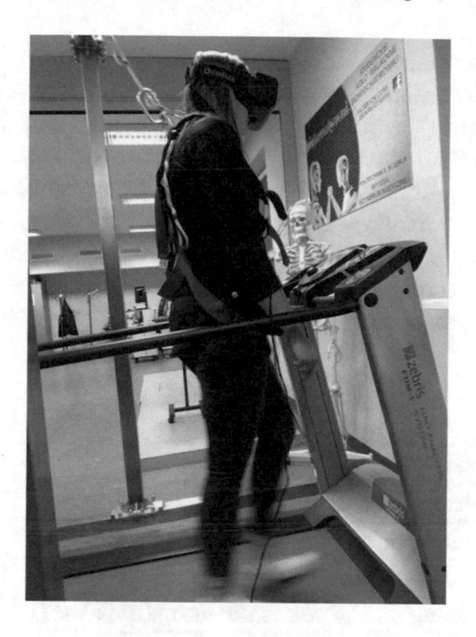

Fig. 1. Test stand.

displayed using the HMD *(head mounted display)* system. To measure the distribution of forces on the ground, Zebris FDM-T treadmill with dedicated software (which could determine the selected kinematic quantities) was used. In addition, a safety platform was used to increase the safety of the subjects. The test stand is shown in Fig. 1. The test was consisted in measurements of the distribution of forces on the ground, during gait at speed of 3.2 km/h in 50-s sections. The test started with a warm-up in the form of a gait on the treadmill for 240 s. At that time, the posture stabilized and the first gait measurement started without the use of virtual reality (NG). In the next step, there was a break in the test, which lasted about 240 s. Then the examined person put HMD glasses on with projection of virtual scenery. In the next step, a 50-s walk in virtual reality (VRG1) was performed. This measurement was repeated 4 times (VRG2, VRG3, VRG4, VRG5), each time with a break of 240 s. During the break, the examined person was stimulated by various stimuli, such as: increasing/decreasing the speed of the treadmill, increasing/decreasing the speed of moving scenery created in the virtual world and by sideways shifting of virtual scenery at a frequency equal to half the frequency of making steps. All this to disperse the examined person and to maximize the destabilization of the posture of this person. Then there was a 30-s break in which the subject stood, followed by the next phase of the study. The course of the study with division into phases is presented in Table 1. With the use of the Zebris-FDM platform, the distribution of forces on the ground was recorded, and with the use of a dedicated treadmill software, the following gait parameters were determined: Cadence (steps per second), Stride time, Stride length, Total double support (period within a gait cycle where both feet have ground contact), Step length, Step time, Pre-swing phase, Swing phase, Single support phase (periot within gait cycle during witch the contralateral foot has no contact with ground), Anterior/Posterior position (shift of intersection point of the course of CoP in butterfly cyclogram display, zero position is the heel strike) and Lateral symmetry (shift of intersection point of the course of CoP in butterfly cyclogram display, a negative value indicates a shift to the left side and a positive value a shift to the right side). For all parameters, mean and standard deviations were calculated, the results were compared with each other. Statistical analysis was carried out using the Matlab and Statistica software.

Table 1. Stages of the study.

Stages	Activity	Duration, s
1	Warm-up - walking on the treadmill	240
2	Walking on the treadmill without VR (NG)	50
3	Break	240
4	Walking on the treadmill with VR (VRG1)	50
5	Break	240
6	Walking on the treadmill with VR (VRG2)	50
7	Break	240
8	Walking on the treadmill with VR (VRG3)	50
9	Break	240
10	Walking on the treadmill with VR (VRG4)	50
11	Break	240
12	Walking on the treadmill with VR (VRG5)	50

3 Results

The Fig. 2 shows the graphs of all measured values. These parameters were determined by the Zebris FDM-T treadmill software, based on the distribution of pressure forces on the ground during 50-s test of the gait. The analysis of the graphs allows to indicate the differences between the averaged values for examined group of measured parameters. In order to examine whether these differences are statistically significant, statistical tests were carried out for the measured values with the division into individual phases of the study. In the first stage, in order to investigate the occurrence of normal distributions, the Shapiro-Wilk test was carried out, and then the Levene test for examination of the homogeneity of variance between samples. In the next step, ANOVA calculations were carried out for gait parameters with normal distribution, and Kruskal-Wallis ANOVA tests were performed for values that did not have normal distributions. The p values obtained with these tests can be found in the Table 2. These values were

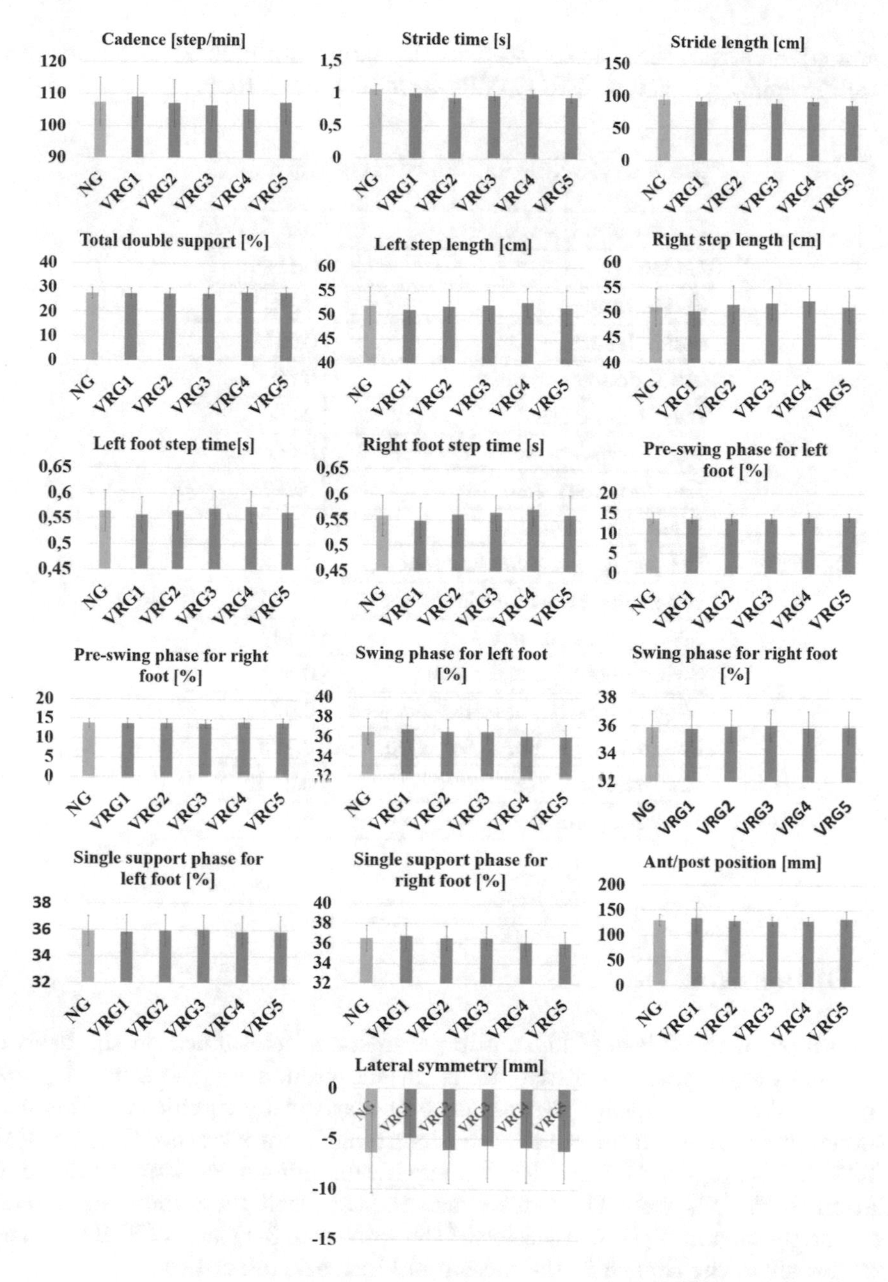

Fig. 2. Analyzed gait parameters.

calculated using the Statistica and Matlab software. The value of p-parameter greater than 0.05 indicates that there are no significant differences between following tests for measured parameters. From Table 2 it can be read that for all measured quantities there are no statistically significant differences between the measurements: NG, VRG1, VRG2, VRG3, VRG4 and VRG5.

Table 2. P-values for individual gait quantities.

Parameters	p(ANOVA)
Cadence	0.1185
Stride time	0.13
Stride length	0.1118
Total double support	0.68
Left step length	0.596
Right step length	0.064
Left foot step time	0.2976
Right foot step time	0.0528
Pre-swing phase for left foot	0.512
Pre-swing phase for right foot	0.885
Swing phase for left foot	0.143
Swing phase for right foot	0.983
Single support phase for left foot	0.994
Single support phase for right foot	0.123
Ant/post position	0.9421
Lateral symmetry	0.3807

4 Discussion

Comparison of the values of kinematic parameters, determined on the basis of measurements of pressure distributions on the ground for the studied group of people, does not indicate the existence of statistically significant differences between these values. In the range of the determined values for the VRG1, VRG2, VRG3, VRG4 and VRG5 studies, no significant differences were observed in relation to the NG test. The results may indicate that for a imposed walking speed on the treadmill, the projection of the scenery in 3D glasses (HMD system) does not affect the change in the measured kinematic quantities.

In studies [3,10], it was observed that all examined kinematic values of gait in VR differ significantly in value, in comparison with those parameters measured for walking without virtual reality. However, it should be noted that these tests did not examine gait on the treadmill, but natural gait. The speed of the gait was not imposed on the subjects, they moved at the speed with which they wanted.

This may means that gait on a treadmill with a predetermined gait speed (by its very nature) is characterized by a higher repeatability compared to natural gait.

On the other hand, the study [6] in which gait on the treadmill in virtual reality and without perturbation of scenery was compared with gait on the treadmill without VR, no statistically significant differences were found between the kinematic parameters of these gaits. The research confirm the results of the carried out studies.

A similar situation can be observed by comparing between the groups the results obtained from the studies VRG1, VRG2, VRG3, VRG4 and VRG5. There were no significant differences in the measured parameters for the group under study. This may indicate a high repeatability in the way of gait on the treadmill at a imposed speed in relation to subsequent measurements. It is worth noting that the break in the research was 240 s, during which attempts were made to disperse the subjects. Despite this, the averaged values of the measured parameters have not changed. In the studies of other authors it can be observed that in gait in VR it is important that the subjects are completely immersed in virtual reality. Because only in this case the gait in VR is a repetitive action [11].

The attempt [12] of rehabilitation with using VR of persons after partial amputation one of the lower limbs, showed no statistically significant changes in the values of gait kinematic parameters measured in this study. In spite of this, a significant improvement in the 2 min walk test was observed and participants of this study was satisfied with effects of this type of rehabilitation.

5 Conclusions

The purpose of this study was to check whether gait on the treadmill in virtual reality is repetitive. Obtained results indicate that there are some differences between gait on the treadmill without VR and gait on the treadmill in VR and between gait (VRG1) and subsequent gait attempts in virtual reality (VRG2, VRG3, VRG4, VRG5). However, considering this overall phenomenon, it can be concluded that gait on the treadmill in VR is repetitive in scope of measurement of kinematic parameters determined by distribution of forces on the ground.

References

1. Corbetta, D., Imeri, F., Gatti, R.: Rehabilitation that incorporates virtual reality is more effective than standard rehabilitation for improving walking speed, balance and mobility after stroke: a systematic review. J. Physiother. **61**, 117–124 (2015)
2. Jurkojć, J.: Balance disturbances coefficient as a new value to assess ability to maintain balance on the basis of FFT curves. Acta Bioeng. Biomech. **20**(1) (2018)
3. Menegoni, F., Albani, G., Bigoni, M., Priano, L., Trotti, C., Galli, M., Mauro, A.: Walking in an immersive virtual reality. Annual Review of Cybertherapy and Telemedicine (2009)

4. Jurkojć, J., Wodarski, P., Bieniek, A., Gzik, M., Michnik, R.: Influence of changing frequency and various sceneries on stabilometric parameters and on the effect of adaptation in an immersive 3D virtual environment. Acta Bioeng. Biomech **19**(3) (2017)
5. Michnik, R., Jurkojć, J., Wodarski, P., Gzik, M., Jochymczyk-Woźniak, K., Bieniek, A.: The influence of frequency of visual disorders on stabilographic parameters. Acta Bioeng. Biomech. **19**(3) (2017)
6. Kastavelis, D., Mukherjee, M., Decker, L., Stergiou, N.: The effect of virtual reality on gait variability. Biomech. Res. Build. (2010)
7. Warren Jr., W.H., Kay, B.A., Zosh, W.D., Duchon, A.P., Sahuc, S.: Optic flow is used to control human walking. Nat. Neurosci. **4**, 213–216 (2001)
8. Meldrum, D., Herdman, S., Moloney, R., Murray, D., Duffy, D.: Effectiveness of conventional versus virtual reality based vestibular rehabilitation in the treatment of dizziness, gait and balance impairment in adults with unilateral peripheral vestibular loss: a randomised controlled trial. BMC Ear Nose Throat Disord. **12**, 3 (2012)
9. Darter, J.B., Wilken, J.M.: Gait training with virtual reality-based real-time feedback: improving gait performance following transfemoral amputation. Phys. Ther. **91**(9), 1385–1394 (2011)
10. Janeh, O.B., Langbehn, E., Steinicke, F., Bruder, G., Gulberti, A., Poetter-Nerger, M.: Walking in virtual reality: effects of manipulated visual self-motion on walking biomechanics. ACM Trans. Appl. Percept. **2**(3), Article no. 1 (2010)
11. Jones, A.J., Swan II, E.J., Singh, G., Reddy, S., Moser, K., Hua, C., Ellis, R.S.: Improvements in visually directed walking in virtual environments cannot be explained by changes in gait alone. In: Proceedings of ACM Symposium on Applied Perception (SAP), pp. 11–16 (2012)
12. D'Angelo, S.M.: Analysis of Amputee Gait using Virtual Reality Rehabilitation Techniques

Three-Dimensional Adults Gait Pattern – Reference Data for Healthy Adults Aged Between 20 and 24

Katarzyna Jochymczyk-Woźniak[1](✉), Katarzyna Nowakowska[1], Robert Michnik[1], Marek Gzik[1], and Dominik Kowalczykowski[2]

[1] Department of Biomechatronics, Faculty of Biomedical Engineering, Silesian University of Technology, Roosevelta 40, Zabrze, Poland
katarzyna.jochymczyk-wozniak@polsl.pl

[2] Students Scientific "BIOKREATYWNI", Faculty of Biomedical Engineering, Silesian University of Technology, Roosevelta 40, Zabrze, Poland
http://www.ib.polsl.pl

Abstract. The research aimed to identify the standards of gait-related kinematic quantities and normalcy indices of the Gillette Gait Index related to adults aged between 20 and 24. In addition, it was decided to compare the standard obtained for adults with the gait standard of children. The study group consisted of 32 healthy adults aged 20 to 24. The tests were performed using the BTS Smart system. The original applications developed in the Matlab environment for all of the children subjected to the tests enabled the identification of the Gillette Gait Index (GGI). Detailed analysis involved a set of sixteen variables describing gait kinematics; the above-named variables are used when creating the GGI. Within the framework of this work, a comparative statistical analysis of 16 parameters constituting the GGI was performed between the groups of adults and children. This analysis showed statistically significant differences for 9 variables. The results obtained in this work indicate that there is the necessity of using adequate gait norm for a particular age. The obtained parameter values of the GGI were also compared with results obtained by other authors. The identified trajectories of the kinematic quantities can be used as normative in relation to adult patients.

Keywords: Gait analysis · Normal pattern · Gillette Gait Index · Kinematics

1 Introduction

Gait is a basic form of human locomotion, which has been proven in the research conducted by Morlock et al. Their research showed that a human being annually covers a distance of 1.5 thousand kilometres making over 2 mln steps [14]. Gait is closely connected with energy expenditure which determines the efficiency of gait. Proper gait is characterized by the endeavour to minimize energy expenditure [12].

E. Tkacz et al. (Eds.): IBE 2018, AISC 925, pp. 169–176, 2019.
https://doi.org/10.1007/978-3-030-15472-1_19

A quantitative analysis of gait based on optical systems is more and more often used in hospitals and outpatient clinics. The analysis of patients' gait takes advantage of implemented standards provided by manufacturers of measuring systems. The manufacturers are rather unwilling to share information on the numerical strength and data of the group on which a certain standard is based. It is advisable that each research centre should use reliable standards. Scientific literature provides standards of adult and child gaits. Authors usually present spatiotemporal parameters of gait rather than kinematic ones [4–7,11,15,17, 19,23,24]. Recently, there have been several indices which describe the gait of tested individuals using one dimensionless numerical value [2,10,13,20–22]. The most popular gait index is the Gillette Gait Index (GGI) developed by Schutte et al. [21], which is based on 16 kinematic as well as spatiotemporal parameters of gait. Normative values of particular parameters for the child group may be found in the following works [1,9,19–21], whereas the values for healthy adults can be found only in one work [3]. Moreover, a part of results presented by these authors considerably diverges from normative values determined for children.

Taking into account the foregoing, this research work aimed to identify the standards of gait-related kinematic quantities and indices of normalcy of the Gillette Gait Index related to adults aged between 20 and 24. In addition, it was decided to compare the obtained gait standard of adults with the children's gait standard.

2 Materials and Methods

Three-planar gait analysis was conducted using an optical measuring system – the BTS Smart consisting of 8 digital cameras work in the infrared range (frequency: 250 Hz). Passive markers were placed on the participant's body according to the Helen Hayes protocol. The test group consisted of 32 healthy individuals (18 women, 14 men) – without any signs of disabilities within the motor organ (age: 24 ± 4, body mass: 68 ± 13 kg, body height: 1.73 ± 7.6 m).

After necessary anthropometric measurements have been carried out (according to the Davis model), individuals subjected to tests covered a measurement path between ten or twenty times walking at their natural speed. The analysis encompassed kinematic values (angles in particular joints of lower limbs and the pelvis position) as well as spatiotemporal parameters. Using the authors' application written in the Matlab environment, the index of gait normalcy – the Gillette Gait Index (GGI) was determined. This index is built on the basis of 16 parameters, such as: percentage of stance phase (time of toe off) (1), walking speed normalized by the length of the lower limb (2), frequency of making steps - cadence (3), mean forward tilt of the pelvis in the sagittal plane (4), range of the pelvic motion in the sagittal plane (5), mean pelvic rotation in the transverse plane (6), minimum flexion of the hip joint in the sagittal plane (7), range of the hip joint motion in the sagittal plane (8), peak abduction of the hip joint in swing (9), mean rotation of the hip joint in stance (10), knee joint flexion at the initial contact with the ground (11), time to the peak knee flexion (12), range of flexion and extension of the knee joint (13), peak dorsiflection in stance (14), peak dorsiflection in swing (15), mean foot progression angle in stance (16).

The obtained gait parameters were compared with normative values presented by the authors of other works [3,18]. The obtained results were referred to normative values determined for a group of 56 healthy children (28 girls, 28 boys, own research), to check the possibility of using common normative data [9]. The age of children was 11.0 ± 3.0, body mass: 41.8 ± 13.2 kg and body height: 1.47 ± 0.17.

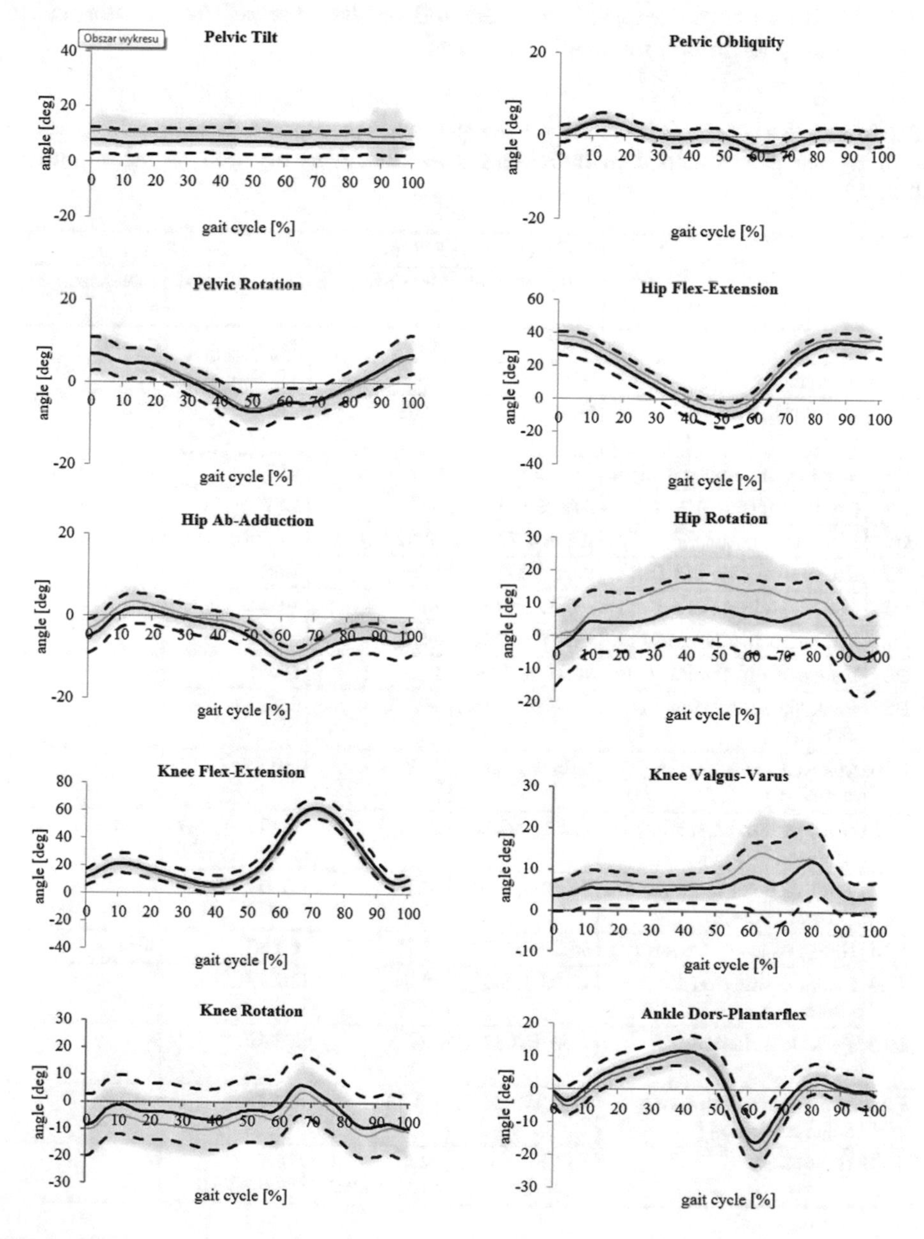

Fig. 1. Pelvis angles and lower limb joints angles during the individual research (n = 32) and for children (n = 56) (the dotted line)

3 Results

In Table 1, the values of the GGI and its individual parameters were juxtaposed with the results obtained for children in own research [9] as well as quantitative variables obtained for adults by the authors of other works [3,18]. Figure 1 shows the trajectories of angles in individual joints of lower limbs as well as the position of the pelvis in three planes. The presented results were referred to normative trajectories determined for healthy children.

Table 1. Values of the GGI and of its 16 constituent parameters in relation to patients with regular gait obtained in individual research and in research by other authors [3,9,18]

		Children	Adults		
		Own research	Own research	Cretual et al.	Pietraszewski et al.
Age (years)		7–17	20–24	22–57	21–23
Group number		56	32	25	17
P1	Time of toe off (% gait cycle)	58.92 ± 1.5	60.68	62.60	65.1
P2	Walking speed/leg length	1.56 ± 0.28	1.31	1.57	1.45 ± 0.18
P3	Cadence (step/s)	2.06 ± 0.24	1.79	1.87	1.84 ± 0.14
P4	Mean pelvic tilt (°)	7.4 ± 4.57	10.22	−5.75	-
P5	Range of pelvic tilt (°)	4.01 ± 1.07	4.09	3.56	1.2
P6	Mean pelvic rotation (°)	-0.88 ± 2.13	0.71	−0.12	-
P7	Minimum hip flexion (°)	-9.75 ± 7.78	−5.58	−11.18	-
P8	Range of hip flexion (°)	46.16 ± 5.84	45.18	44.72	45.5
P9	Peak hip abduction in swing (°)	-11.58 ± 2.26	−9.83	6.73	-
P10	Mean hip rotation in stance (°)	3.68 ± 7.69	7.42	−0.52	-
P11	Knee flexion at initial contact (°)	11.51 ± 4.41	11.12	9.42	-
P12	Time of peak knee flexion (% gait cycle)	70.46 ± 1.16	70.18	72.91	-
P13	Range of knee flexion (°)	58.88 ± 3.74	58.82	60.30	56.1
P14	Peak dorsiflexion in stance (°)	13.29 ± 3.64	12.04	15.97	-
P15	Peak dorsiflexion in swing (°)	4.69 ± 3.42	3.46	10.41	-
P16	Mean foot progression angle in stance (°)	-11.17 ± 4.68	−8.46	1.84	-
GGI (min-max)		15.71 (7.46-30)	12.97 (6.04-25.34)	15.7 (6.9-33.8)	-

4 Discussion

This work presents the normative gait data for adults aged 20–24, which was compared with the normative data of healthy children (Fig. 1). The obtained normative trajectories may be used for the assessment of the gait of patients with various dysfunctions of the motor organ [8]. More and more often, quantitative indices of gait, which are based on a defined group of parameters, are used for the evaluation of locomotor functions in patients. The most popular index is the Gillette Gait Index whose mathematical algorithm is based on 16 spatiotemporal and kinematic parameters of gait. In this work, the comparative statistical analysis of the GGI values was performed regarding the analyzed groups.

Normalcy of the distribution of the analyzed variables obtained from own investigations was verified using the Shapiro-Wilk test. The variables of normal distribution were marked with an asterisk in Table 1 ($p \geq 0.05$).

Statistical differences between the GGI parameters for the analyzed groups (adults, children) were determined by means of the Student's t-test or the Mann

Table 2. Results of the Shapiro-Wilk test, student's t-test and Mann-Whitney u-test

		Shapiro-Wilk test		Student's t-test/the Mann-Whitney u-test
		Adults	Children	p
P1	Time of toe off (% gait cycle)	0.07*	0.91*	<0.05*
P2	Walking speed/leg length	0.02	0.31*	<0.05*
P3	Cadence (step/s)	0.65*	0.21*	<0.05*
P4	Mean pelvic tilt (°)	0.04	0.35*	0.009*
P5	Range of pelvic tilt (°)	0.00	0.43*	0.192
P6	Mean pelvic rotation (°)	0.20*	0.00	0.004*
P7	Minimum hip flexion (°)	0.10*	0.51*	0.015*
P8	Range of hip flexion (°)	0.06*	0.21*	0.404
P9	Peak hip abduction in swing (°)	0.20*	0.46*	0.001*
P10	Mean hip rotation in stance (°)	0.02	0.97*	0.071
P11	Knee flexion at initial contact (°)	0.62*	0.60*	0.665
P12	Time of knee peak flexion (% gait cycle)	0.01	0.71*	0.422
P13	Range of knee flexion (°)	0.81*	0.27*	0.954
P14	Peak dorsiflexion in stance (°)	0.58*	0.23*	0.131
P15	Peak dorsiflexion in swing (°)	0.00	0.81*	0.023*
P16	Mean foot progression angle in stance (°)	0.06*	0.25*	0.013*

*For the Shapiro-Wilk test: $p = 0.05$.

*For the student's t-test and Mann-Whitney u-test: $p = 0.05$.

Whitney u-test, depending on the obtained normalcy of variables distribution (Table 2).

Statistically significant differences between the group of children and the group of adults were observed for the following parameters: time of toe off (P1), walking speed/leg length (P2), cadence (P3), mean pelvic tilt (P4), mean pelvic rotation (P6), minimum hip flexion (P7), peak hip abduction in swing (P9), peak dorsiflexion in swing (P15), mean foot progression angle in stance (P16). The differences between groups for parameters P1-P3 may be determined by the differences in anthropometric values of test groups, including first of all the differences in height and length of lower limbs. Despite the fact that the Student's t-test showed statistical differences between groups for the parameter of mean pelvic rotation, it can be observed that the mean angle of rotation is similar and the difference resulted only from the direction of rotation (right/left). However, greater abduction of the hip joint in swing in the children's group results from the lower position of the body mass centre during gait.

Discrepancies of the parameters regarding the kinematics of the ankle joint (P15, P16) between analyzed groups are connected both with the length of the foot and the consequence of everyday footwear. A considerable part of the adult group was made up by women, who informed about the fact of frequent wearing high heel shoes, which undoubtedly has impact on the shortening of the Achilles tendon and thus the decrease of the foot dorsiflection [16].

5 Summary

Within the framework of this research, the normative gait data of adults aged 20–24 was developed. The conducted statistical analysis showed that the gait of healthy adults is different from the children's gait. The variables which differentiated the children's gait and adult gait most were the spatiotemporal gait parameters, the parameters characterizing the pelvic motion in the sagittal plane, the motion of the hip joint in the sagittal and frontal planes as well as the motion of the ankle joint in the sagittal and transverse planes. No significant differences were observed in the area of the knee joint. The conducted investigations clearly indicate that in the gait assessment of patients with motor organ dysfunctions it is necessary to use normative gait data adequate to a given age.

References

1. Assi, A., Ghanem, I., Lavaste, F., Skalli, W.: Gait analysis in children and uncertainty assessment for Davis protocol and Gillette Gait Index. Gait Posture **30**(1), 22–26 (2009)
2. Baker, R., McGinley, J.L., Schwartz, M.H., Beynon, S., Rozumalski, A., Graham, H.K., Tirosh, O.: The gait profile score and movement analysis profile. Gait Posture **30**(3), 265–269 (2009)
3. Cretual, A., Bervet, K., Ballaz, L.: Gillette Gait Index in adults. Gait Posture **32**(3), 307–310 (2010)

4. Dusing, S., Thorpe, D.: A normative sample of temporal and spatial gait parameters in children using the GAITRite electronic walkay. Gait Posture **25**(1), 135–139 (2007)
5. Ganley, K.J., Powers, C.M.: Gait kinematics and kinetics of 7-year-old children: a comparison to adults using age-specific anthropometric data. Gait Posture **21**(2), 141–145 (2005)
6. Hillman, S.J., Stansfield, B.W., Richardson, A.M., Robb, J.E.: Development of temporal and distance parameters of gait in normal children. Gait Posture **29**(1), 81–85 (2009)
7. Holm, I., Teter, A.T., Fredriksen, P.M., Vollestad, N.: A normative sample of gait and hopping on one leg parameters in children 7–12 years of age. Gait Posture **29**(2), 317–321 (2009)
8. Jochymczyk-Woźniak, K., Nowakowska, K., Michnik, R., Konopelska, A., Luszawski, J., Mandera, M.: Assessment of locomotor functions of patients suffering from cerebral palsy qualified to treat by different methods. In: Gzik, M., Tkacz, E., Paszenda, Z., Piętka, E. (eds.) Innovation in biomedical engineering. Advances in Intelligent System and Computing, vol. 623, pp. 225–233. Springer, Cham (2018)
9. Jochymczyk-Woźniak, K., Nowakowska, K., Michnik, R., Gzik, M., Wodarski, P., Gorwa, J., Janoska, P.: Three-dimensional children gait pattern - reference data for healthy children aged between 7 and 17. In: Piętka, E., Badura, P., Kawa, J., Wieclawek, W. (eds.) Information Technologies in Medicine 5th International Conference. Advances in Intelligent System and Computing, vol. 762, pp. 589–601. Springer, Cham (2018)
10. McMulkin, M.L., MacWilliams, B.A.: Application of the Gillette Gait Index, Gait Deviation Index and Gait Profile Score to multiple clinical pediatric populations. Gait Posture **41**(2), 608–612 (2015)
11. Michnik, R., Jochymczyk-Woźniak, K., Kopyta, I. (eds.): Use of engineering methods in gait analysis of children with cerebral palsy. Wyd. Politechniki Śląskiej, pp. 98–158 (2016). ISBN: 978-83-7880-398-0 (in Polish)
12. Michnik, R., Nowakowska, K., Jurkojć, J., Jochymczyk-Woźniak, K., Kopyta, I.: Motor functions assessment method based on energy changes in gait cycle. Acta Bioeng. Biomech. **19**(4), 63–75 (2017). https://doi.org/10.5277/ABB-00894-2017-03
13. Molloy, A., McDowell, B.C., Kerr, C., Cosgrove, A.P.: Further evidence of validity of the Gait Deviation Index. Gait Posture **31**, 479–482 (2010)
14. Morlock, M., Schneider, E., Bluhm, A., Vollmer, M.A., Bergmann, G., Muller, V., Honl, M.: Duration and frequency of everyday activities in total hip patients. J. Biomech. **34**(7), 873–881 (2011)
15. Nowakowska, K., Michnik, R., Jochymczyk-Woźniak, K., Jurkojć, J., Mandera, M., Kopyta, I.: Application of gait index assessment to monitor the treatment progress in patients with cerebral palsy. In: Piętka, E., Badura, P., Kawa, J., Wieclawek, W. (eds.) Information Technologies in Medicine 5th International Conference. Advances in Intelligent System and Computing, vol. 472(2), pp. 75–85. Springer, Cham (2016)
16. Opila-Correia, K.A.: Kinematics of high-heeled gait. Arch. Phys. Med. Rehabil. **71**(5), 304–309 (1990)
17. Pierce, R., Orendurff, M., Thomas, S.S.: Gait parameters norms for children ages 6–14. Gait Posture **16**(Suppl. 1), 53–54 (2002)
18. Pietraszewski, B., Winiarski, S., Jaroszczuk, S.: Three-dimensional human gait pattern - reference data for normal men. Acta of Bioeng. Biomech. **14**(3), 9–16 (2002)

19. Pinzone, O., Schwartz, M.H., Thomason, P., Baker, R.: The comparison of normative reference data from different gait. Gait Posture **40**(2), 286–290 (2014)
20. Romei, R., Galli, M., Motta, F., Schwartz, M., Crivellini, M.: Use of the normalcy index for the evaluation of gait pathology. Gait Posture **19**(1), 85–90 (2004)
21. Schutte, L.M., Narayanan, U., Stout, J.L., Selber, P., Gage, J.R., Schwartz, M.H.: An index for quantifying deviations from normal gait. Gait Posture **11**(1), 25–31 (2000)
22. Schwartz, M., Rozumalski, A.: The gait deviation index: a new comprehensive index of gait pathology. Gait Posture **28**(3), 351–357 (2008)
23. Stansfield, B.W., Hillman, S.J., Hazlewood, M.E., Robb, J.E.: Regression analysis of gait parameters with speed in normal children walking at self-selected speeds. Gait Posture **23**(3), 288–294 (2006)
24. Steinwender, G., Saraph, V., Scheiber, S., Zwick, E.B., Witz, Ch., Hackl, K.: Intrasubject repeatability of gait analysis data in normal and spastic. Child. Clin. Biomech. **15**(2), 134–139 (2000)

Application of Construction Solutions of Biped Walking Robots in Designing a Prosthetic Foot

Adam Gramala[1(✉)], Paweł Drapikowski[1], Adam M. Pogorzała[2], and Tomasz Walczak[3]

[1] Institute of Control, Robotics and Information Engineering, Poznan University of Technology, Piotrowo 3a, 60-965 Poznan, Poland
adamgramal@wp.pl
[2] Hipolit Cegielski State College of Higher Education in Gniezno, Wyszynskiego 38, 62-200 Gniezno, Poland
[3] Institute of Applied Mechanics, Poznan University of Technology, Jana Pawla II 24, 60-965 Poznan, Poland

Abstract. The recreation of the human gait pattern has been a challenge for biped robot constructors for years. A foot and ankle are important structural components of a biped robot, while in the case of amputation their proper design is a decisive factor in the recreation of the correct gait stereotype in a person with a prosthesis. The paper presents design solutions of selected biped robots' foot models. The chosen foot models' mechanic operation is similar to the biomechanics of the human foot. The work is the review of prosthetic feet dynamics as well. The design of a prosthetic foot begins with the analysis of the biomechanics of a healthy human gait. The gait tests of a healthy person lead to determining correct gait determinants. Determining the angles of flexion of the ankle during walking also shows imperfections of current prostheses and the possibility of eliminating errors in subsequent designs. In addition, research allows to determine the correct trajectory of movement of key points of the body and the value of their accelerations while walking. Gait analysis was performed using the BTS Smart system. When designing a prosthetic foot, it is important to ensure the stability of the gait by positioning a prosthetic foot's centre of gravity and elasticity as well as taking into account an appropriate ankle drive, including dorsiflexion of the foot, which is often overlooked in design concepts and whose lack can influence the patient's greater susceptibility to falls. A common feature for prosthetic and robotic feet is to ensure optimal static and dynamic balance. The main innovation that is possible to implement when designing prosthetic feet is the use of construction solutions of drives used in walking robots.

Keywords: Biped robots · Prosthetic foot · Design · Gait

E. Tkacz et al. (Eds.): IBE 2018, AISC 925, pp. 177–189, 2019.
https://doi.org/10.1007/978-3-030-15472-1_20

1 Introduction

Robots' bipedal way of movement has been a challenge for designers and constructors for years. Following the development of such constructions as, for example, Honda's Asimo [8], it can be noticed how robots are approaching the unsurpassed walking pattern observed in humans. The development of biped robots results from several factors: miniaturization with simultaneous increase in the power of drives, development of batteries of increased energy density with a high instantaneous current output and development of miniature sensors and energy-saving computing units of high power. The development of robots is accelerated by service-related and military applications.

An important construction element of a biped robot is a foot and ankle. These elements play a markedly more important role than in case of multi-legged walking robots. The latter robots due to the multi-support gait, despite a moving ankle lack but equipped with a force sensor, provide adequate friction force. An example of such a construction is the Messor robot built at Poznan University of Technology [3].

A foot also plays a very important role in the construction of a lower limb prosthesis. Regarding the level of amputation, prostheses can be divided into those that must include knee and ankle joints (above-knee amputations) and those requiring reconstruction of mobility at the level of the anatomical ankle (below-knee amputations). The reasons for amputation are diverse. They may be caused by injuries or various diseases as well as developmental defects in the form of deficiencies and losses of limbs. In the case of foot amputation, it is necessary to provide the patient with a prosthesis in such a way as to restore the correct gait pattern and enable him/her to return to physical activity and ensure full independence.

Amputation-related problems can be considered in many aspects. The most important factor that largely determines the patient's ability to return to pre-amputation activity includes numerous medical issues. These are, for instance, the level of selective amputation which influences the way of providing the patient with a prosthesis (a short stump may cause difficulties in stabilizing a prosthesis funnel), the cause of amputation, and diseases, in particular, vascular diseases. In the case of limb amputation, a very important element is the correct formation of a stump and then adapting it to prosthetic fitting. Improper preparation of a stump may greatly limit prosthetic possibilities or completely prevent them. In case skin and other soft tissues are not properly formed, it may lead to abrasions or intertrigo on the stump. A separate problem is the situation in which a prosthesis funnel "slides" from the stump. This most often occurs in patients with an improperly formed and irregularly-shaped stump. Another factor affecting the patient's ability to function is the type of prosthesis used. The above-mentioned factors should be considered in technical and economic aspects. However, when choosing the type of prosthesis, due to the kind of mechanism applied, also medical aspects should be considered taking into account the patient's ability of using it.

Robots and men differ in the way they achieve stability. The human brain must "learn" to adapt to the conditions in which the body part is replaced by a mechanical component - a prosthesis. The whole robot relies on sensors and vision systems. Therefore, the process of learning how to walk in humans looks different than in robots. In humans the process of learning to walk is based on rehabilitation and re-education as well as on the creation of muscle habits. Whereas, in robots the process is based on an algorithm. Here opens an opportunity for prosthesis designers. Using analogy, it is possible to teach a prosthesis to adapt to changing conditions based on data obtained from sensors. The use of neural networks in a prosthetic foot should facilitate the process of achieving stability by humans. Therefore, it is important to use BTS gait analysis systems when developing robot feet. This will enable to model motion as precisely as possible. Thanks to the development of a learning algorithm, a prosthetic foot should more accurately detect changes in the inclination angle of the ground or the detection of walking on stairs. Disciplines such as robotics and bionics are penetrating each other more and more. The exchange of information between scientists in these fields and related areas will further improve bionic foot prostheses. In both cases, the aim is to design and produce a construction that in terms of gait parameters will match the mobility of a healthy person. Correct prosthesis implementation in amputees - especially in working age - allows them to find a job, be socially integrated and reduces the risk of frustration and depression that may occur in people who have not been adequately taken care of. Investing in a good quality prosthesis is beneficial for all parties not only for economic reasons, but also empirical ones - in the first case, service providers (society) are not burdened with the costs of allowances and disability benefits, while in the second case the patient feels needed and fully appreciated. The aim of the future project is to create a bionic prosthetic foot, i.e. one that will allow make it possible to move in various phases of the gait. Thanks to this, the sufferer will not only be able to use it as a support but also walk in a manner similar to the natural walk of a healthy person.

Today's reality promotes interdisciplinary design teams. Cooperation in the field of developing robotic and robot feet may be beneficial for specialists both in robotics and bionics. Thanks to this, it will be possible to reduce design costs and, above all, increase the quality of both constructions.

2 Biped Robot Feet

The human foot is a static part of the lower limb motor organ. It fulfills two important functions: support and body movement in relation to the ground [11]. A biped robot foot has a similar role. However, its shape and construction of the ankle can be very different. Functionally, the ankle joint is an integral part of the foot and the mutual interaction of these elements makes this complex fully efficient and complementary. The foot itself is usually a monolithic element of diverse shape and size, the task of which is to ensure contact with the ground avoiding the risk of slipping. The exception to this rule is the SoftFoot [15] adapted to move on uneven terrain.

The RunBot robot [12] does not have a rotary ankle and is not able to maintain its vertical position in a static manner. However, the purpose of the construction and control system was first of all to obtain dynamic stability and not static one. Most robots' feet have one or two degrees of freedom. One degree of freedom is sufficient to achieve static and dynamic stability, as exemplified by the Cassie robot made by Agility Robotics [1]. Moving in a way similar to human gait requires an ankle joint with two degrees of freedom. The number of robots is increasing. The best known are Assimo [8], Atlas [4], HRP-4C, HRP-2 [10].

An interesting foot solution was used in the construction of the DURUS robot [5]. During heel contact with the ground energy is accumulated and then released during tearing-off the toes from the ground.

3 Prosthetic Feet

Prosthetic feet are one of the most important elements of lower limb prostheses. It is up to them to maintain the user's body balance and the proper functioning of the other components of a prosthesis. The main manufacturers of prosthetic feet are Ottobock, Osuur and Blatchford. In terms of properties there are static feet, dynamic feet with no mobility in the ankle joint and dynamic active feet with a moving ankle joint.

The most similar to the solutions used in biped robots are dynamic feet with a moving ankle joint. The occurrence of rotational movement increases the naturalness of the gait and, above all, stability during standing. Due to the need to reduce the energy consumption of prostheses, manufacturers use prosthetic feet whose executive element regulates the angular speed of rotation in the ankle by means of damping force. In the most expensive prostheses this parameter is controlled in real time depending on the speed of the gait, slope or ground type. The above solutions are featured by the Ottobock Meridium prosthetic foot (Fig. 1) [14]. The foot adapts to the user, which significantly increases the naturalness of gait phases and does not subject joints to excessive overload. The element responsible for changing the angular speed of the ankle's rotation is a hydraulic damper controlled by electrovalves. The user can use an application to determine the damping factor and to select a method of activity, such as cycling or moving in high-heeled shoes. Despite the fact that thanks to technical solutions used the Meridium foot is among the most advanced ones, unfortunately it also has some drawbacks. A negative feature, which is noticed by engineers and its users, is the lack of water resistance (waterproofing at IP45 level makes it impossible to use in rainy conditions). Due to a significant reduction in the elasticity of carbon composite, which consists only of one composite element and due to the vulnerability of hydraulics to be damaged the foot is not used in intense physical activity, including running. This is a kind of disappointment for this model's users. However, it is worth noting that typical running prostheses are made of carbon composite only. They are characteristically profiled and have no drive. Another example of an active prosthetic foot is the ProPrio Foot

Fig. 1. Ottobock Meridium prosthetic foot [14]

by Ossur [13,17,18]. It has a flexible composite spring and its calculation unit controls the electric motor responsible for rotation in the ankle joint. This system is based on information from the implanted sensor inside the patient's muscles. Thanks to this, it is even easier to adapt the prosthesis to changing environmental conditions. The authors did not find any patients in Poland who had such a prosthesis.

A prosthetic foot which is the most similar to robotic feet is BiOM Ankle (Fig. 2), currently being implemented for production [9]. Its creator is Dr. Hugh Herr, director of Biomechatronics at MIT. Herr is passionate about mountain climbing. In his youth during one of his mountain treks he suffered frostbite of both feet and underwent bilateral amputation below the knee joint. This prompted him to create a prosthesis that in its assumption and structure would reproduce biomechanics of the human foot.

The most important advantage of the BiOM Ankle prosthetic foot is that it emulates functions and power of lost muscles and tendons. The system located inside calculates real-time torque in the ankle joint and balances it with the active drive. The BiOM Ankle prosthetic foot allows bending in the propulsion phase, unlike in prostheses with hydraulic systems. In increases gait speed and makes it more comfortable.

4 Biomechanical Parameters of the Gait

The design of a prosthetic foot begins with the analysis of the biomechanics of a healthy human gait. The presented results are based on own research conducted

Fig. 2. BiOM Ankle prosthetic foot [7]

in the Institute of Applied Mechanics at Poznan University of Technology on the BTS Smart motion analysis system. The system is based on the operation of passive markers reflecting infrared light. It consists of six cameras emitting and recording infrared radiation, two dynamometric platforms and a workstation with software that enables the processing of data collected by cameras [6]. There are other methods of gait evaluation like Inertial Sensors Wavelets [18]. Gait tests of a healthy person lead to specifying correct gait determinants. Thanks to this, it is possible to design a prosthesis that provides correct plots of the reaction forces of the ground, and which also, to the least possible extent, affects the movement of the hips. Determining angles of flexion of the ankle during walking also shows imperfections of current prostheses and the possibility of eliminating errors in subsequent designs. The use of the correct walking pattern is also important in the design of biped robots. Thanks to this, it is possible to determine the "ground truth" to which a robot will strive during the walk. Such tests allow to determine the correct trajectory of movement of key points of the body and the value of their acceleration during walking. The trajectory determination using BTS Smart system may also be more precisely calculated, where the forward kinematics for the leg is used to define quantitative measures of the manipulability and workspace in a sagittal plane by the inverse kinematics algorithm [19].

In Figs. 3 and 4 is presented the knee flexion (plot1), the ankle dorsiflexion and plantar flexion (plot 2) and ground reaction forces (plot3) which are shown as well as a purple vectors on the visualization of the gait (selected frame) on the left side of those figures. Determination of flexion angles was based on the SFTR system. In this system the hypertrophy in the knee is marked as a minus, so every dorsiflexion will be show as a negative angle and every plantar flexion will be show as a positive angle. On both of the generated plots in SMART analyzer a

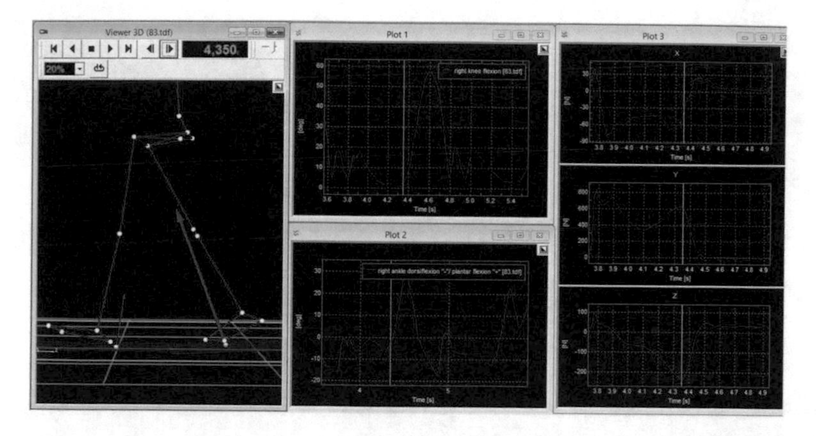

Fig. 3. Analysis of the step of a person after one-sided amputation in the middle of the left thigh for the right foot at the heel detachment from the ground (Plot 1- knee flexion, Plot 2- ankle flexion, Plot 3- ground reaction forces- X- lateral, Y- vertical, Z- posterior)

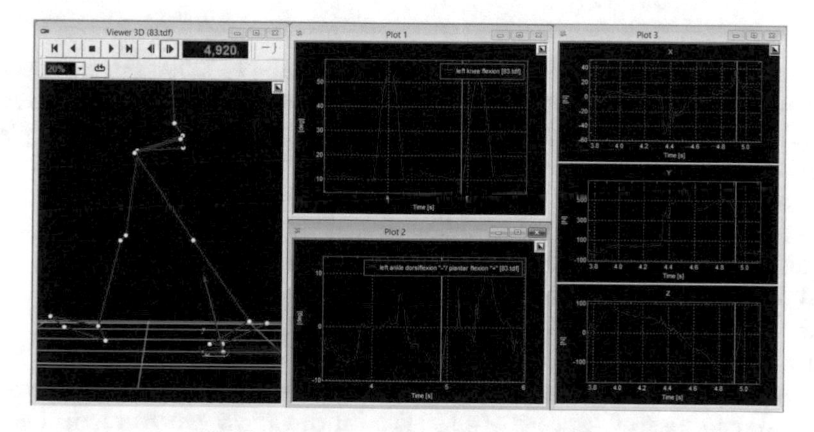

Fig. 4. Analysis of the step of a person after one-sided amputation in the middle of the left thigh for the left foot at the heel detachment from the ground (Plot 1- knee flexion, Plot 2- ankle flexion, Plot 3- ground reaction forces- X- lateral, Y- vertical, Z- posterior)

white line is presented as a moment of detachment of the heel from the ground, which is a reference point. For better illustration of the results, Fig. 5 shows a comparison the ankle dorsiflexion (plot 1), plantar flexion (plot 2) and ground reaction forces (plot 3) of the step of a person after one-sided amputation in the middle of the left thigh for both limbs.

The analysis of the angle of bending in the ankle joint and bending in the knee joint presented in Figs. 3 and 4 on the example of a person after one-sided amputation of the left lower limb at the thigh level leads to the conclusion that to ensure a more natural gait an element allowing the transition from dorsiflexion

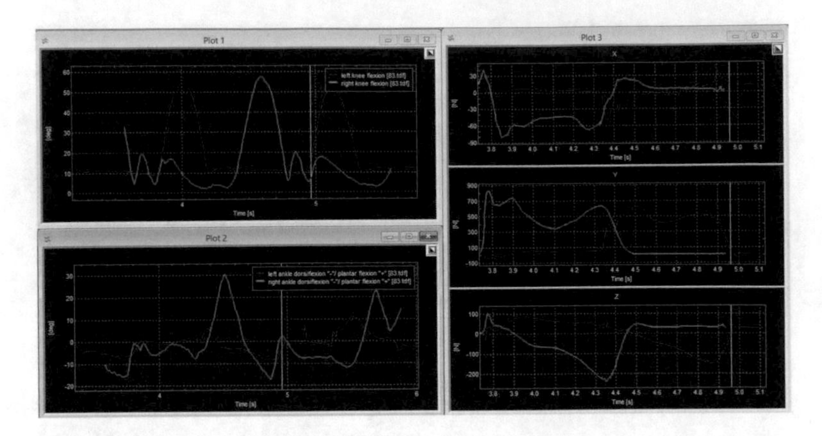

Fig. 5. Comparison of results of the step of a person after one-sided amputation in the middle of the left thigh (red- amputated leg, green- not amputated leg, Plot 1- knee flexion, Plot 2- ankle flexion, Plot 3- ground reaction forces- X- lateral, Y- vertical, Z- posterior).

to plantar flexion should be used in a foot prosthesis. The gait results were compiled for both lower limbs in the push-off phase. The right lower limb shown in Fig. 3 during the gait behaves correctly, i.e. during the push-off phase plantar flexion takes place. The prosthesis illustrated in Fig. 4, which does not have an active element, thus prevents plantar flexion of the foot in this gait phase. As a consequence, it causes less dynamic gait and its asymmetry. This asymmetry may also be cause due to the lack of flexion in the knee prostheses during the initial phase of contact with the ground.

Another important factor that causes a significant gait asymmetry is the smaller vertical component of the reaction of the ground forces for the limb in the prosthesis. The person with amputated leg puts more strain on the preserved limb. This can be seen especially on plot 3 on the Fig. 5, where vertical component of the ground reaction forces presented as a y-axis. On this plot the most important is a peak on the heal strike for the non-amputated leg which is 20% higher than the amputated leg. For a non-amputated limb, the maximum value is 738N, howewer for the limb with a prosthesis it is only 620N. Consequently, this causes overloading of the preserved limb, which may result in pathological changes at a later time.

Therefore, in order to ensure maximum naturalness of the gait, active elements should be used in a prosthesis which allow plantar flexion of the foot. However, this is not being currently used - the BiOM prosthesis is an exception. In order to ensure greater stability, currently in prosthetics prostheses are used which in the swing phase are set in dorsiflexion. This provides less dynamics of walking, but does not require the use of drives. This approach coincides with the direction of work on the foot in biped robots. Robots such as Asimo, Marlo, Pal Robotics, and Atlas do not straighten the knee joint during walking, thus maintaining constant dorsiflexion in the ankle joint. This reduces deflection of the centre of gravity of the system during walking, which significantly increases the robot's balance.

5 Ensuring Gait Stability - Location of the Center of Gravity and Elasticity of a Prosthetic Foot

When designing prostheses and robotic feet, it is important to place the centre of gravity of the entire system in the individual phases of the gait. This ensures a stable walk both on flat and uneven surfaces, on ramps and when moving up the stairs. The balance of man and robot depends on the position of the centre of gravity of the body in the support quadrilateral. Man controls stability by means of sight, vestibular system, support surface and musculoskeletal system [16]. After amputation these functions are disturbed. The human brain, after properly conducted and long-lasting neurorehabilitation, is able to adapt to new conditions using adaptive abilities of the preserved parts of the body and other systems. It is therefore able to control the position of the body in space during walking and in this way to prevent falls. The specific compensation mechanism is complicated in operation and requires cooperation of many systems. Additionally, it can influence overloading of compensating structures. The main goal in the design of prostheses is to reduce the impact of the used prosthesis on healthy tissues. The above assumptions were taken into account by the authors during the design and implementation of their idea of a prosthetic foot. The creators used a composite prosthetic foot which absorbs the heel impact on the ground thus reducing the negative impact of the prosthesis on the patient's other parts of the body and systems.

The use of elastic elements also results from the need to ensure adequate fatigue strength. Daily, a human averages about 5–7 thousand steps [2]. This value depends, among others, on physical activity and the type of work performed. Research shows that the number of steps varies greatly depending on the place of residence. For example, US residents perform on average 5,117 steps, while the Swiss as many as 9,650 steps. These values show that every year a statistical person takes between 1.8 million and 3.5 million steps. In Poland, a modular below-knee prosthesis is refunded once every three years (Ministry of Health, 2017). The above situation requires an assumption that the minimum durability of all elements is 5.5 million steps. Therefore, a prosthetic foot should primarily be resistant to fatigue wear, which requires proper selection of materials. A foot prosthesis should be made of materials that will absorb the shock of the heel against the ground, so it should be resistant to fatigue components. Elastic materials, such as carbon composite, allow energy to be stored during heel contact with the ground and then release it during the toe-off phase. Difficult atmospheric conditions, solar radiation, humidity and temperature changes determine the use of corrosion-resistant material that will have constant properties and geometric dimensions regardless of the presence of water, low or high ambient temperature.

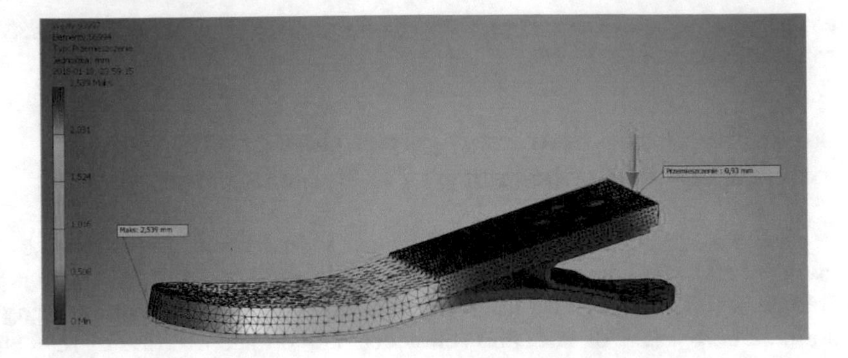

Fig. 6. MES simulation of the own concept of a prosthetic foot during heel impact on the ground

Figure 6 shows the simulation of heel impact on the ground. Fixed binding was applied at the point of contact of the foot during the step. It was assumed that the pressure exerted on the foot is 1100N. The yield strength of carbon composite was taken at 900 MPa. Forefoot is not bound. Thanks to the shape of the foot corresponding to the human foot, satisfactory deflection parameters were obtained. It is 0.93 mm in the heel part. As a consequence, it will allow full use of the elastic properties of carbon composite. At the same time, it will not cause disturbances when walking. In order to ensure greater stability, a cut on the heel and forefoot was also used, which will provide greater comfort when walking on an uneven surface.

6 Ankle Joint Drives

Most amputees use prosthetic feet without elements responsible for controlling the flexion angle in the ankle joint. This is mainly due to economic reasons as well as the need to provide sufficient power for at least day-long operation of a prosthesis. Current trends in the design of prostheses indicate blending of robotics and biomechanics. Constructions tested so far and based on controlling the bending angle by means of changing the damping force are compared with constructions based on active electric drives. The use of motors in prostheses significantly increases the naturalness of walking. However, this has also potentially negative consequences. First of all, the size and weight of a prosthesis increases. It is therefore important to use motors or actuators with high torque and low mass.

As part of the work on the creation of our own concept of a prosthetic foot shown in Fig. 7, we set out to use an element that allows dorsiflexion of the foot and which also maintains it during the swing phase. This is important because, as results from conversations with users of prostheses, the main cause of falls, which most often occur in the "toe-off" phase (critical instability) is tripping over protruding elements such as stones or uneven terrain. Initially, in order to obtain dorsiflexion of the foot, an electromagnetic actuator was used which acted analogously to the tibialis anterior muscle. Torsion springs were designed to be used on axles, the role of which is analogous to tendons. This solution is cheap. However, it does not provide adaptation of flexion during the step by the absence of a damping element. Therefore, it was decided that in the development phase of the project, the possibility of adjusting the angle of flexion in the joint as well as adjusting the angular speed of rotation would be accompanied by a magnetorheological damper placed parallel to the upper composite component of the sole, which will be connected to the joint by a carbon composite spring. This concept coincides with biomechanical modelling and is based on the previously mentioned robotic foot constructions.

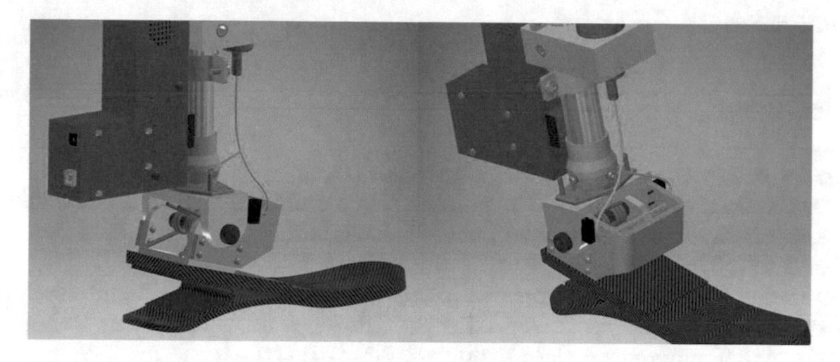

Fig. 7. Own concept of a prosthetic foot with possible pronation and supination in the ankle joint

7 Conclusions

The analysis of the available sources as well as own concepts and research show that the main similarity of a robotic and prosthetic foot is the beginning of the design process. In both cases, the natural gait of a human being is used as a model. It is therefore possible to use common models of gait dynamics and kinematics for designing desired trajectories of movement, bending angles in joints, etc. In order to save money, it is also worth investing in motion analysis systems that can be used in both fields.

Prosthetic feet differ from robotic feet primarily in the fact that in the case of robots there is no occurrence of overloading in the joints, intertrigo on a stump or its abrasions, as well as excessive sweating. Proper design of a robot excludes susceptibility to fatigue damage. Moreover, a robot inside does not have such varied mechanical elements as man. In turn, for human tissues a completely different coefficient of elasticity of materials is important, in comparison with natural tissues such as bones, muscles or ligaments, which affects changes in the body during use.

A common feature for prosthetic and robotic feet is to ensure optimal static and dynamic balance. Having knowledge about the position of all segments of the prosthesis, it is not necessary to implant sensors in the muscles, as in the case of Ossura. At the same time, low mass makes it possible to use DC motors with belt transmissions analogous to the BiOM prosthetic foot without the risk of mechanical damage to the motor shaft or bolt due to occurring overloads.

The main innovation that is possible to implement when designing prosthetic feet is the use of construction solutions of drives used in walking robots. This will be ensured by the aforementioned transition from dorsiflexion of the foot, which is necessary to ensure the naturalness of movement and which physiologically occurs in the human from the footflat to heel off phase, to plantar flexion which occurs in the push off phase. As a result, it will increase the dynamics of the gait and relieve the hip joint.

References

1. Agility Robotics. http://www.agilityrobotics.com
2. Bassett, D.R., Wyatt, H.R., Thompson, H., Peters, J.C., Hill, J.O.: Pedometer-measured physical activity and health behaviors in U.S. adults. Med. Sci. Sports Exerc. **42**(10), 1819–1825 (2010)
3. Belter, D., Walas, K.: A compact walking robot-flexible research and development platform. In: Szewczyk, R. (ed.) Recent Advances in Automation, Robotics and Measuring Techniques, vol. 267, pp. 343–352 (2014)
4. Boston Dynamics. https://www.bostondynamics.com/atlas
5. Engadget. https://www.engadget.com/2016/07/12/georgia-tech-robotdurus-robot-human-walking/
6. Gramala, A., Pogorzala, A.M., Walczak, T., Drapikowski, P.: Wykorzystanie programów obliczeniowych typu CAD w projektowaniu stopy protezowej, Rehabilitacja w praktyce (2018)
7. Harris, D.L.: BiOM may go public, but not before 2016, Bedford bionics company. https://www.bizjournals.com/boston/blog/startups/2014/12/bedford-bionics-maker-biom-may-go-public-but-not.html
8. Honda. http://asimo.honda.com/asimo-specs/
9. Infinite Technologies. https://www.infinitetech.org/biom-anklefoot/
10. Kajita, S., et al.: Cybernetic human HRP-4C: a humanoid robot with human-like proportions. In: Pradalier, C., Siegwart, R., Hirzinger, G. (eds.) Robotics Research. Springer Tracts in Advanced Robotics, vol. 70. Springer, Heidelberg (2011)
11. Kasperczyk, T.: Wady postawy ciała - diagnostyka i leczenie. KASPER S.C, Kraków (2004)
12. Manoonpong, P., Woergoetter, F.: Efference copies in neural control of dynamic biped walking. Rob. Auton. Syst. **57**(11), 1140–1153 (2009)
13. Ossur. https://www.ossur.com/
14. Ottobock. https://www.ottobockus.com/
15. Piazza, C., et al.: Toward an adaptive foot for natural walking. In: IEEE-RAS 16th International Conference on Humanoid Robots (Humanoids), Cancun, pp. 1204–1210 (2016)
16. Souchard, P.: Fizjoterapeutyczna metoda globalnych wzorcow posturalnych, Edra Urban and Partner (2014)

17. Sofge, E.: Brain-Controlled Bionic Legs Are Finally Here (2015). https://www.popsci.com/brain-controlled-bionic-legs-arehere-no-really
18. Glowinski, S., Blazejewski, A., Krzyzynski, T.: Human gait feature detection using inertial sensors wavelets. Wearable Rob.: Chall. Trends Biosyst. Biorob. **16**, 397–401 (2017)
19. Glowinski, S., Krzyzynski, T.: An inverse kinematic algorithm for the human leg. J. Theor. Appl. Mech. **16**, 53–61 (2016)

Height of the Countermovement Vertical Jump Determined Based on the Measurements Coming from the Motion Capture System

Jakub Krzysztof Grabski(✉), Tomasz Walczak, Martyna Michałowska, Marta Szczetyńska, and Patrycja Pastusiak

Institute of Applied Mechanics, Faculty of Mechanical Engineering and Management, Poznan University of Technology, Jana Pawła II 24, 60-965 Poznań, Poland
jakub.grabski@put.poznan.pl

Abstract. A simple and common method for assessing the jumping ability is determination of the jump height during the vertical countermovement jump. However there are different techniques of measurements in biomechanics, e.g. using force plates or accelerometers. In this study the center of jumper's body is determined based on the motion capture system measurements and Clauser or Zatziorsky formulas. Then the jump height can be obtained in a simple way determining the take-off instant and the peak position instant. In such a case the flight height can be defined as a difference between the position of the center of jumper's body at these two instants. The results are compared with the jump height obtained based on the force plates measurements, after double integration of the motion equation.

Keywords: Vertical jump · Countermovement jump · Flight height · Motion capture system

1 Introduction

Vertical jump can be a simple method for assessing jumping ability. Because of that the method is widely described in the literature. There are three basic types of vertical jumps: countermovement jump (CMJ), squat jump and drop jump. In this paper the CMJ is studied.

Different techniques of measurements can be applied in biomechanical studies [1], e.g. force plates [2–4], electromyography [5,6] or dynamometers [7,8]. In this study the measurements coming from the motion capture system [9,10] are used in order to estimate the center of jumper's position (vertical component). Based on this estimation the height of the vertical jump can be determined.

Determination of center of mass for human body is a very important issue in biomechanics. Knowing of the center of mass during different human movements can be valuable for physicians, athletes and their trainers. However it is

E. Tkacz et al. (Eds.): IBE 2018, AISC 925, pp. 190–199, 2019.
https://doi.org/10.1007/978-3-030-15472-1_21

a very difficult task. A well-known and common method in biomechanics for determining the center of mass for human mass is division of the human body into a several segments. Then calculating the center of mass for each segments (using previously defined coefficients) the center of mass for the whole body can be determined. This approach is examined in this paper and compared with the results obtained based on the measurements of the ground reaction force [11] for determining the height of the countermovement vertical jump. It is also well-known that in modeling and simulation of human movement individual parameters of human body should be taken into account. The aim of this study is to show that application of the same coefficients for each person for calculating the center of mass of human body quite significant errors can be obtained in results.

2 Vertical Countermovement Jump

The gravitational force and the ground reaction force (GRF) act on the jumper body. The air resistance is negligible during the CMJ. In such a case the motion equation for the center of mass the jumper body (COM) takes the form

$$ma_y(t) = R_y(t) - mg, \tag{1}$$

where m is the jumper mass, $R_y(t)$ is the GRF and g is the gravitational acceleration.

Thus the acceleration in the vertical direction is given by

$$a_y(t) = \frac{R(t)}{m} - g. \tag{2}$$

Based on the above equation the vertical velocity and vertical position of the COM can be obtained. For this purpose any method for numerical integration is needed, e.g. the trapezoidal rule.

There are five different phases which can be distinguished during the CMJ. These phases are illustrated in Fig. 1: stationary phase (Fig. 1a), downward movement phase (Fig. 1b), pushoff phase (Fig. 1c), flight phase (Fig. 1d) and landing phase (Fig. 1e). The curves of the GRF, vertical velocity and vertical position of the COM is shown in Fig. 2.

The first phase is the stationary phase (Fig. 1a). During this phase the jumper is standing upright. The GRF is equal to the jumper's body weight and it is constant. Because of that the initial vertical acceleration and vertical velocity of the COM are equal to zero. The vertical position of the COM can be also assumed to be zero initially.

The initiation of the second phase (Fig. 1b) is related to beginning of any movement performed by the jumper. In the phase (which is called the downward movement phase) the resultant force acting on the COM and the acceleration are negative at the beginning of the phase. They are initially decreasing. In the second part of the phase the resultant force and acceleration are increasing and they become positive in the end of the phase. The velocity of the COM is negative

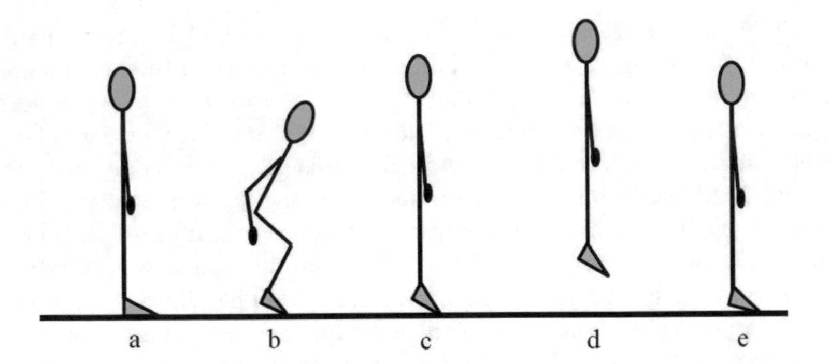

Fig. 1. Five phases of the countermovement vertical jump: the stationary phase (a), the downward movement phase (b), the pushoff phase (c), the flight phase (d), the landing phase (e)

and the phase ends when it equals to zero. During the downward movement phase the position of the COM is decreasing and it achieves a minimal value in the end of the phase.

The next phase is the pushoff phase (Fig. 1c). At the beginning of this phase the resultant force acting on the COM is slightly increasing. Then it achieves a maximal value (of the pushoff phase). It is equal to the jumper's body weight in the end of the phase. In the first part of the phase the velocity of the COM is increasing. Then it achieves a maximal values and it is slightly decreasing in the end of the pushoff phase. The position of the COM is increasing. In the end of

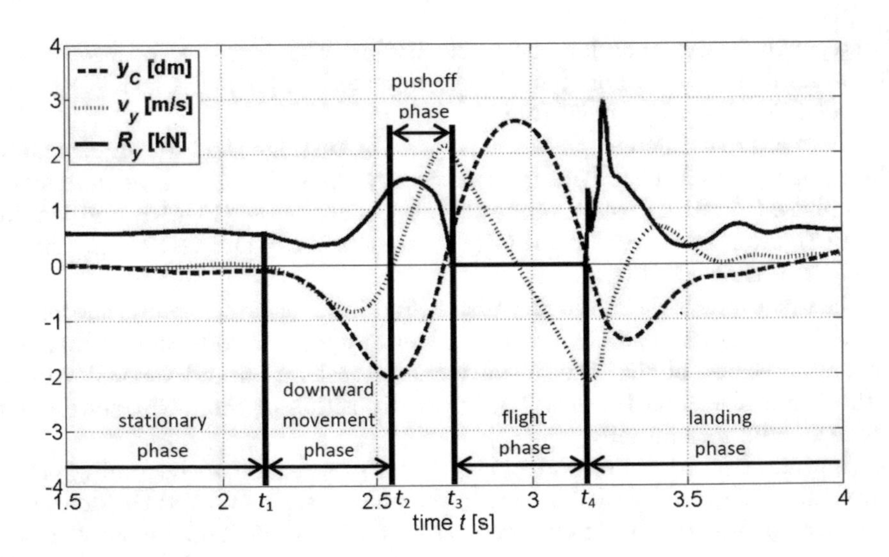

Fig. 2. The ground reaction force, vertical velocity and vertical position of the jumper's center of mass during the countermovement vertical jump

the phase it is a little grater than zero. It means that the position of the COM in the end of the pushoff phase is a little greater than initially in the stationary phase.

During the fourth phase, which is called the flight phase (Fig. 1d), the GRF is equal to zero. Thus only the gravity force acts on the COM and the resultant force is equal to the body weight. The velocity of the COM is decreasing linearly. It is positive in the first part and negative in the second part of the flight phase. Because of that the position of the COM is increasing in the first part of this phase, then it achieves a maximal values and in the end it is decreasing.

The last phase is the landing phase (Fig. 1e). There is an irregular peak in the GRF curve during this phase. After the phase the GRF is equal to the jumper's body weight. Thus the resultant force acting on the jumper's body equals zero and because of that the velocity and position are equal to zero too. The jumper is standing upright again.

3 Determination of the Flight Height

In this paper the position of the COM is obtained from the well-known works of Clauser et al. [12] and Zatsiorky et al. [13].

It is assumed that two dimensional movements are considered. Furthermore in the present study the jump was assumed to be symmetrical with respect to the sagittal plane. It is considered in this study in the sagittal plane. In this approach the human body is divided into 14 segments: head, trunk, 2 upper arms, 2 forearms, 2 hands, 2 thighs, 2 shanks, 2 feet.

In order to determine the center of mass for each segment the ends of these segments can be determined using the motion capture system. In this study the points presented in Fig. 3 were used.

The positions of the center of mass for each segment can be calculated based on the coefficients presented in Table 1 (using Clauser's or Zatsiorky's formulas).

Table 1. Position of the center of mass for each segment as a percentage of the segment length [14]

i	Segment	Clauser *et al.* l_{pi}	Zatsiorsky *et al.* l_{pi}
1	Head	46.6%	50.0%
2	Trunk	38.0%	44.5%
3	Arm	51.3%	45.0%
4	Forearm	39.0%	42.7%
5	Hand	48.0%	37.0%
6	Thigh	37.2%	45.5%
7	Shank	37.1%	40.5%
8	Foot	44.9%	44.1%

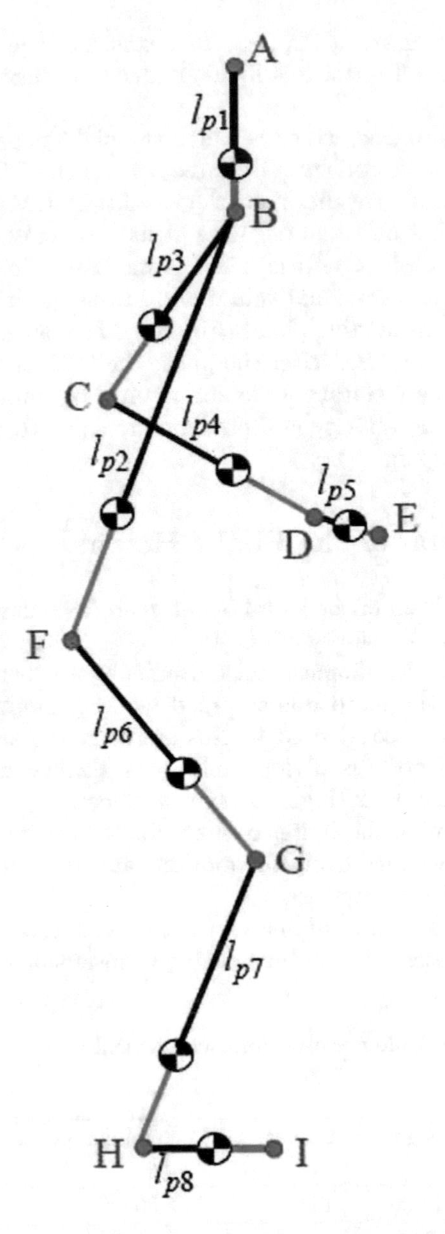

Fig. 3. Model of human body with characteristic points used in this study to determine the center of mass for each segment

Thus the center of mass for each segments can be expressed by the following formula

$$(X_i, Y_i) = (X_{\mathrm{M}} + l_{pi}\,(X_{\mathrm{N}} - X_{\mathrm{M}})\,, Y_{\mathrm{M}} + l_{pi}\,(Y_{\mathrm{N}} - Y_{\mathrm{M}}))\,, \tag{3}$$

where M is the beginning point of the segment vector and N is the beginning point of the segment vector. The beginning of the segment vector is always the point located closer to the top of the head.

When the centers of mass for each segment are determined the COM can be calculated using the coefficients presented in Table 2 and the following formula:

$$(X_{\mathrm{C}}, Y_{\mathrm{C}}) = \left(\sum_{i=1}^{2} m_{pi} X_i + 2 \sum_{i=3}^{8} m_{pi} X_i, \sum_{i=1}^{2} m_{pi} Y_i + 2 \sum_{i=3}^{8} m_{pi} Y_i \right), \tag{4}$$

where m_{pi} is the relative mass of the i-th segment expressed by

$$m_{pi} = \frac{m_i}{M}. \tag{5}$$

In the above equation m_i is the mass of the i-th segment and M is the mass of the whole body.

Table 2. Relative mass of each segment as a percentage of the body weight [14]

i	Segment	Clauser *et al.* m_{pi}	Zatsiorsky *et al.* m_{pi}
1	Head	7.3%	6.940%
2	Trunk	50.7%	43.457%
3	Arm	2.6%	1.707%
4	Forearm	1.6%	1.625%
5	Hand	0.7%	0.614%
6	Thigh	10.3%	14.165%
7	Shank	4.3%	4.330%
8	Foot	1.5%	1.371%

Based on this estimation of the vertical position of the COM the height of the vertical countermovement jump can be defined as

$$h = y_{\mathrm{C}} \left(t_{\mathrm{peak}} \right) - y_{\mathrm{C}} \left(t_3 \right), \tag{6}$$

where $y_{\mathrm{C}} (t_{\mathrm{peak}})$ is peak of the vertical position of the COM and $y_{\mathrm{C}} (t_3)$ denotes the vertical position of the COM at the take-off (see t_3 in Fig. 2).

4 Results

In experiments two volunteers took part. Each person performed 100 jumps. The persons jumps every time with the same position at the same place. In order to avoid muscles fatigue after each jump the jumper had a rest. The measurements were done using BTS Smart-DX system with 6 infrared cameras. The markers were placed on the following antropometric points which refer to the points of the model presented in Fig. 3. The antropometric points are: vertex (point A), acromion (point B), radiale (point C), stylion (point D), daktylion (point E), symphsion (point F), tibiale (point G), pternion (point H), akropodin (point I). Simultaneously the GRF was recorded using two AMTI force plates.

In Fig. 4 one can find comparison between the position of the center of mass y_C calculated using the Clauser's, Zatsiorky's formulas and based on the GRF measurements (by double numerical integration) for persons A (Fig. 4a) and B (Fig. 4b), respectively. It can be observed that both for person A and B the vertical position of the COM using the Clauser's or Zatsiorky's formulas are quite close to each other. For person B these two estimations of the vertical position are also quite close to the curve obtained from the GRF measurements and double numerical integration of the motion equation during the flight phase (near the peak value of the curve). During the downward movement and pushoff phases the difference between these results are bigger but for determining the jump height crucial is the flight phase. For person A the differences between the curves during the flight phase are also bigger.

Table 3 presents comparison of the average jump height obtained for persons A and B during all performed jumps. The observations presented for Fig. 4 are confirmed in the results shown in Table 3. The differences between the curves are significant during the flight phase of the countermovement vertical jump because for this phase the jump height is calculated. One can notice that differences between the results obtained using Clauser's or Zatsiorsky's formulas and the results obtained based on the GRF measurements are much smaller for person B (for which the smaller differences during the flight phase in the previously presented curves are obtained).

Table 3. Comparison of the average jump height obtained for persons A and B based on the Clauser's or Zatsiorsky's formulas or the GRF measurements

Person	Clauser's formulas	Zatsiorky's formulas	GRF measurements
A	0.2264 ± 0.0319 m	0.2281 ± 0.0322 m	0.2098 ± 0.0306 m
B	0.2021 ± 0.0258 m	0.2052 ± 0.0256 m	0.2003 ± 0.0228 m

Two relative difference between the results obtained from the Clauser's or Zatsiorsky's formulas and the GRF measurements are defined as follows:

$$\delta^{\mathrm{C}} = \left| \frac{h^{\mathrm{C}} - h^{\mathrm{GRF}}}{h^{\mathrm{GRF}}} \right|, \tag{7}$$

$$\delta^{\mathrm{Z}} = \left| \frac{h^{\mathrm{Z}} - h^{\mathrm{GRF}}}{h^{\mathrm{GRF}}} \right|, \tag{8}$$

where h^{C}, h^{Z}, h^{GRF} denote the height of the vertical CMJ obtained based on the Clauser's formula, Zatsiorsky's formula and GRF measurements, respectively. In

Table 4 one can find the average values of these relative differences. It can noticed that in both cases the higher values of the differences are obtained for person A.

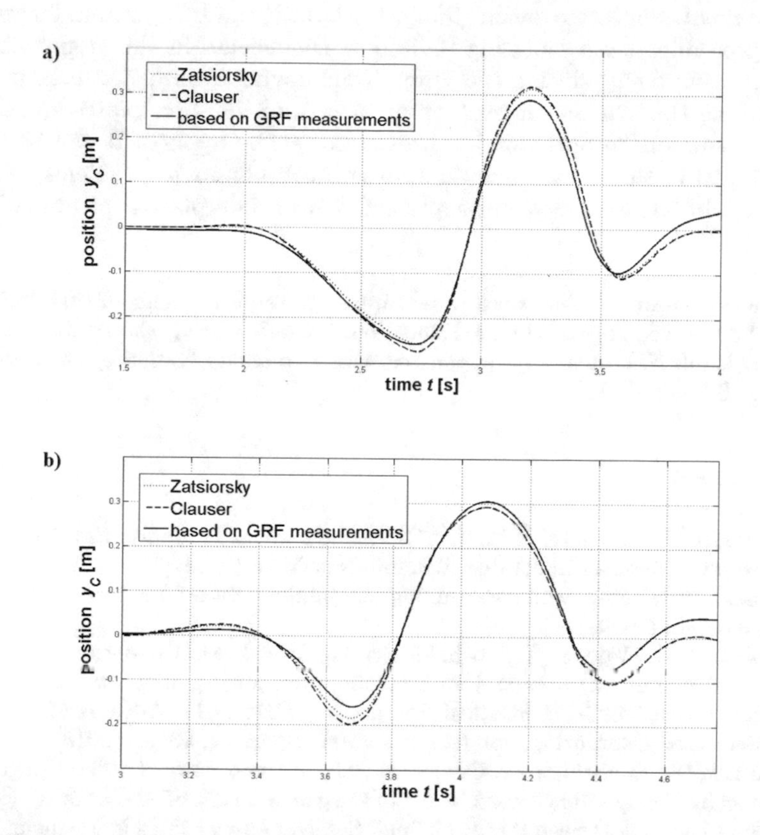

Fig. 4. Comparison between the position of the center of mass calculated using the Clauser's, Zatsiorky's formulas and based on the GRF measurements for persons A (a) and B (b), respectively

Table 4. Comparison of the average relative difference between the results obtained from the Clauser's or Zatsiorsky's formulas and the ground reaction force measurements for persons A and B

Person	δ^{C}	δ^{Z}
A	8.52%	9.22%
B	5.52%	5.54%

5 Conclusions

The presented results show that using the same coefficients for each person in order to calculate the center of human body based on the approach in which the whole body is divided into some segments can generate quite big errors in the results, especially for calculating any special precise parameters, e.g. the height of the vertical countermovement jump. However it can be a ground for establish a new procedure for calculating individual coefficients in this approach based on the measurements during the jump. Then having a serious of measurements coming from the motion capture system and from the force plates and defining a fitness function an optimization procedure can be applied. It can be done in future works in this area. Such an approach can eliminate problems related to individual differences between people and appropriate placement of markers on human body.

Acknowledgements. The work was supported by the grant 02/21/DSPB/3513 funded by the Ministry of Higher Education, Poland. During the realization of this work Dr. Jakub K. Grabski was supported with scholarship funded by the Foundation for Polish Science (FNP).

References

1. Payton, C.J., Bartlett, R.M. (eds.): Biomechanical Evaluation of Movement in Sport and Exercise. Routledge, London, New York (2008)
2. Cross, R.: Standing, walking, running and jumping on a force plate. Am. J. Phys. **67**, 304–309 (1999)
3. Grabski, J.K., Walczak, T., Michałowska, M., Cieślak, M.: Gender recognition using artificial neural networks and data coming from force plates. In: Gzik, M., et al. (eds.) Innovations in Biomedical Engineering, IIBE 2017. Advances in Intelligent Systems and Computing, vol. 623, pp. 53–60. Springer, Cham (2018)
4. Walczak, T., Grabski, J.K., Grajewska, M., Michałowska, M.: The recognition of human by the dynamic determinants of the gait with use of ANN. In: Awrejcewicz, J. (ed.) Dynamical Systems: Modelling. Springer Proceedings in Mathematics and Statistics, vol. 181, pp. 375–385. Springer, Cham (2016)
5. Merletti, R., Parker, P. (eds.): Electromyography. Physiology, Engineering and Noninvasive Applications. Wiley, Hoboken (2004)
6. Grabski, J.K., Kazimierczuk, S., Walczak, T.: Analysis of the electromyographic signal during rehabilitation exercises of the knee joint. Vibr. Phys. Syst. **26**, 79–86 (2014)
7. Nathan, R.H.: A dynamometer for biomechanical use. J. Biomed. Eng. **1**(2), 83–88 (1979)
8. Grygorowicz, M., Michałowska, M., Walczak, T., Owen, A., Grabski, J.K., Pyda, A., Piontek, T., Kotwicki, T.: Discussion about different cut-off values of conventional hamstring-to-quadriceps ratio used in hamstring injury prediction among professional male football players. PLoS ONE **12**(2), e0188974 (2017)
9. Corazza, S., Mündermann, L., Chaudhari, A.M., Demattio, T., Cobelli, C., Andriacchi, T.P.: A markerless motion capture system to study musculoskeletal biomechanics: visual hull and simulated annealing approach. Ann. Biomed. Eng. **34**, 1019–1029 (2006)

10. Walczak, T., Grabski, J.K., Grajewska, M., Michałowska, M.: Application of artificial neural networks in man's gait recognition. In: Kleiber, M., et al. (eds.) Advances in Mechanics: Theoretical, Computational and Interdisciplinary Issues. Proceedings of the 3rd Polish Congress of Mechanics (PCM) and 21st International Conference on Computer Methods in Mechanics (CMM), pp. 591–594. CRC Press, Taylor & Francis Group, London (2016)
11. Grabski, J.K., Walczak, T., Michałowska, M., Pastusiak, P., Szczetyńska, M.: On different methods for calculating the flight height in the vertical countermovement jump analysis. In: Arkusz, K., et al. (eds.) Biomechanics in Medicine and Biology, BIOMECHANICS 2018. Advances in Intelligent Systems and Computing, vol. 831, pp. 242–251. Springer, Cham (2019)
12. Clauser, C.E., McConville, J.T., Young, J.W.: Weight, volume and center of mass of segments of the human body. AMRL Technical report 69-70. Wright Patterson Air Force Base, Ohio (NTIS No. AD-710-622.) (1969)
13. Zatsiorsky, V.M., Seluyanov, V.N., Chugunova, L.G.: Methods of determining mass-inertial characteristics of human body segments. In: Chemyi, G.G., Regirer, S.A. (eds.) Contemporary Problems of Biomechanics, pp. 272–291. CRC Press, Massachusetts (1990)
14. Tejszerska, D., Świtoński, E. (eds.): Biomechanika inżynierska. Zagadnienia wybrane. Laboratorium, Gliwice (2004). (in Polish)

Analysis of an Impact of Hemodynamic Parameters in Relation to Variable Morphometric Features of the Middle Cerebral Artery (MCA)

Marta Sobkowiak[1(✉)], Wojciech Wolański[1], Mikołaj Zimny[2], Marek Gzik[1], and Wojciech Kaspera[3]

[1] Department of Biomechatronics, Faculty of Biomedical Engineering, Silesian University of Technology, ul. F. D. Roosevelta 40, 41-800 Zabrze, Poland
{marta.sobkowiak,wojciech.wolanski,marek.gzik}@polsl.pl
[2] Department and Clinical Division of Neurosurgery, Silesian University of Medicine in Katowice, Voivodeship Specialist Hospital no. 5, Students' Research Society, Plac Medyków 1, 41-200 Sosnowiec, Poland
zimny.mikolaj@gmail.com
[3] Department and Clinical Division of Neurosurgery, Silesian University of Medicine in Katowice, Voivodeship Specialist Hospital no. 5, Plac Medyków 1, 41-200 Sosnowiec, Poland
wkaspera@sum.edu.pl

Abstract. The article-related research work aimed to investigate the effect of the morphometric features of the middle cerebral artery (MCA) on values of hemodynamic parameters present on the artery wall. The research-related tests were performed using the ANSYS® CFX software programme. The objective of the tests involved the development of a parametric model representing the morphometric features of a vessel. The model was based on geometric models of middle cerebral arteries (MCAs) obtained on the basis of angiotomography of patients. Morphometric features subjected to analysis included the radius of the artery inlet and that of the outlet as well as the angle of bifurcation.

Information about values of artery radiuses was necessary to determine the angle of bifurcation. Test simulations involved the use of information about blood flow velocity obtained using a transcranial Doppler ultrasound examination with transcranial colour-coded duplex sonography (TCCS) of a given patient. Hemodynamic parameters subjected to analysis included pressure and wall shear stress (WSS). The simulation results enabled the correlation of the MCA morphometric features with blood flow hemodynamic parameters and revealed the mutual influence of the aforesaid factors.

Keywords: Computational Fluid Dynamics (CFD) · Blood flow · Artery · CT · Murray's law · Velocity of blood

E. Tkacz et al. (Eds.): IBE 2018, AISC 925, pp. 200–209, 2019.
https://doi.org/10.1007/978-3-030-15472-1_22

1 Introduction

Morphological parameters can give us information about the risk of aneurysms. This parameters are strongly related with hemodynamic parameters. There are many research-related publications [1,2,8,9] concerning the search for mechanical reasons for aneurysm formation. Researchers point to high wall stress shear (WSS) as the primary initiator of lesions. Other research teams verified the effect of arterial blood pressure on the velocity of blood flow in vessels [4]. Based on a specific case study, the aforesaid tests revealed that an increase in pressure translates into an increase in arterial blood flow velocity. Other works presented the effect of vascular morphometric features on hemodynamic factors [7]. Team [6] performed multi-phase simulations of blood flow through an aorta with coarctation. The tests revealed that the correlation between radiuses of the arterial inlet and outlets (Murray's law) and the angle of bifurcation affected both arterial blood pressure and the WSS. The primary objective of the study discussed in this article involved the analysis of the sensitivity of hemodynamic parameters using variable morphometric features of the MCA.

2 Methods

The above-named analysis of sensitivity was performed using geometric models of patients' arteries obtained in computed tomography (CT). The obtainment of three-dimensional models of vessels involved the use of the Mimics software programme developed by the Materialise company on the basis of two-dimensional CT images (Fig. 1).

Resolution of CT scans is 512×512 px, mean value of pixel size is 0.41 mm. There were used 40 slices to generate each model. In the simulations, the arteries of six patients were considered at the risk of aneurysm formation (2 males, 4 females, average age - 40 years). Blood flow simulations were performed on three different models: real - segmented in Mimics software (Materialise) and two parametric models - designed in Inventor (Autodesk). Substitute diameters of vessels were calculated on the basis of an area identified in the place of a cross-section subjected to analysis. A subsequent step involved the definition of boundary conditions (arterial inlet and outlets) and their midlines. The above-presented model was subjected to discretisation. In addition, an inflationary layer was defined (Fig. 2). The inflationary layer was composed of six coatings, the first of which was 0.0075 mm in height. Afterwards, the model was exported to the Ansys CFX software programme.

The simulations of blood flow involved the use of parameters specified in reference publications [2,3,5]:

- molar mass - 18,02 kg/kmol,
- density - 1050 kg/m^3,
- thermal capacity - 4181.7 J/kg$*$K,
- viscosity - 0.0035 Pa$*$s.

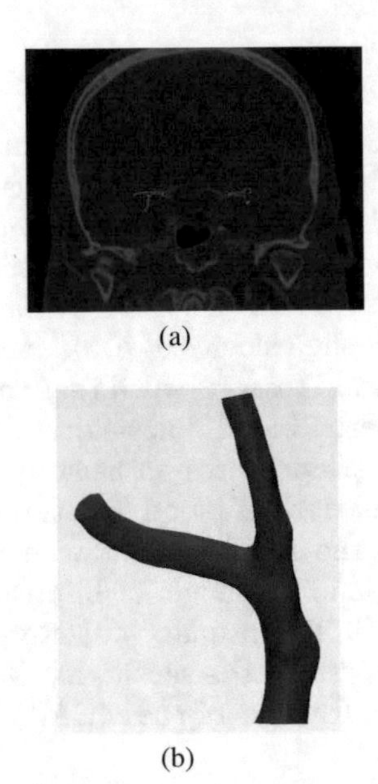

(a)

(b)

Fig. 1. Geometric model of the artery obtained on the basis of two-dimensional CT images using the Mimics software programme: (a) location of the middle cerebral artery, (b) model of the artery prepared for further analysis

The velocity of blood flow used in the simulations was constant $V_s = 0.45\,\mathrm{m/s}$ [2]. In this assumption it was possible to compare the results of simulation process (WSS and pressure). The diameters of arteries, viscosity coefficient and the velocity of blood flow through the MCA indicated the Reynolds number value below 1000 (formula 1).

$$Re = \frac{\upsilon \cdot l \cdot \rho}{\mu} \tag{1}$$

where

υ - fluid flow velocity,
l - linear dimension of the vessel, perpendicular to velocity (v),
μ - fluid viscosity,
ρ - fluid density.

In view of the foregoing, the laminar type of flow was adopted. Blood was treated as non-Newtonian fluid. Simulations involved the application of the Bird Carreau model. Figure 3 presents the formulated model of the vessel with marked boundary conditions. At the inlet, a blood velocity of 0.45 m/s was applied. A pressure of 0 Pa was applied at the outlets. The liquid flowing through the vessel was given a body temperature of 37 °C.

The correlation between both radiuses of the arterial inlet and outlets is described by Murray' law (2):

$$r_0^3 = r_1^3 + r_2^3 \tag{2}$$

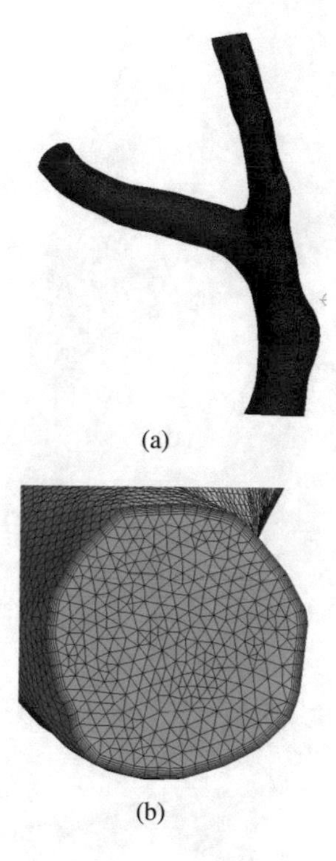

Fig. 2. Discrete model of the MCA: (a) volumetric mesh, (b) paramural (inflationary) layer

where

r_0 - radius of the inlet branch of the vessel,
r_1, r_2 - radiuses of the outlet branches of the vessel.

The values of arterial radiuses were used to calculate a theoretical angle of bifurcation. The calculation was performed using the following formulas:

$$cos\phi_1 = (r_0^4 + r_1^4 + r_2^4)(2r_0^2 r_1^2)^{-1} \tag{3}$$

$$cos\phi_2 = (r_0^4 + r_2^4 + r_1^4)(2r_0^2 r_1^2)^{-1} \tag{4}$$

The above-presented data were used to develop the parametric models of arteries of selected patients. The parametric models were developed using the Inventor software programme. Variable parameters in the models were geometric features, i.e. radiuses and the angle of bifurcation. For each case two parametric models were created, i.e. one with the real angle of bifurcation and the other one with the theoretical angle of bifurcation. The real angle was measured on the artery model using the 3-matic software (Materialize), whereas the theoretical angle was calculated on the basis of dependences numbered 3 and 4. The adopted mechanical properties of blood and the simulation settings were the same as in the cases of real geometric models.

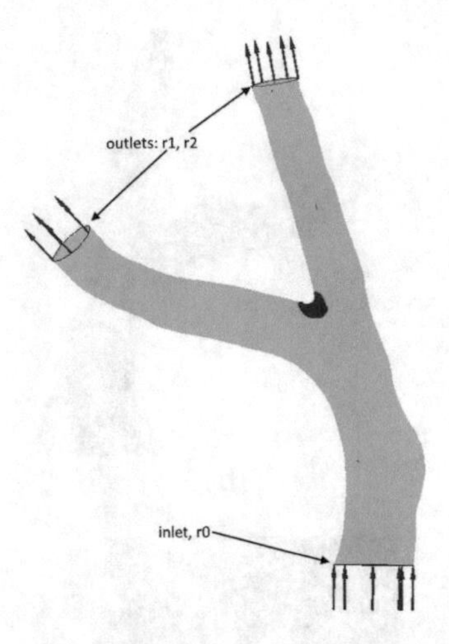

Fig. 3. Model of the artery with marked boundary conditions and subarea of bifurcation angle (red coloured)

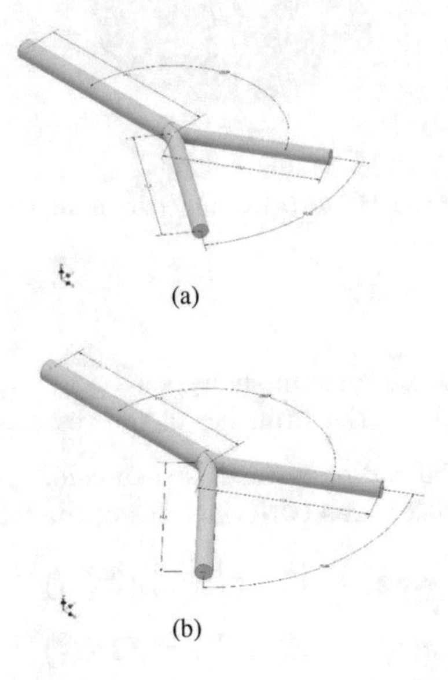

Fig. 4. Examples of parametric models of arteries: (a) real angle of bifurcation, (b) theoretical angle of bifurcation

When creating the parametric models of arteries, it was assumed that the lengths of the inlet and outlet branches will be 10 times longer than the length of the radius of a given branch (10*r_n, Fig. 4).

3 Results

The simulations of blood flow through the real and parametric models of the MCA were used to perform the analysis of hemodynamic parameters. The morphometric features of the MCA in individual patients and the distribution of blood flow velocity at inlets and outlets are presented in Table 1. Figure 5 presents maps of the WSS of the right MCA of patient P2 along with the WSS distribution in the parametric model.

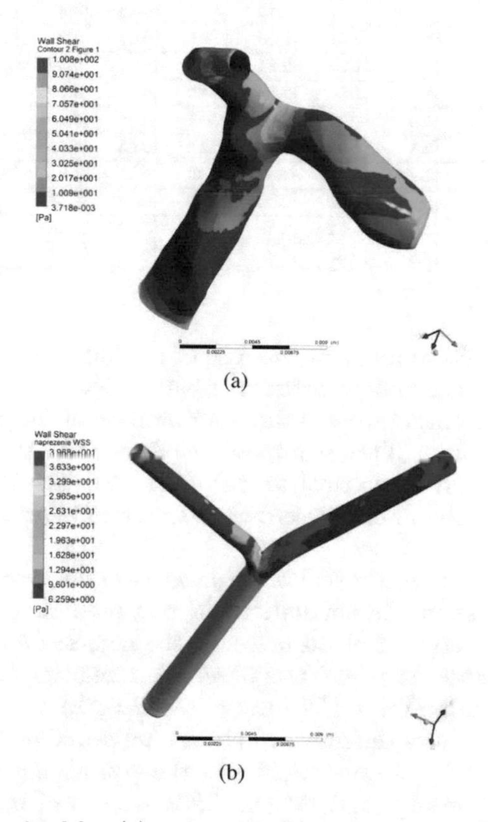

Fig. 5. WSS field obtained for: (a) real model, (b) parametric model with the measured angle of bifurcation

The simulation results obtained for the selected case (Fig. 5) demonstrated that formulas 3 and 4, which were used to determine the theoretical angle of bifurcation, enabled the obtainment of the lower values of the WSS among all three models.

The distribution of blood velocity in real models of the MCA revealed that, locally (in the bifurcation area), values of velocity were lower in relation to the theoretical angle of patient P2. The global values of velocity were higher in

Table 1. List of morphometric features and velocity distribution at the inlets and outlets of the MCA

Patient	Angle [°]		Radius [mm]			Velocity distribution [m/s]					
						Measured angle, local values			Theoretical angle, local values		
	Measured	Theoretical	Inlet radius r0	Outlet radius r1	Outlet radius r2	Inlet radius r0	Outlet radius r1	Outlet radius r2	Inlet radius r0	Outlet radius r1	Outlet radius r2
P1_R	65.62	76.6	2.82	2.21	2.29	0.45	0.55	0.57	0.45	0.56	0.56
P1_L	88.85	75	2.89	1.83	2.61	0.45	0.54	0.61	0.45	0.55	0.61
P2_R	87.6	102.2	3.25	2.42	3.18	0.45	0.42	0.5	0.45	0.43	0.49
P2_L	66.61	18	3.85	2.51	2.95	0.45	0.67	0.71	0.45	0.68	0.69
P3_R	115.99	67.2	2.82	1.73	2.51	0.45	0.53	0.59	0.45	0.54	0.65
P3_L	71.54	158	2.86	1.5	2.26	0.45	0.7	0.78	0.45	0.71	0.74
P4_R	95.79	160.6	3.05	1.89	2.02	0.45	0.82	0.84	0.45	0.77	0.85
P4_L	101.63	162	3.3	1.93	2.2	0.45	0.83	0.85	0.45	0.84	0.8
P5_R	68.52	160.7	3.22	1.9	2.22	0.45	0.85	0.82	0.45	0.77	0.79
P5_L	64.76	170	3.15	1.26	2.24	0.45	0.87	1.07	0.45	1.01	0.76
P6_R	69.15	77.3	2.48	1.81	2.13	0.45	0.54	0.58	0.45	0.54	0.57
P6_L	88.1	106	2.5	2.07	2.42	0.45	0.54	0.58	0.45	0.42	0.45

relation to the theoretical angle in the case of the left artery of patient P3 and P5 as well as in relation to both arteries of patient P4.

The research work also involved the confrontation of the results of the simulation of blood flow through the parametric models of the MCA with the results of the real models and theoretical models (Table 2). The compared values of hemodynamic parameters were present at the point of bifurcation of each model of the MCA.

The analysis of the results revealed the presence of the lower values of pressure and wall shear stress in the simulations of real models, i.e. obtained in CT tests of patients. In terms of blood pressure, the highest difference between the theoretical and parametric model was observed in relation to the left artery of patient P3. The comparison of the results (see Table 2) obtained in relation to parametric results reveals differences between values of pressure and those of the WSS in relation to the measured and theoretical angles. In the cases of the left side of patients P2 and P3 and both arteries of patients P4 and P5, it was possible to observe significantly lower results of the pressure present on the artery wall in relation to the theoretical angle of bifurcation. In relation to the above-named patients, the average difference in pressure between the real and theoretical angle amounted to 339 Pa. Lower WSS values were observed in the right artery of patients P2, P4 and P6, the left artery of patient P3 and in both arteries of patient P5. In relation to the above-named patients, the average difference in the WSS between the real and theoretical angle amounted to approximately 10 Pa.

Table 2. Comparison of parametric model results in relation to anatomical and theoretical angles

Patient	Pressure [Pa]			WSS [Pa]		
	Parametric model	Theoretical model	Real model	Parametric model	Theoretical model	Real model
P1_R	634	666	413	44.03	44.04	47.35
P1_L	606	573	525	45.36	47.35	41.85
P2_R	441	461	314	39.31	36.45	40.91
P2_L	760	489	287	59	83	54.81
P3_R	587	662	570	35.21	52.21	7.26
P3_L	1057	803	230	52.17	47.41	43.8
P4_R	1178	765	-	56.21	51.32	-
P4_L	1241	708	-	48.49	53.1	-
P5_R	1134	876	-	59.96	38.03	-
P5_L	1522	1215	-	91.22	65.58	-
P6_R	672	664	-	49.13	46.52	-
P6_L	455	409	-	38	54.68	-

4 Analysis of Results

The research work involved the analysis of the morphometric features of the artery on the hemodynamics of blood flow and, in particular, the effect of the angle of bifurcation. The research-related analysis involved two types of models, i.e. real and parametric with the real and theoretical angle of bifurcation determined using formulas numbered 3 and 4. The obtained simulation results concerning both models (Figs. 6 and 7) revealed that the lower values of pressure and those of the WSS were related to the real model. However, taking into consideration only the parametric model with the theoretical and real angle of bifurcation, it can be seen that lower values occurred in the case of the theoretical angle.

The performed tests demonstrated the usability of the development of parametric models on the basis of anatomical parameters of the MCA. The models enable the performance of any modifications of the vessel and, as a result, make it possible to perform the analysis of the artery wall sensitivity to the hemodynamic parameters of blood.

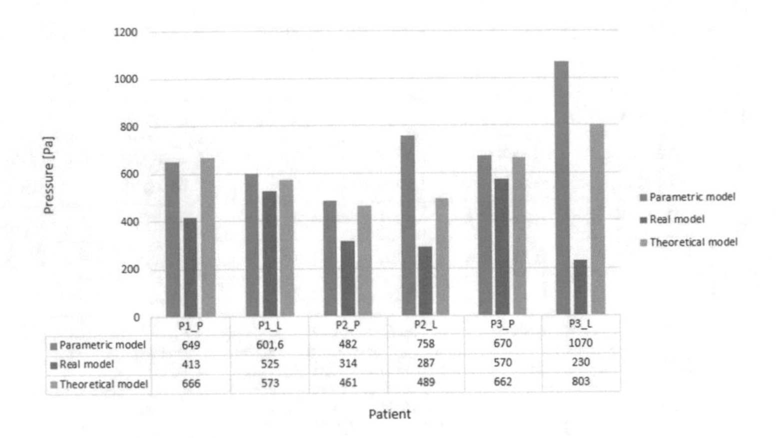

Fig. 6. Pressure measurement results in relation to various artery models

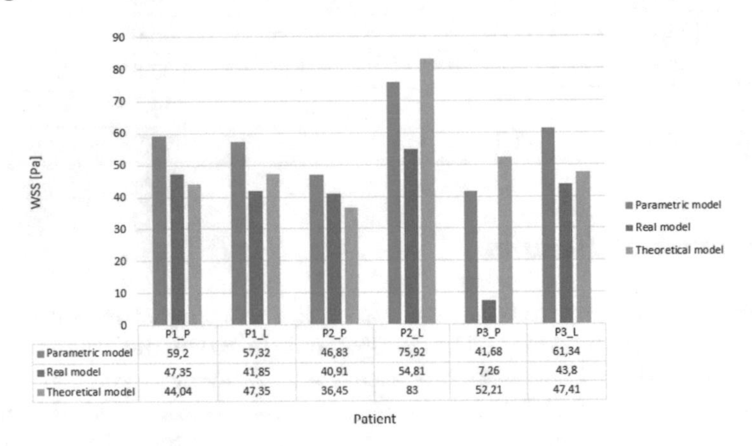

Fig. 7. WSS results in relation to various artery models

5 Conclusions

The research work resulted in the development of the parametric model of the middle cerebral artery (MCA). The geometric features of the above-named model can be adjusted (modified) to corresponding morphometric features of the artery. In addition, the model enables the adjustment of geometry to theoretical values of geometric parameters, i.e. radiuses of artery branches and the angle of artery bifurcation. The model developed in the above-presented manner makes it possible to analyse the sensitivity of hemodynamic parameters using variable morphometric features of the middle cerebral artery.

The above-presented simulation results concerning the real and parametric models of the middle cerebral artery (MCA) and the comparison of the aforesaid results enabled a more extensive analysis of the effect of morphometric features on the hemodynamics of blood flow. The developed parametric model makes it possible to analyse not only the anatomical but also theoretical morphometric parameters of the artery. However, the above-presented simulation results are an introduction to further model tests performed using a larger number of patients. Only a sufficiently large test group and the analysis of sensitivity performed for such a group will make it possible to draw objective conclusions concerning the effect of morphometric features on the hemodynamics of the middle cerebral artery. For this reason, the subsequent stage of tests will involve the analysis of the sensitivity of hemodynamic parameters when using variable morphometric features of the MCA in relation to a group of 50 patients. The above-named tests will be performed in accordance with the presented methodology using the above-presented parametric model of the middle cerebral artery. The research work presented in this article has demonstrated that the aforesaid model constitutes a good tool enabling the performance of such an analysis.

Acknowledgements. The study was supported by the research grant StrategMed 2/269760/1/NCBR/2015 of the National Centre for Research and Development (NCBR).

References

1. Ferruzzi, J., Vorp, D.A., Humphrey, J.D.: On constitutive descriptors of the biaxial mechanical behavior of human abdominal aorta and aneurysms. J. Roy. Soc. Interface **8**, 435–450 (2011)
2. Fung, Y.C.: Biomechanics: Mechanical Properties of Living Tissues, 2nd edn. Springer, New York (1993)
3. Holzapfel, G.A., Ogden, R.W.: Constitutive modelling of arteries. Proc. Roy. Soc. A: Math. Phys. Eng. Sci. **466**(2118), 1551–1597 (2010)
4. Jiang, J., Strother, C.: Computational fluid dynamics simulations of intracranial aneurysms at varying heart rates: a "patient-specific" study. J. Biomech. Eng. **131**, 091001 (2009)
5. Lasheras, J.C.: The biomechanics of arterial aneurysms. Ann. Rev. Fluid Mech. **39**, 293–319 (2007)
6. Melka, B., Gracka, M., Adamczyk, W., Rojczyk, M., Golda, A., Nowak, A.J., Białecki, R.A., Ostrowski, Z.: Multiphase simulation of blood flow within main thoracic arteries of 8-year-old child with coarctation of the aorta. In: Heat and Mass Transfer (2017). https://doi.org/10.1007/s00231-017-2136-y
7. Murray, C.D.: The physiological principle of minimum work. I. The vascular system and the cost of blood volume. Department of Biology, Bryn Mawr College (1926)
8. Wolański, W., Gzik-Zroska, B., Joszko, K., Kawlewska, E., Sobkowiak, M., Gzik, M., Kaspera, W.: Impact of vessel mechanical properties on hemodynamic parameters of blood flow. In: Gzik, M., Tkacz, E., Paszenda, Z., Piętka, E. (eds.) Proceedings of the Innovations in Biomedical Engineering, IiBE 2017, pp. 271–278. Springer, Cham (2018)
9. Wolański, W., Gzik-Zroska, B., Joszko, K., Gzik, M., Sołtan, D.: Numerical analysis of blood flow through artery with elastic wall of vessel. In: Gzik, M., Tkacz, E., Paszenda, Z., Piętka, E. (eds.) Proceedings of the Innovations in Biomedical Engineering, IiBE 2017, pp. 193–200. Springer, Cham (2017)

Resultant Reactions in the Hip, Knee and the L5-S1 Joint During the Back Squat with Variation External Loads

Paulina Szyszka[1(✉)], Robert Michnik[2], Katarzyna Nowakowska[2], Adam Czaplicki[1], and Jarosław Sacharuk[1]

[1] Faculty of Physical Education and Sport in Biała Podlaska, Academy of Physical Education in Warsaw, Akademicka 2, 21-500 Biała Podlaska, Poland
paulina.szyszka@awf-bp.edu.pl

[2] Department of Biomechatronics, Faculty of Biomedical Engineering, Silesian University of Technology, Roosevelta 40, 41-800 Zabrze, Poland
http://www.awf-bp.edu.pl

Abstract. The research work aimed to identify and analyse reactions in the joints of the lower limb as well as in the intervertebral joints of the lumbar spine when lifting weights and making a squat. The determination of loads of the skeletal-muscular system involved the performance of simulations in the AnyBody software programme. Input data used in the research-related tests came from the tests concerning the kinematics of weightlifting by a weightlifter. The use of mathematical modelling and static optimisation methods made it possible to identify resultant responses in the joints in individual positions of a squat made under various external loads. The highest resultant responses in all of the joints subjected to analysis were identified at an angle of 135° in knee joints. As a result, it is important to pay attention during training not to stop moving during an exercise at the above-named angle.

Keywords: Mathematical modelling · Squat · Anybody Modeling System · Reactions in joint · Weightlifting

1 Introduction

A squat is an exercise performed in a closed kinematic chain commonly used in training programmes in various sports events [9]. The above-named exercise is also frequently recommended and applied in rehabilitation, e.g. after the reconstruction of the sacral ligament [17]. The squat is an exercise involving many joints [10] activating approximately 200 muscles and characterised by an easy technique [20]. The squat can be modified in a variety of ways. Squats can be made with additional objects, e.g. usually a barbell or, less frequently, a kettle or dumbbells [9]. The most popular types of squats are front squats and back squats. Depending on the position of a barbell, squats can be further subdivided

E. Tkacz et al. (Eds.): IBE 2018, AISC 925, pp. 210–217, 2019.
https://doi.org/10.1007/978-3-030-15472-1_23

into low and high bar squats. In relation to the angular position in the knee joint at the final stage of the exercise, back squats can also be divided into semi-squats and full squats [12,15]. There is no agreement as to the effectiveness of squats in relation to the value of an angle in the knee joint in the final position of the exercise. According to Escamilla [6], it is more favourable for healthy athletes to make a semi-squat. In turn, Hartmann et. al [11] and Contreras [3] indicate that, as regards the shaping of muscles acting during squats it is necessary to perform the entire movement during the squat. A more clear division among scientists is the one concerning assessments of differences between front and back squats. Both in training and rehabilitation, coaches and therapists more often recommend the back squat. However, research results published by various authors indicate greater advantages related to the making of front squats [24]. In turn, Gullet [10] refers to slight differences in the technique of both squat types, yet a noticeable difference is the fact that back squats enable the lifting of heavier loads. Related tests revealed that back squats are characterised by a high trunk inclination angle, a lower bend angle in the knee joints and higher angular velocity of the lumbar spine in comparison with front squats [15]. Specialist strength training, e.g. weightlifting involves other types of squats enabling the strengthening of the entire body, i.e. overhead squats [1]. Subject-related reference publications often emphasize that anthropometry and ranges of movements are some of the most important variables affecting the making of a squat with a barbell [23]. Another factor affecting the kinematics of the movements and loads in the joints when making a squat is an increase in the external load. Kim [15] posed a hypothesis, according to which an increase in external load when making a squat is accompanied by a change in the angle in the lumbar spine. The above-named hypothesis was not confirmed by related verification tests. In turn, Walsha et al. [26] indicated that a change in load from 60% RM (RM - one-rep maximum) to 80% RM leads to significant hyperextension in the lumbar spine. List, Gulay and Lorenzetti [16] indicated in their work that a change in load increases the inclination in the lumbar spine. Specialist publications often referred to the analysis of kinematics [5,9,19] during the making of a squat. Tests were performed using the EMG [2,5,10]. However, issues concerning the determination of responses in joints and muscular strength moments were addressed to a lesser degree [7,8]. Because of numerous disputes unsettled in available reference publications, this work aimed to determine interior reaction in hip and knee joints as well as in the intervertebral joint L5-S1 when making back squat. Analyses also involved various terminal squat-making positions determined by an angle in the knee joint with variation external loads.

2 Materials and Methods

The test materials resulted from kinematic analyses of 4 squats made by a weightlifter (body weight: 72 kg, body height: 189 cm). The first stage involved tests of the kinematics of a squat in 4 configurations, i.e. without external load, with a load of 20 kg, 70 kg and 150 kg, i.e. the athlete's weightlifting life record in the squat. The tests were conducted using a Vicon optical system provided with 8 IR cameras and 2 AMTI dynamographic platforms. Reflective markers were placed on the athlete's body in accordance with the Plug-in-Gait model. The determination of the response system in the joints of the lower limbs and intervertebral forces in the lumbar spine involved the use of the static optimisation method, i.e. a non-invasive method enabling the assessment of loads in the skeletal-muscular system. The above-named method has been successfully used in the assessments of both athletes [4,13] and patients suffering from chronic diseases of the locomotor system [18]. The kinematic data obtained in the tests were used to simulate a squat in the AnyBody Modeling System environment. The simulation involved the use of the model of the entire body (Standing Model) composed of 69 rigid bodies corresponding to segments of the body or individual osseous elements. The muscular system was represented by more than 1000 active force elements corresponding to muscular actions. A simple model of the muscle was adopted. The model took into consideration intra abdominal pressure. The model was scaled taking into consideration the height, weight and the percentage of the fatty tissue of a test participant (ScalingLengthMassFatExt method). This model was validated by the authors of other work [22]. The recorded kinematic parameters were used to appropriately adjust the position of the body weight centre in the model. The simulations were made for 6 values of an angle in the knee joint (90°, 100°, 105°, 115°, 125° and 135°) in each of the analysed configurations, i.e. without external load and with an external load of 20 kg, 70 kg and 150 kg (Fig. 1). The lifted load was replaced with two concentrated forces affecting the athlete's arms vertically, in areas corresponding to the contact between the barbell and the shoulder girdle. The analysis involved resultant responses in the hip and knee as well as in the intervertebral joint L5-S1. The determined values of response in the knees were normalised in relation to the athlete's body weight (BW).

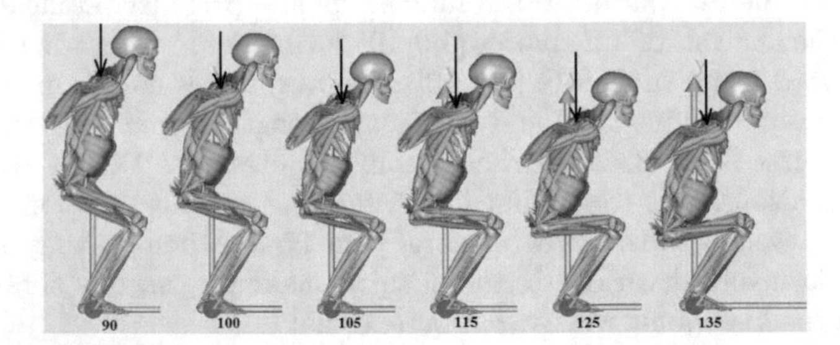

Fig. 1. Angular positions in the knee joint subjected to analysis in the Anybody Modeling System software programme

3 Results

Table 1 contains the results of the kinematic analyses of the squat made without external load and the squat made under a load of 20 kg, 70 kg and 150 kg. The test results revealed that the highest values of responses in the knee joint were observed in relation to an angle of 115° (9.3 N/BW) when lifting the highest external load. An angle of 125° and 135° in the knee joint was accompanied by the lowest load values in the knee joint (Fig. 2). Another correlation was observed in relation to the hip joint, where an increase in an angle in the knee joint was accompanied by higher resultant values of responses in the hip joint. The highest response values with reference to the hip joint were observed in relation to an angle of 135° in the knee joint. Under the highest external load they were 12 times higher than the athlete's weight (Fig. 3). Along with an increase in the value of an angle in the knee joint, resultant responses in the intervertebral joint L5-S1 grew up to an angle of 125°. The highest rate of an increase in response was observed in relation to the highest external load (Fig. 4).

Table 1. *Mean* ± *SD* of the kinematic analyses of the squat made under changing external load

Knee joint angle [°]	Hip joint angle [°]	Ankle joint angle [°]	Trunk inclination angle [°]	Shoulder joint angle [°]	Elbow joint angle [°]	Body centre of mass displacement [cm]
90	65 ± 2.3	41 ± 1.2	28.9 ± 0.2	35.6 ± 0.2	128.8 ± 0.3	2.4 ± 0.03
100	74 ± 2.2	40 ± 1.7	29.2 ± 0.4	35.3 ± 0.4	128.9 ± 0.3	2.6 ± 0.02
105	77 ± 2.5	37 ± 1.5	29.2 ± 0.3	35.3 ± 0.3	128.8 ± 0.4	2.7 ± 0.03
115	88 ± 3.2	35 ± 1.1	29.6 ± 0.5	35.1 ± 0.4	128.7 ± 0.4	2.9 ± 0.04
125	97 ± 3.6	35 ± 1.3	29.8 ± 0.4	35.6 ± 0.2	128.6 ± 0.3	3.2 ± 0.05
135	105 ± 4.1	32 ± 1.2	30.1 ± 0.3	35.9 ± 0.2	128.6 ± 0.4	3.2 ± 0.05

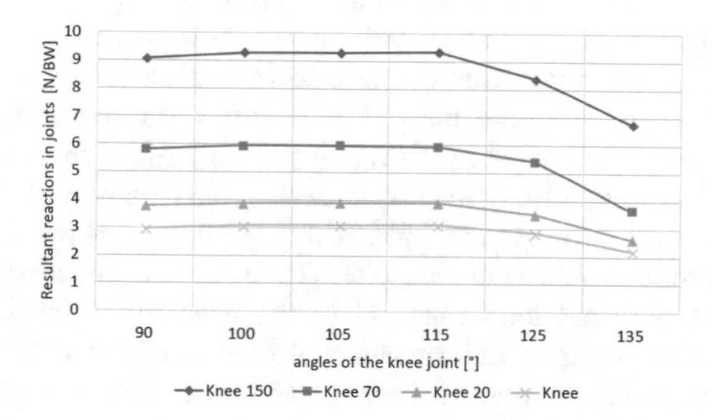

Fig. 2. Resultant values in the knee joint in relation to the angular position in the knee joint and external load

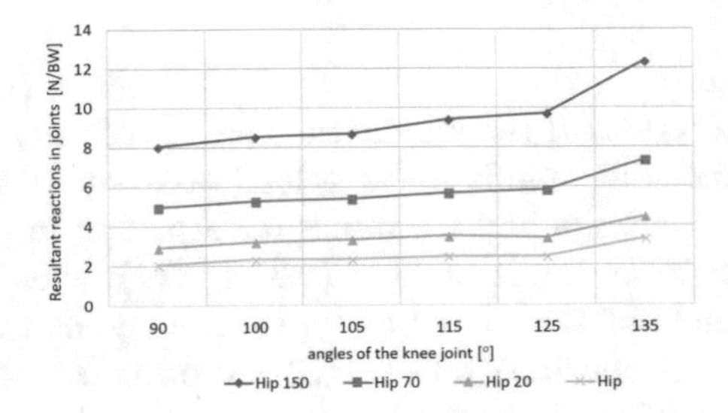

Fig. 3. Resultant values in the hip joint in relation to the angular position in the knee joint and external load

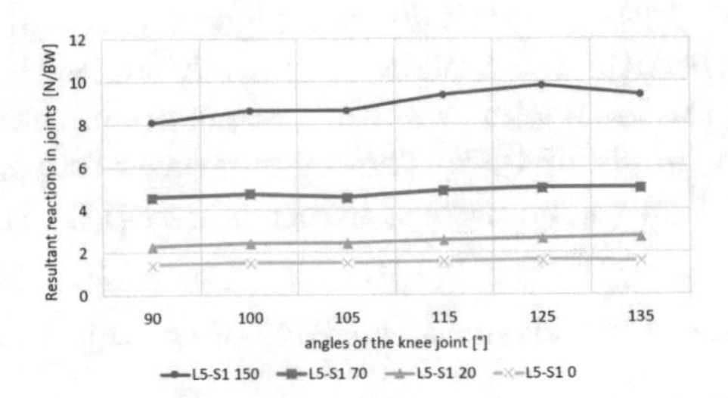

Fig. 4. Resultant values in the intervertebral joint L5-S1 in relation to the angular position in the knee joint and external load

4 Discussion

The research work aimed to assess resultant responses in the joints when making a squat, i.e. the most common exercise in weightlifters' training [25]. The study involved the analysis of a squat made without external load as well as under a load of 20 kg, 70 kg and 150 kg. The method based on the mathematical modelling of the human locomotor system is increasingly often used to identity loads of the skeletal-muscular system e.g. in sport [4,13] or in daily life activities [21]. The analyses of kinematic variables of motion, recorded using a Vicon system, did not reveal significant changes in the values of angles in individual joints under increasing external load. The results obtained in the research are consistent with those presented by Kim [15] yet they differ from those presented by Walsh et al. [26], List, Gulay and Lorenzetti [16]. Walsh et al. [26] stated that a change in load from 60% RM to 80% RM resulted in the significant hyperextension of the lumbar spine, whereas List, Gulay, and Lorenzetti [16] concluded that a change in external load resulted in the greater inclination of the lumbar spine. The above-named differences could be attributed to the selection of test group participants. Similar to tests presented in this study, Kim [15] performed tests involving professional athletes. The remaining tests involved groups of athletes less experienced at strength training. Differences could result from the repeatability of kinetic patterns, which are higher among athletes representing a higher level of training. By analysing resultant responses in intervertebral joint

L5S1, in hip and knee joints in 6 various squat terminal positions, the authors of this study aimed to identify the positions exerting the highest loading. The most frequent injuries occurring in force training include injuries of the lower limb and those of the lumbar spine [14]. A survey involving 140 respondents, i.e. Polish weightlifters revealed that as many as 45% of the athletes suffered from pain-related ailments in the lumbar spine, whereas 40% complained about knee-related problems. Kim [15] indicated that the excessive inclination of the trunk could damage the spine or result in pain ailments. In individual tests the angle of the athlete's trunk inclination amounted to approximately ok 30°, which was consistent with the data presented by Russell and Phillips [24]; the aforesaid angle being lower than that presented by Kim [15]. The analyses performed by the authors of this study indicated that the highest resultant responses of inter-vertebral joint L5S1 accompanied a knee joint bend angle of 125°. The higher external load causes higher values of resultant responses. As regards changes in the depth of the squat, it was ascertained that responses in joints were the same if external load was not applied. At the highest external load, i.e. 150 kg, it was possible to observe a higher rate of increments of resultant responses in joints in relation to a knee joint bend angle of 125°. The results of the analyses confirmed data presented by Flanagan [8] who indicated that the moment of force in the hip joint grew in a linear manner along with an increase in external load. Similarly, an increase in external load led to an increase in values of resultant responses in the hip joint. An increase in the depth of a squat translated into increased responses in the hip joint. The fastest rate of increased resultant responses was observed within the angular position range of 125° to 135°. The analyses performed by the authors revealed that an increase in the depth of the squat was accompanied by an increase in resultant responses in the knee joint; the highest values were obtained in relation to an angle of 115°. An increase in the depth of the squat, after obtaining an angle of above 115° in the knee joint, was accompanied by a decrease in the values of resultant responses in the knee joint. Regrettably, the authors of the study have not found data presenting the value of response in the knee joint in the above-named angular positions when making the squat. However, on the basis of various research apparatuses, some authors state that making the full squat is more convenient, whereas the others claim that making semi-squats is more favourable. Escamilla [6] emphasized that the training of the squat within the knee joint bending functional range of 0° to 50° can be favourable for patients undergoing knee rehabilitation as knee forces are minimum within the functional range. The activity of the quadriceps muscle, hamstring tendon and the gastrocnemius muscle increased along with an increase in the knee bending, which is crucial in relation to athletes with healthy knees making semi-squats within the knee joint bending range of 0° to 100°. It was also revealed that the semi-squat was not harmful for healthy knees. In addition, the author emphasized that the full squat increases the risk of damage to the ligamentous apparatus within the knee joint and the meniscus. Hartmann et al. [11] stated that the depth of squats was an important variable related to the effective training of squats. Contreras [3] indicated that, as regards the shaping of the muscles active during squats, when making squats it is necessary to perform movements within the entire range. In addition, in spite of the fact that front squats require lower load, they provide similar training stimulus as back squats. In view of the foregoing, when making squats it is necessary to take into consideration both the depth of squats and the weight of loads used in training. Depending on the purpose of an exercise, it is necessary to take optimum terminal positions of the squat and to avoid holding the positions exposed to heaviest loads, i.e. stopping the movement in such positions.

5 Conclusion

The research-related tests and simulation revealed that the highest resultant responses in knee joints during squats accompanied a knee bend angle of 115°. In turn, the highest resultant responses in the hip joint and in the intervertebral joints L5-S1 were observed in relation to a knee joint bend angle of 135°. The subsequent stage of the research will involve the analysis of individual constituent forces of responses (compressive, shear and translator forces) in the joints of the lower limb and in the intervertebral joints of the lumbar spine when making squats under additional external load.

References

1. Aspe, R.R., Swinton, P.: Electromyographic and kinematic comparison of the back squat and overhead squat. J. Strength Conditioning Res. **28**(10), 2827–2836 (2014)
2. Bryanton, M.A., Carey, J.P., Kennedy, M.D., Chiu, L.Z.F.: Quadriceps effort during squat exercise depends on hip extensor muscle strategy. Sports Biomech. **14**(1), 122–38 (2015)
3. Contreras, B., Vigotsky, A.D., Schoenfeld, B.J., Beardsley, C., Cronin, J.A.: Comparison of gluteus maximus, biceps femoris, and vastus lateralis electromyography amplitude in the parallel, full, and front squat variations in resistance-trained females. J. Appl. Biomech. **32**(1), 16–22 (2016)
4. Czaplicki, A., Sacewicz, T., Jaszczuk, J.: Identification and performance optimization of a mae-geri kick. Model. Eng. **11**(42), 99–104 (2011). (in Polish)
5. Dionisio, V.C., Almeida, G.L., Duarte, M., Hirata, R.P.: Kinematic, kinetic and EMG patterns during downward squatting. J. Electromyogr. Kinesiol. **18**, 134–143 (2008)
6. Escamilla, R.F.: Knee biomechanics of the dynamic squat exercise. Med. Sci. Sports Exerc. **33**, 127–139 (2001)
7. Flores, V., Becker, J., Burkhardt, E., Cotter, J.: Knee kinematics during squats of varying loads and depths in recreationally trained females. J. Strength Conditioning Res. (2018). https://doi.org/10.1519/JSC.0000000000002509
8. Flanagan, S.P., Salem, G.J.: Lower extremity joint kinetic response to external resistance variations. J. Appl. Biomech. **24**, 58–68 (2008)
9. Glassbrook, D.J., Brown, S.R., Helms, E.R., Duncan, J.S., Strey, A.G.: The high-bar and low-bar back-squats: a biomechanical analysis. J. Strength Conditioning Res. (2017). https://doi.org/10.1519/JSC.0000000000001836

10. Gullett, J.C., Tillman, M.D., Gutierrez, G.M., Chow, J.W.: A biomechanical comparison of back and front squats in healthy trained individuals. J. Strength Conditioning Res. **23**(1), 284–292 (2009)
11. Hartmann, H., Wirth, K., Klusemann, M., Dalic, J., Matuschek, C., Schmidtbleicher, D.: Influence of squatting depth on jumping performance. J. Strength Conditioning Res. **26**, 3243–3261 (2012)
12. Hegedus, J.: Enciclopedia de la musculacióndeportiva. Editorial Stadium, Buenos Aires (1987)
13. Jurkojć, J., Michnik, R., Czapla, K.: Mathematical modelling as a tool to assessment of loads in volleyball player's shoulder joint during spike. J. Sports Sci. **35**(12), 1179–1186 (2017)
14. Kerr, Z.Y., Collins, C.L., Comstock, R.D.: Epidemiology of weight training-related injuries presenting to United States emergency departments, 1990 to 2007. Am. J. Sports Med. **38**, 765–771 (2010)
15. Kim, J.: Lower body kinematic comparisons between front and back squats in response to loads. BSU master's theses and projects, Item 5 (2014)
16. List, R., Gülay, T., Lorenzetti, S.: Kinematic of the trunk and the spine during unrestiricted and restricted squats. In: Proceedings of International Symposium on Biomechanics in Sports, vol. 28, p. 1 (2010)
17. Lutz, G.E., Palmitier, R.A., An, K.N., Chao, E.Y.: Comparison of tibiofemoral joint forces during open kinetic chain and closed kinetic chain exercises. J. Bone Joint Surg. Am. **75**(5), 732–739 (1993)
18. Michnik, R., Jurkojć, J., Pauk, J.: Identification of muscles forces during gait of children with foot disabilities. Mechanika **6**(80), 48–51 (2009)
19. Miletello, W.M., Beam, J.C., Zachary, C.: A biomechanical analysis of the squat between competitive collegiate, competitive high school, and novice powerlifters. J. Strength Conditioning Res. **23**(5), 1611–1617 (2009)
20. Nisell, R., Ekholm, J.: Joint load during the parallel squat in powerlifting and force analysis of in vivo bilateral quadriceps tendon rupture. Scand. J. Sports Sci. **8**, 63–70 (1986)
21. Nowakowska, K., Gzik, M., Michnik, R., Myśliwiec, A., Jurkojć, J., Suchoń, S., Burkacki, M.: The loads acting on lumbar spine during sitting down and standing up. In: Gzik, M., Tkacz, E., Paszenda, Z., Piętka, E. (eds.) Innovations in Biomedical Engineering. Advances in Intelligent Systems and Computing, vol. 526, pp. 169–176. Springer, Cham (2017)
22. Rasmussen, J., de Zee, M., Carbes, S.: Validation of a biomechanical model of the lumbar spine. In: Congress XXII of the International Society of Biomechanics (2009)
23. Rippetoe, M.: Starting Strength Basic Barbell Training, 3rd edn. The Aasgaard Company, Wichita Falls (2011)
24. Russell, P.J., Phillips, S.J.: A preliminary comparison of front and back squat exercises. Res. Q. Exerc. Sport **60**(3), 201–208 (1989)
25. Szyszka, P., Jaszczuk, J., Sacharuk, J., Parnicki, F., Czaplicki, A.: Relationship between muscle torque and performance in special and specific exercises in young weightlifters. Pol. J. Sport Tourism **23**, 127–132 (2016)
26. Walsh, J.C., Quinlan, J.F., Staplenton, R., FitzPatrick, D.P., McCormack, D.: Three dimensional motion analysis of the lumbar spine during free weight lift training. Am. J. Sports Med. **35**(6), 927–932 (2007)

The Impact of Virtual Reality on Ranges of COP Motions During Gait

Piotr Wodarski(✉), Jacek Jurkojć, Marek Gzik, Andrzej Bieniek, Miłosz Chrzan, and Robert Michnik

Department of Biomechatronics, Faculty of Biomedical Engineering, Silesian University of Technology, Gliwice, Poland
piotr.wodarski@polsl.pl

Abstract. An essential aspect of rehabilitation is the maintenance of balance both when standing and in gait. Clinical practice sees equipment enabling the dynamic evaluation of posture stability involving the use of virtual reality. The use of virtual reality in rehabilitation supporting processes and, particularly, with respect to the improvement of motor activity and ability to balance the body, requires that the above-named technology should be tested in terms of its effect on humans. The objective of the tests was to assess changes in the ranges of movements of the COP during gait on treadmill. The assessment was based on measurements of the distribution of pressure on the ground in a system enabling the projection of sceneries created by means of virtual reality. The tests were performed using two different gait velocities as well as matched and mismatched velocity of the scenery motion in relation to actual treadmill motion conditions. The tests involved 32 individuals (26 females and 6 males) aged 23 (SD 1,8). The investigations consisted of 6 stages, in which a person provided with a safety harness walked on a treadmill and was subjected to measurements focused on the distribution of forces exerted by feet on the ground during gait. The tests justified a conclusion that the projection of a moving scenery performed using an HMD headset increased ranges of movements of the COP on the ground during gait on a treadmill. The mismatch between the treadmill velocity and the velocity displayed in virtual reality changed the range of the COP displacement in the AP direction only in the cases of the treadmill velocity changes. The test results constitute the first stage of research on the impact of virtual reality on the stability of human gait.

Keywords: Dynamic stability · Gait · Virtual reality · Balance

1 Introduction

The use of Virtual Reality Technology in medical support engineering systems is becoming increasingly approved by physicians and physiotherapists [1,2]. The number of commercially available devices as regards pre-operative planning and

E. Tkacz et al. (Eds.): IBE 2018, AISC 925, pp. 218–232, 2019.
https://doi.org/10.1007/978-3-030-15472-1_24

rehabilitation support has grown significantly [3,4]. The above-named devices range from medical simulators to equipment supporting diagnostic and therapeutic processes [5]. The development of devices supporting therapeutic processes makes patients rehabilitation more attractive, motivating them to exercise more and recover sooner [1].

An essential aspect of rehabilitation is the maintenance of balance, both when standing and during gait [6,7]. Stability plays an important role in human locomotion. It is often defined as the ability to maintain balance which can be disturbed as a result of disorders caused by anomalies related to the nervous system (local stability) or triggered by external stimuli (global stability) [8,9]. Defects in the locomotor system or neurological disorders are frequently manifested by motor or stability-related perturbations [10]. Early detection of the above-named perturbations may accelerate the diagnosis of the reason for such perturbations and result in the immediate commencement of required treatment [11]. Widely defined detection and diagnostics are aided by stabilographic equipment, where vision disorders are triggered by virtual reality and human motor activity is evaluated using stabilographic platforms. Investigations concerned with the aforesaid areas were conducted by Kessler [12], McAndrew [13,14], Jurkojć [15] and Michnik [16,17]. Related developed tests, enabling the assessment of the ability to balance the body, utilise special indicators, e.g. BRU [12] or w15 and w30 [16,17]. Values of these indicators make it possible to identify reactions to visual stimuli. An improper response can imply the improper functioning of the nervous system, which, in turn, could provide information about the progression of selected conditions in patients [18].

Clinical practice also utilises equipment enabling the virtual reality-based dynamic assessment of posture stability [18]. The principle of operation of such equipment involves the projection of an interesting scenario, e.g. in the form of a game, during gait on the treadmill or in the Cave system [19]. Measurements of angular values and displacements are performed using optical motion analysis systems, whereas pressure distribution measurements are performed using sensors installed in the measuring treadmill [19,20].

The analysis of data obtained during the tests of body stability in dynamic conditions enables the identification of dynamic stability considered in the local and global respect. Local dynamic stability is the resilience of the locomotor system to the infinitesimally small (local) perturbations that occur naturally during walking [8,9], and global dynamic stability is the capacity of the system to respond to larger perturbations, such as tripping or slipping [8,9].

The assessment of dynamic stability and gait variability can be performed using analyses of trajectories of selected points and the stability theory by Lyapunov [8,9] or by analysing the area of support [18,21,22]. The first method is based on criteria of systems stability and, among other things, is used in tests concerned with the prevention of falls in the elderly [9]. The second method assumes that the greater the displacement of the centre of mass on the base or of the centre of pressure in the ML and AP directions, the higher the probability of a fall of a person subjected to a test [21,22]. The area of stability depends not only on the plane of support but also on the velocity and acceleration of individual segments of the body [23]. Both during tests involving a standing position and during gate it is important that displacement of the centre of mass should not be big, as increased displacement raise the probability of a fall of a person subjected to a test. The determination of the precise location of the centre of mass is often unfeasible if performed using optical and accelerometric measurement methods [23]. It is significantly easier to measure the position of the COP (Centre of Pressure).

The use of Virtual Reality Technology in rehabilitation-aiding processes, particularly as regards the improvement of motor activity and ability to balance the body requires the determination of the impact of the VR on a human [16,24]. It is possible to identify changes in kinematic parameters and to determine maximum displacement of the COP triggered by the projection of a scenery created using virtual reality.

2 The Aim of Study

The tests aimed to assess changes in ranges of the COP movements during gait on a treadmill. The assessment was based on measurements concerning the distribution of pressure on the ground using a system enabling the projection of sceneries created using virtual reality. The tests were performed using two different gait velocities as well as matched and mismatched velocity of scenery motion in relation to actual treadmill motion conditions.

3 Testing Methodology

3.1 Measuring Station

The tests were performed using a 3D graphic projection system, i.e. HMD Oculus DK2 headset and a WinFDM-T measurement treadmill manufactured by Zebris. In addition, to increase the safety of a test participant against falling, a safety suspension (safety harness) was used. The design of the safety harness did not affect the movements of a person subjected to a test. The test station and a test participant are presented in Fig. 1.

The HMD system was provided with a graphic application in the form of a forest track moving with the same velocity as the treadmill. The graphic scenery is presented in Fig. 2.

3.2 Test Group

The test group consisted of 32 individuals (26 females and 6 males) aged 23 (SD 1.8), being of an average height of 170 cm (SD 8.7) and an average weight of 63 kg (SD 15.2). Each test participant declared that had not suffered any injury in the lower limbs and did not suffer from any motor system or balance-related ailments. This study was previously approved by the Ethics in Research Committee of the Academy of Physical Education in Katowice (protocol number 11/2015).

Fig. 1. Test station and a test participant

Fig. 2. Graphic scenery used in the 3D application

3.3 Tests

To obtain the previously assumed objective it was necessary to perform tests consisting of 6 measurements. At each stage a test participant walked on the treadmill. The participant was wearing a fastened safety harness. The harness did not impede any movements.

The tests were performed using two treadmill velocities:

v1 = 3.2 km/h

v2 = 4 km/h

Velocity v2 was by 25% higher than velocity v1.

At the first stage, a participant walked on the treadmill at velocity v1. After the stabilisation of the body posture and the information expressed by the participant that they had become accustomed to the pre-set treadmill velocity, a 50-s long measurement of ground response (NGv1) was measured. After 50 s the treadmill velocity was increased by 25% to v2. Next, after the stabilisation of the body posture of the participant another 50-s long measurement of ground response (NGv2) was performed.

At the subsequent stage the test participant put on the HMD headset displaying pre-designed graphic scenery. The motion of the graphic scenery was synchronised with the motion of the treadmill. Afterwards, similar to the previous part of the test, the test participant walked on the treadmill at velocity v1 for 50 s (VRGv1) and velocity v2 for 50 s (VRGv2).

The final stage of the tests involved the assessment of the impact of the mismatch of the scenery motion in relation to the treadmill velocity on the dynamic variability of human body. Initially, the participant walked at a velocity of 4 km/h, whereas the scenery moved at a velocity of 3.2 km/h (VRtG). In the final measurement the person walked on the treadmill at a velocity of 3.2 km/h, whereas the scenery moved at a velocity of 4 km/h (VRGsG).

Test stages along with velocities of the scenery and those of the treadmill are presented in Table 1.

Table 1. Test stages

Test stage	Velocity of treadmill	Velocity of scenery	Research designation
1	3.2 km/h	-	NGv1
2	4 km/h	-	NGv2
3	3.2 km/h	3.2 km/h	VRGv1
4	4 km/h	4 km/h	VRGv2
5	4 km/h	3.2 km/h	VRtG
6	3.2 km/h	4 km/h	VRGsG

3.4 Analysis of Data

Every recorded measurement was 50 s in duration. The recording of the distribution of the pressure exerted by feet on the ground was performed at a frequency of 100 Hz. The dedicated software of the WinFDM-T treadmill enabled the determination of ranges of movement of the COP in the ML and AP directions and the determination of the following kinematic analyses: step width (distance between left and right heel), Ant/Post position (shift forward/backwards of the intersection point of the course of the COP in the butterfly diagram, the zero position is the heel strike) and Lateral symmetry (shift to left/right of the intersection point of the course of the COP in the butterfly diagram, a negative value indicates a shift to the left side and a positive value indicates a shift to right side, the zero position is the centre of the presentation).

The first stage of the analysis involved each test and each person and was concerned with the determination of the COP shifts in the ML and AP directions in the treadmill coordinate system (without taking into consideration the gain in distance related to the motion of the treadmill). Afterwards, in order to remove the COP shift constant constituents and only analyse the COP positions of the standing person in relation to the initial position, the coordinates of the initial test position (the COP positions related to a time of 0 s) were subtracted from determined values.

The subsequent stage involved the analysis of the virtual reality effect on changes in maximum shifts of the COP in the AP and ML directions. Next, software developed by the authors in the Matlab environment was used to determine the ranges of the COP shifts in the ML and AP directions in relation to each participant and tests NGv1, NGv2, VRGv1 and VRGv2. Afterwards, the results in relation to each person were averaged and the standard deviation and the scatter of values were calculated. In addition, the mean values of Step width, Ant/Post position and Lateral symmetry were calculated in relation to the above tests.

The subsequent stage involved the assessment of the impact of the mismatch of the graphic on changes in the range of the COP shifts in the ML and AP directions in relation to VRtG and VRGsG. The test results were compared with the results obtained in relation to NGv1. Next, the results in relation to the entire group were averaged and the standard deviation and the scatter of values were calculated. Similar to the previous test, the mean values of Step width, Ant/Post position and Lateral symmetry were calculated in relation to VRtG and VRGsG.

All of the calculated results were compared in diagrams. In order to indicate statistically relevant differences, the occurrence of normal distributions was subjected to the Shapiro-Wilk test, whereas the homogeneity of variance was subjected to the Levene test. The subsequent stage involved the determination of the presence of statistically significant differences between individual parameters. The above-named determination was performed using parametric and non-parametric ANOVA tests and post-hoc tests.

4 Test Results

The results of statistical analyses and comparisons of the means are presented in diagrams in a form clarified in Fig. 3. Figures 4 and 5 present examples of trajectories of the COP displacements in the ML and AP in relation to a randomly selected measurement after the correction of the treadmill motion.

Figure 6 contains a butterfly diagram concerning the above-presented trajectories and describing the COP displacement in the ML and AP directions during gait. Figure 7 presents maximum ranges of the COP movement in the ML direction in relation to NGv1, VRGv1, NGV2 and VRGv2 tests (averaged for all test participants).

Figure 8 presents maximum ranges of the COP movement in the AP direction during gait in relation to NGv1, VRGv1, NGV2 and VRGv2 tests (averaged for all test participants) (after the correction of the treadmill motion).

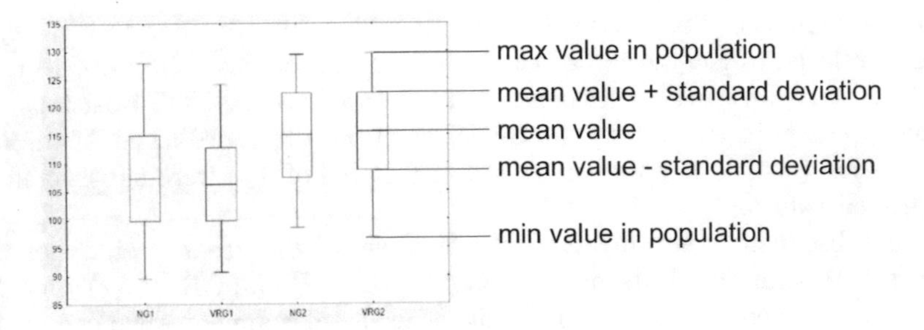

Fig. 3. Description of diagrams

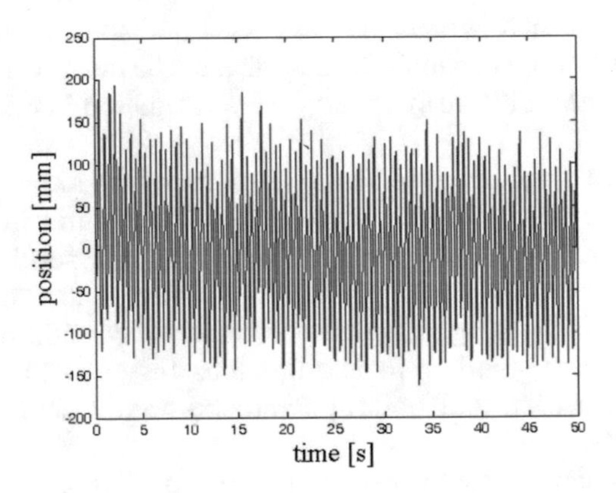

Fig. 4. Example of a diagram of the COP displacement in the AP direction during gait after the correction of treadmill velocity

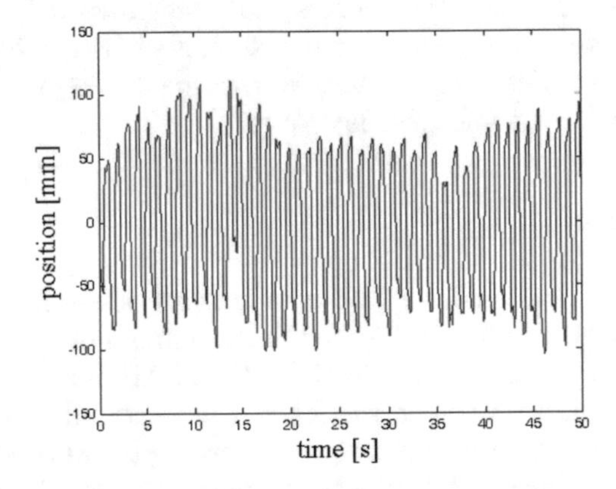

Fig. 5. Example of a diagram of the COP displacement in the ML direction during gait

The next stage, following the previously adopted methodology, involved the determination of the mean step width in relation to the tests. The results are presented in Fig. 9.

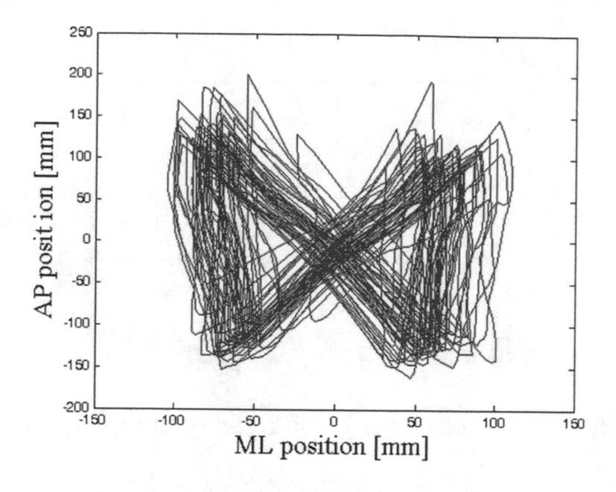

Fig. 6. Example of a butterfly diagram

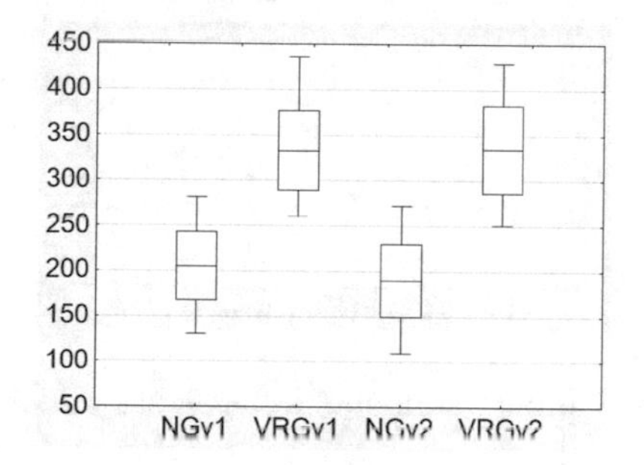

Fig. 7. Maximum ranges of the COP movement in the ML direction [mm] during gait in relation to NGv1, VRGv1, NGV2 and VRGv2 tests, averaged for all test participants.

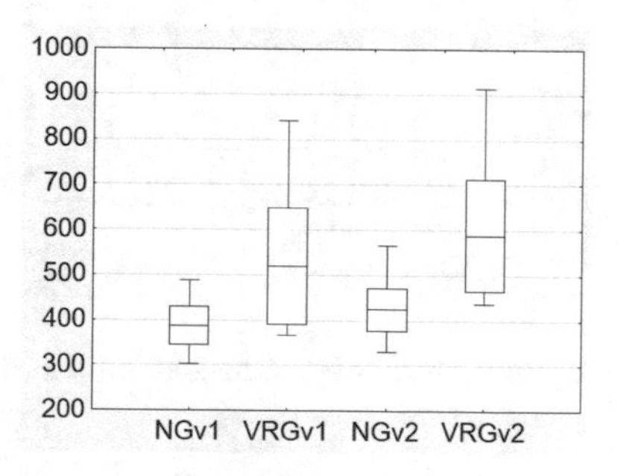

Fig. 8. Maximum ranges of the COP movement in the AP direction [mm] during gait in relation to NGv1, VRGv1, NGV2 and VRGv2 tests (averaged for all test participants)

The tests also involved the determination of the mean values of Lateral symmetry and Ant/Post position parameters. The results are presented in Figs. 10 and 11.

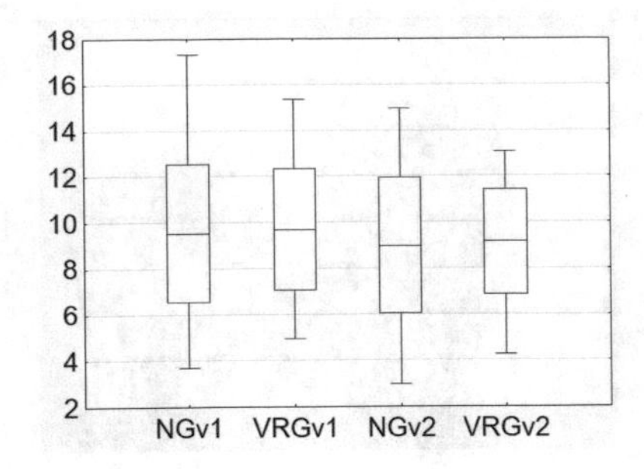

Fig. 9. Step width [cm] results averaged for the entire group of test participants

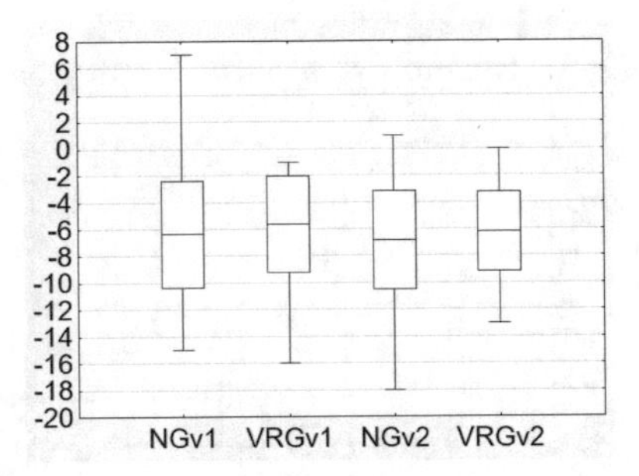

Fig. 10. Lateral symmetry [mm] parameter values averaged for the entire group of test participants

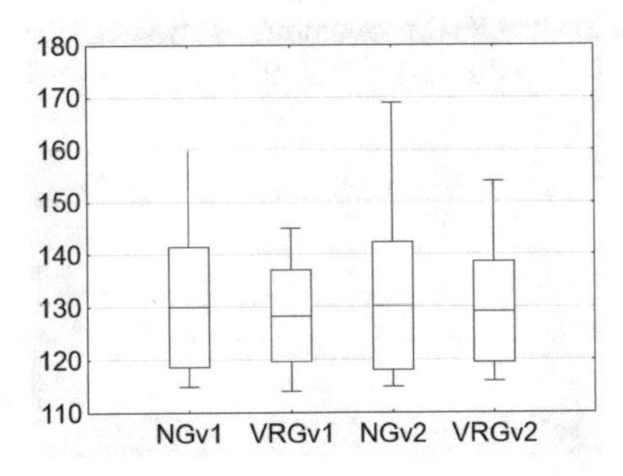

Fig. 11. Ant/Post position [mm] parameter values averaged for the entire group of test participants

The next stage involved the assessment of the effect of the mismatch of the scenery in relation to the treadmill velocity on the COP displacements. The comparative results (averaged for the entire group of participants) of maximum ranges of the COP displacements in relation to NGv1, VRtG and VRGsG tests in the ML direction are presented in Fig. 12, whereas the displacements in the AP direction are presented in Fig. 13.

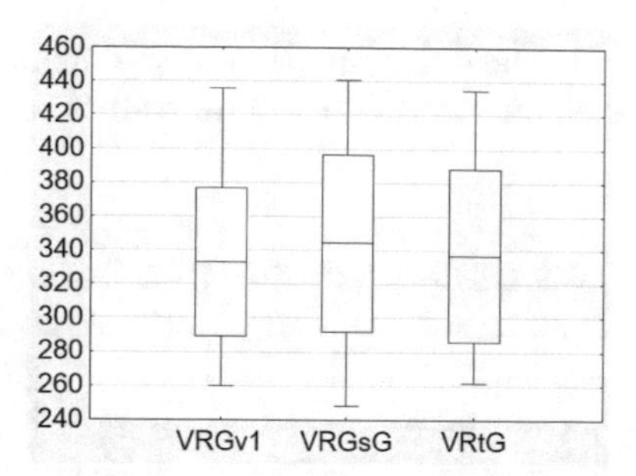

Fig. 12. Maximum ranges of the COP displacement [mm] in relation to NGv1, VRtG and VRGsG tests in the ML direction, averaged for the entire group of test participants

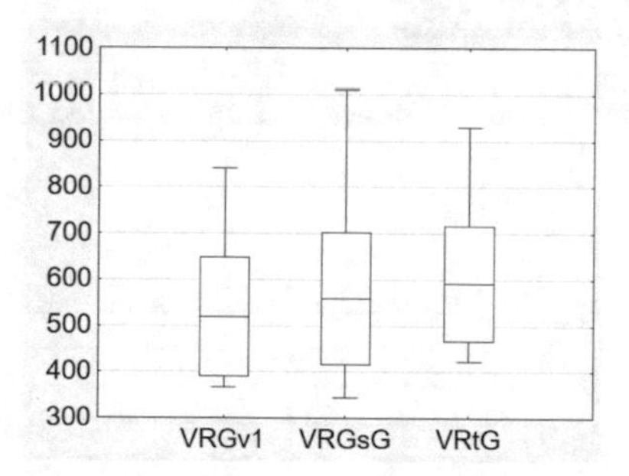

Fig. 13. Maximum ranges of the COP displacement [mm] in relation to NGv1, VRtG, and VRGsG tests in the AP direction, averaged for the entire group of test participants

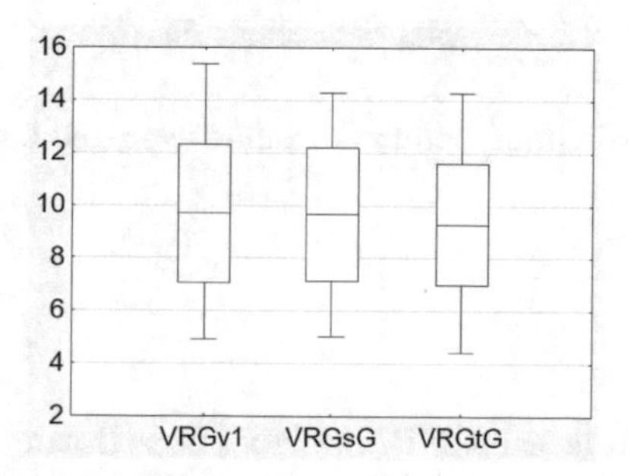

Fig. 14. Mean step width [cm] results in relation to NGv1, VRtG and VRGsG tests, averaged for the entire group of test participants

The next stage, following the previously adopted methodology, involved the determination of the mean step width in relation to NGv1, VRtG and VRGsG tests. The results are presented in Fig. 14.

The final stage involved the determination of the mean values of Lateral symmetry and Ant/Post position parameters. The results are presented in Figs. 15 and 16.

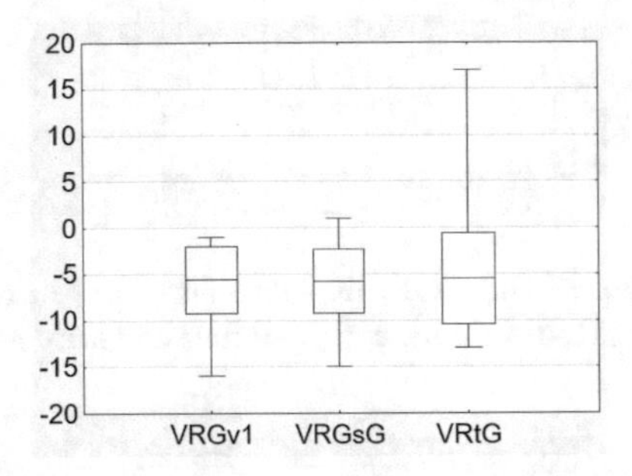

Fig. 15. Lateral symmetry [mm] parameter values averaged for the entire group of test participants

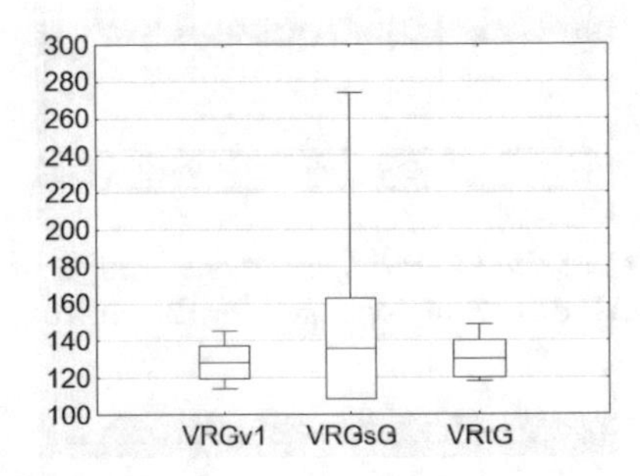

Fig. 16. Ant/Post position [mm] parameter values averaged for the entire group of test participants

5 Discussion

The diagrams concerning the COP displacements during gait on the treadmill reveal oscillations around the initial position. When analysing the diagrams it can be seen that the frequency of the above-named oscillations is by twice higher in relation to the AP direction (Fig. 4) than for the ML direction (Fig. 5). The butterfly diagram (Fig. 6) presents the COP displacement areas during measurements. The central part of the diagram presents intersections of movement trajectories. In numerical terms in relation to the diagram centre and in relation to the heel position, these trajectories describe further analysed parameters of Lateral symmetry (distance between the COP and the centre in the ML direction) and Ant/Post position (distance between the COP from the heel in the AP direction) [14].

The first part of data analysis includes the analysis of the effect of virtual reality on changes in the range of the COP movement. Figure 7 reveals a statistically significant increase in the mean value of the range of the COP movement in the ML direction in relation to the VRGv1 test in comparison with NGv1 ($p < 0.0001$) and in relation to VRGv2 in comparison with NGv2 ($p < 0.0001$). Similarly to ranges in the COP movement in the AP direction, it is possible to observe an increase in the mean value in relation to the VRGv1 test in comparison with NGv1 ($p < 0.001$) and in relation to VRGv2 in comparison with NGv2 ($p < 0.001$). The increase in value both in the AP and in the ML directions indicates that, as regards both analysed velocities, i.e., v1 and v2, after putting on the HMD headset and the projection of 3D scenery, the area of the COP displacements increased during gait.

Figure 9 reveals that, as regards the above-named measurements, the step width did not increase ($p = 0.594$ test-t for VRGv1 and NGv1 and $p = 0.535$ test-t for VRGv2 and NGv2). The lack of a change in the step width combined with an increase in the range of movements in the ML direction indicates a change in the manner of gait. For instance, the manner of taking steps might have changed or the range of displacement of the centre of mass (COM) in the ML direction might have increased, which was observed by Bierbaum [25]. The above-named changes can indicate the greater susceptibility of the system to a fall, thus indicating a decrease in the stability range [25,26].

The analysis of Lateral symmetry parameters presented in Fig. 10 reveals that the symmetry of COP displacements in the ML direction was not changed by the impact of virtual reality ($p = 0.144$ for v1 and $p = 0.227$ for v2). Similarly, the values of Ant/Post position parameters (Fig. 11) do not indicate the existence of statistically relevant changes ($p = 0.239$ for v1 and $p = 0.971$ for v2). The foregoing indicates that the wearing of the HMD headset displaying virtual reality did not lead to a change in the body posture of a test participant walking on the treadmill significantly enough to trigger a change in the central point of the COP intersection in butterfly diagrams (the body did not lean in the ML and AP directions). The results confirmed observations reported in publications of other authors [26].

The subsequent stage involved the assessment of changes in the COP trajectories triggered by the matching and the mismatch of treadmill velocity to conditions present in the scenery displayed by the HMD headset. To this end, Fig. 12 was used to compare the maximum COP displacements in the ML direction in relation to the VRGv1, VRGsG and VRtG tests. The statistical analysis performed using the ANOVA tests did not reveal differences between the mean values of this parameter ($p = 0.638$). A slightly different situation was observed in relation to the ranges of the COP movement in the AP direction (Fig. 13). A statistically relevant difference was observed between the value in relation to the VRtG and VRGv1 tests was observed (ANOVA post-hoc, $p = 0.0015$). However, no differences were observed between VRGv1 and VRGsG (ANOVA post-hoc, $p = 0.1162$). The results also revealed the lack of differences in the ranges of the COP displacements in the ML direction in relation to the test where the scenery velocity increased by 25% (VRGsG) and in relation to the test where the treadmill velocity increased by 25% (VRtG) in comparison with velocity v1 in the VRGv1 test. As regards the AP direction, the differences in the average ranges of the COP displacements between VRGv1 and VRtG could be ascribed to the change in the treadmill velocity and, consequently, the velocity of gait, which was confirmed by tests performed by England [27] in relation to the effect of gait velocity on the treadmill on the ranges of the COP and COM displacements.

As regards, the VRGv1, VRGsG and VRtG tests, no statistically significant differences between the mean values of Step width (Fig. 14, ANOVA KW $p = 0.756$), Lateral symmetry (Fig. 15, ANOVA $p = 0.598$) and Ant/Post position (Fig. 16, ANOVA KW $p = 0.470$) parameters were observed. The foregoing indicates the lack of changes in the support plane width as well as the lack of changes in the area of the intersection of the COP movement trajectory in the central part of butterfly diagrams in relation to test participants.

6 Conclusions

The tests justified the conclusion that the projection of moving scenery displayed using the HMD headset increases the ranges of the COP movements during gait on the treadmill. The lack of an increase in the area of support in the ML direction combined with an increase in the range of the COP movement in the above-named direction indicates a decrease in the distance between the maximum COP deflection and the end of the plane of support, which could result in the fall of a test participant. The mismatch between the treadmill velocity and the velocity of displayed virtual reality scenery only affected the range of the COP displacements in the AP direction (in the cases where the treadmill velocity was changed). The conclusions were confirmed by test results available in reference publications indicating that changes in the treadmill velocity are accompanied by a change in gait variability in the AP direction. The results constitute the first stage of tests concerning the impact of virtual stability on human gait variability and can be verified using other stability assessment methods.

References

1. da Costa, R.M.E.M., de Carvalho, L.A.V.: The acceptance of virtual reality devices for cognitive rehabilitation: a report of positive results with schizophrenia. Comput. Methods Programs Biomed. **73**, 173–182 (2004)
2. Cikajlo, I., Matjačić, Z.: Advantages of visual reality technology in rehabilitation of people with neuromuscular disorders. In: Recent Advances in Biomedical Engineering, pp. 301–320 (2009). ISBN 978-953-307-004-9
3. Kusz, D., Wojciechowski, P., Cieliński, Ł., Iwaniak, A., Jurkojć, J., Gąsiorek, D.: Stress distribution around a TKR implant: are lab results consistent with observational studies? Acta Bioeng. Biomech. **10**(4), 21–26 (2008)

4. Wolański, W., Larysz, D., Gzik, M., Kawlewska, E.: Modeling and biomechanical analysis of craniosynostosis correction with the use of finite element method. Int. J. Numer. Meth. Biomed. Eng. **29**, 916–925 (2013)
5. Cooper, R.A., Dicianno, B.E., Brewer, B., LoPresti, E., Ding, D., Simpson, R., et al.: A perspective on intelligent devices and environments in medical rehabilitation. Med. Eng. Phys. **30**, 1387–1398 (2008)
6. Duque, G., Boersma, D., Loza-Diaz, G., Hassan, S., Suarez, H., et al.: Effects of balance training using a virtual-reality system in older fallers. Clin. Interv. Aging **8**, 257–263 (2013)
7. Griffin, H.J., Greenlaw, R., Limousin, P., Bhatia, K., Quinn, N.P., Jahanshahi, M.: The effect of real and virtual visual cues on walking in Parkinson's disease. J. Neurol. **258**, 991–1000 (2011)
8. Full, R.J., Kubow, T., Schmitt, J., Holmes, P., Koditschek, D.E.: Quantifying dynamic stability and maneuverability in legged ocomotion. Integr. Comp. Biol. **42**(1), 149–157 (2002)
9. Dingwell, J.B., Marin, L.C.: Kinematic variability and local dynamic stability of upper body motions when walking at different speeds. J. Biomech. **39**, 444–452 (2006)
10. Kang, H.G., Dingwell, J.B.: Effects of walking speed, strength and range of motion on gait stability in healthy older adults. J. Biomech. **41**, 2899–2905 (2008)
11. Winter, D.A., Patla, A.E., Frank, J.S., Walt, S.E.: Biomechanical walking pattern changes in the fit and healthy elderly. Phys. Ther. **70**(6), 340–347 (1990)
12. Kessler, N., Gananca, M.M., Gananca, C.F., Gananca, F.F., Lopes, S.C., Serra, A.P., Caovilla, H., et al.: Balance rehabilitation unit (BRUTM) posturography in relapsing-remitting multiple sclerosis. Arq. Neuropsiquiatr. **69**(3), 485–490 (2011)
13. McAndrew, P.M., Dingwell, J.B., Wilken, J.M.: Walking variability during continuous pseudo-random oscillations of the support surface and visual field. J. Biomech. **43**(8), 1470–1475 (2010)
14. McAndrew, P.M., Wilken, J.M., Dingwell, J.B.: Dynamic stability of human walking in visually and mechanically destabilizing environments. J. Biomech. **44**(4), 644–649 (2010)
15. Jurkojć, J., Wodarski, P., Bieniek, A., Gzik, M., Michnik, R.: Influence of changing frequency and various sceneries on stabilometric parameters and on the effect of adaptation in an immersive 3D virtual environment. Acta Bioeng. Biomech. **19**(3), 129–137 (2017)
16. Michnik, R., Jurkojć, J., Wodarski, P., Gzik, M., Jochymczyk-Woźniak, K., Bieniek, A.: The influence of frequency of visual disorders on stabilographic parameters. Acta Bioeng. Biomech. **18**(1), 25–33 (2016)
17. Michnik, R., Jurkojć, J., Wodarski, P., Gzik, M., Bieniek, A.: The influence of the scenery and the amplitude of visual disturbances in the virtual reality on the maintaining the balance. Arch. Budo **10**(1), 133–140 (2014). ISSN 1643-8698
18. Caderby, T., Yiou, E., Peyrot, N., Begon, M., Dalleau, G.: Influence of gait speed on the control of mediolateral dynamic stability during gait initiation. J. Biomech. **47**, 417–423 (2014)
19. Keshner, E.A., Kenyon, R.V.: The influence of an immersive virtual environment on the segmental organization of postural stabilizing responses. J. Vestib. Res. **10**, 207–219 (2000)
20. Cho, K.H., Lee, K.J., Song, C.H.: Virtual-reality balance training with a video-game system improves dynamic balance in chronic stroke patients. Exp. Med. NCBI **228**, 69–74 (2012)

21. Hof, L., van Bockel, R.M., Schoppen, T., Postema, K.: Control of lateral balance in walking experimental findings in normal subjects and above-knee amputees. Gait Posture **25**, 250–258 (2007)
22. McAndrew, P.M., Wilken, J.M., Dingwell, J.B.: Dynamic margins of stability during human walking in destabilizing environments. J. Biomech. **45**, 1053–1059 (2012)
23. Hof, A.L., Gazendam, M.G.J., Sinke, W.E.: The condition for dynamic stability. J. Biomech. **38**, 1–8 (2005)
24. Jurkojć, J., Wodarski, P., Bieniek, A.: Influence of changing frequency and various sceneries on stabilometric parameters and on the effect of adaptation in an immersive 3D virtual environment. Acta Bioeng. Biomech. **19**(3), 129–137 (2017)
25. Bierbaum, S., Peper, A., Karamanidis, K., Arampatzis, A.: Adaptational responses in dynamic stability during disturbed walking in the elderly. J. Biomech. **43**, 2362–2368 (2010)
26. Sloot, L.H., van der Krogt, M.M., Harlaar, J.: Effects of adding a virtual reality environment to different modes of treadmill walking. Gait Posture **39**, 939–945 (2014)
27. England, S.A., Granata, K.P.: The influence of gait speed on local dynamic stability of walking. Gait Posture **25**, 172–178 (2007)

Can We Learn from Professional Dancers Safe Landing? Kinematic and Dynamic Analysis of the 'grand pas de chat' Performed by Female and Male Dancer

Joanna Gorwa[1](✉), Anna Fryzowicz[1], Robert Michnik[2], Jacek Jurkojć[2], Jarosław Kabaciński[1], Katarzyna Jochymczyk-Woźniak[2], and Lechosław B. Dworak[3]

[1] Department of Biomechanics, Faculty of Physical Education, Sport and Rehabilitation, Poznań University of Physical Education, Królowej Jadwigi 27/39, 61-871 Poznań, Poland
sarbi@poczta.onet.pl

[2] Department of Biomechatronics, Faculty of Biomedical Engineering, Silesian University of Technology, Roosevelta 40, 41-800 Zabrze, Poland

[3] Chair of Bionics, University of Arts, Marcinkowskiego 29, 61-745 Poznan, Poland
http://www.awf.poznan.pl

Abstract. PURPOSE: The aim of the study was to assess and compare kinematic and dynamic aspects of 'grand pas de chat' dancing element performed by ballet dancers: a female and a male. METHOD: Tests were carried out with the use of the motion analysis system and the force plate. RESULTS: The research enabled kinematic and dynamic analysis of elements performed by dancers. The results made it possible to indicate when, during dancers' movement, the greatest loads in their body occur. Also differences in kinematics of landing performed by female and male were observed. CONCLUSION: According to directives on technique and esthetics required in classical ballet by Vaganova [1] female rather than male performance of landing phase after 'grand pas de chat' was done safer. For amateurs an instruction on safe landing preserving sufficient amortisation should be incorporated.

Keywords: Classical dance · Professional dancers · Kinematic analysis · Ground reaction forces · Safety

1 Introduction

Dance is a form of art and as such has a positive influence on health and well-being among amateurs of different age and social status [2–5] and may be employed in rehabilitation process [6,7]. On the other hand, dance has a lot in common with professional sport. At the beginning of the 1970s, the research

E. Tkacz et al. (Eds.): IBE 2018, AISC 925, pp. 233–240, 2019.
https://doi.org/10.1007/978-3-030-15472-1_25

of Nicholas, Grossman, and Hershman [8] revealed surprising facts about the scale of loads occurring in professional dance. According to these authors, similarly to American football, professional dance is at the top of the list of the most demanding physical exercises which exert the greatest loads on the body. In preliminary study [9] the jump 'grand pas de chat' – a jump imitating a delicate, graceful jump of a cat – belonging to the family of the so called 'big jumps' of classical ballet, generated the highest vertical component of ground reaction forces (vGRF) (7.47 BW for the female dancer and 9.43 BW for the male dancer). On the other hand in normal gait vGRF values reach 1.16-1.17 BW [10]. Additionally even with moderate vGRF values internal forces occurring in landing after the jump may be extremely high [11]. According to several authors [1,12–15] correct technique that allows for proper shock absorption is a factor that significantly reduces the risk of injury. Since dance become a form of rehabilitation, remedy or just attractive form of spending free time an emphasis should be put on its safe practice. It is crucial for amateurs to keep correct dance technique to preserve appropriate amortisation especially in the light of mentioned above vGRF values occurring in landing after the jump.

The aim of this research was to assess the kinematic and dynamic aspects of classical dance movement called 'grand pas de chat' performed by female and male dancer. Additionally the comparison between female and male performance was carried out to verify who can land safer preserving the aesthetics and correctness crucial in classical ballet.

2 Materials and Methods

Participants
Two professional ballet dancers training in Classical Ballet Theatre with no history of low back or lower limb pathology participated in this study (a female and a male). The age, dance training experience, body height and body mass were respectively: 20 yr, 10 yr, 1.69 m, 52 kg for woman and 19 yr, 9 yr, 1.86 m, 71 kg for man.

Biomechanical Instrumentation
Kinematic values were computed with the use of Ariel Performance Analysis System (APAS) (Ariel Dynamics Inc., USA) and GRFs were measured with the use of Kistler force plate (Kistler Gruppe, Switzerland) at a sampling rate 1000 Hz. The motion of tested participants was recorded by four Basler infrared digital cameras with recording frequency of 200 Hz (Basler AG, Germany). Ten reflective markers were placed over specific anatomical landmarks of the pelvis and right lower limb: the head of second metatarsal bone, the calcaneus, the lateral malleolus, the femur lateral epicondyle, the femur great trochanter, right and left anterior superior iliac spines, L5S1 vertebrae and two markers on wands on lateral side of the lower leg and thigh.

Experimental Protocol
Each participant, after few-minutes warm up, performed barefoot the 'grand pas de chat' movement. The performance was repeated several times until three

technically correct jumps with landing within the force platform were recorded. Afterwards the jump with the highest maximum vGRF, indicating the highest loads acting on joints, was selected for further analysis. The participants were blinded to the aim of this research i.e. to perform the jump in the safest way, merely they were asked to do it technically correct. The correctness of performed jumps was evaluated by choreographer present during the tests.

Data Analysis
APAS software was used to process images recorded by cameras and the location of markers positioned on dancers' bodies was determined. GRFs acting during the landing phase were processed with the use of self-prepared software. The cameras were arranged in a manner that allowed the authors to precisely determine the kinematics of the pelvis and the right lower limb during the landing phase. The number and location of markers made it possible to determine the positions of right lower limb joint centers and then the relative angular motion of particular lower limb segments and the pelvis. The calculations were performed with the use of proprietary software developed in the Matlab environment.

Ethics
Before testing both dancers were familiarized with the testing protocol and provided informed consent to participate in this study. The experiment was approved by local ethics committee.

3 Results

Elaborated methodology enabled determination of kinematic and dynamic quantities describing dancers' motion. Figure 1 presents the trajectories of joint centers and successive positions of lower limb segments determined during the trials. The line graphs of vGRF during landing are shown in Fig. 2A. The peak vGRF obtained by female and male dancers were similar: 4.92 BW and 4.75 BW, respectively.

The analysed movements also show significant range of motion (ROM) of the pelvis both in the sagittal and horizontal planes during the landing phase (Figs. 2B and C respectively). The line graphs of lower limb joint angles in the sagittal plane show that they are more or less similar in shape in both participants, but differ in terms of the ROM. This concerns especially hip joint angle (Fig. 2D) which ROM accounted for 76° for the female and 62° for the male. In case of knee joint (Fig. 2E), ROM was 32° for the female and 23° for the male. Ankle joint ROM (Fig. 2F) in the sagittal plane was 87° for the female, and 52° for the male. Figure 3 presents foot angle in relation to the ground in the sagittal plane determining whether a dancer stands on tiptoe or on the whole foot. In each figure the moment when the vGRF is maximal is marked.

4 Discussion

The results made it possible to analyse technical quality and precisely describe 'grand pas de chat' movement carried out by both female and male dancers.

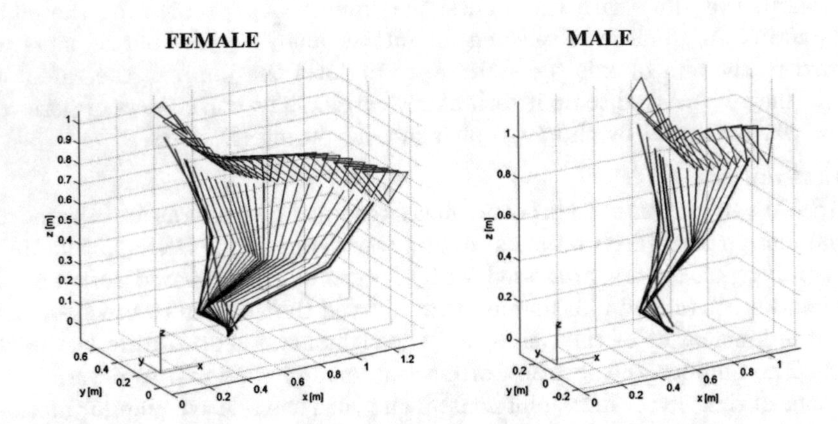

Fig. 1. Trajectories of joint centers and successive positions of lower limb segments during the landing phase of the 'grand pas de chat' movement

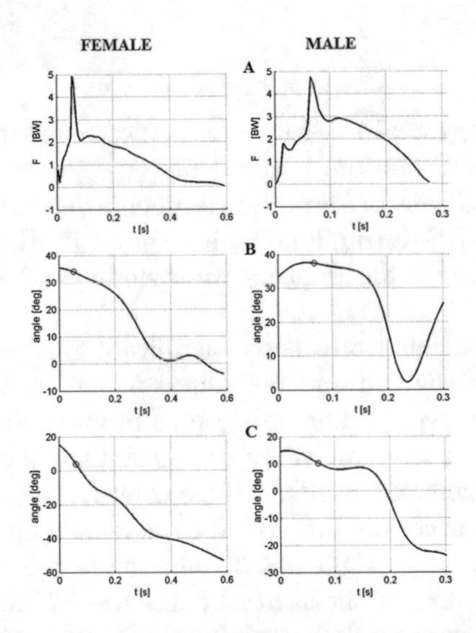

Fig. 2. Line graphs of (A) vGRF, (B) pelvis tilt angle in sagittal plane, (C) pelvis rotation angle in horizontal plane, (D) hip flexion-extension joint angle, (E) knee flexion-extension joint angle, (F) foot plantar-dorsal flexion joint angle. The mark indicates the moment of peak vGRF. Positive values mean: anterior tilt, external rotation, flexion or dorsal flexion. Negative values mean: posterior tilt, internal rotation, extension or plantar flexion

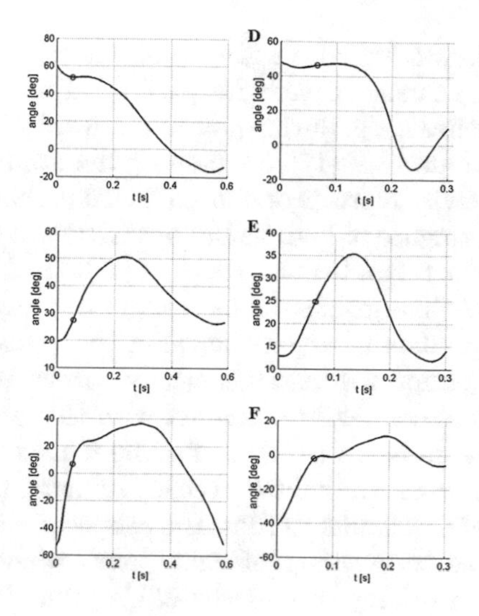

Fig. 2. (*continued*)

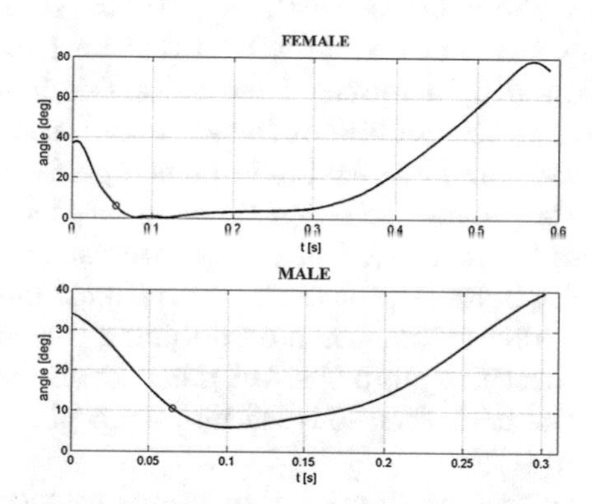

Fig. 3. Line graphs of foot angle in relation to the ground in the sagittal plane. The mark indicates the moment of peak vGRF

Interestingly, the landing phase duration in case of classical ballet movements is very short. High peak values of vGRF during the landing phase that follows a ballet jump creates the greatest risk of injury, particularly for the biokinematic chain of the lower limb [12,13,16]. Determination of the moment of maximal vGRF can be crucial to estimate loads in the skeletal system. When analysing the entire chain taking part in the final impact phase of the jump, the values of 4.75 and 4.92 BW can be observed not only in the foot of the dancer, but also in all sections of this chain. One may presume, that such high values of vGRF will exert serious loads on the joint surfaces, bones, muscles, tendons and ligaments of the supporting limb.

Kinematic analysis was based on graphs of angles of pelvis as well as hip, knee and ankle joints. In most movements performed in everyday life the pelvis assumes horizontal position, e.g. during gait the forward tilt of the pelvis remains at the level of several degrees [17]. In case of the studied movements, great anterior tilt of the pelvis, more pronounced in man than woman, during the landing phase in the presence of high values of vGRF will cause overloads to the lumbar spine, and further, result in its damage [18].

After analysing the line graphs of foot angle in relation to the ground in the sagittal plane, as well as knee joint angle in the sagittal plane, one can say that the landing was performed in a technically correct way. In both cases the landing begins with toes coming into contact with the ground (the tilt angle of foot in relation to the ground was 37° – for the female dancer, and 35° – for the male dancer) and when the vGRF reaches its maximum, the feet are in a more horizontal position. Thanks to that, the center of mass is transferred from toes and metatarsus to the vicinity of ankle joint, which decreases the forces generated by muscles working around the ankle joint. Nevertheless in female performance greater ROM of the tilt angle of foot in relation to the ground in landing phase enhance the potential of the foot to act as a shock absorber. At the moment of landing, the knee is slightly bent (20° for female and 13° for male dancer) and it is followed by a further bending of this joint which reaches its maximum more or less in the middle of the support phase, finally followed by extension. Greater maximum knee flexion in female performance provides more efficient shock absorption thanks to eccentric quadriceps activation.

Such a structure of movement of the joints proves that the movement was performed correctly. Vaganova [1], the author of the most important book on the principles of classical ballet so far, when describing a correct landing technique after a jump, states: ‘after the jump the feet must touch the floor first with the toe, then softly with the heel, then lowered into demi-plié. After this the knees should be straightened’ [1].

Attention should be paid to the fact that jumps performed by a male differ in their technique. It should be obvious, since it is sex that determines the role played by the artist in ballet [1]. Women are required to move precisely controlling the body balance on a support base that is reduced to a minimum (such a reduced surface of support can be observed while standing on the tips of all toes during ‘relevé en pointe’) and to show gentleness and ‘ethereality’ in their jumps. Men mostly work on entire foot surfaces or on the balls of feet in the so called ‘demi - pointe relevé’, and the jumps they perform are very expressive. During such a dynamic exercise the coordination of movement of lower and upper limbs is crucial. Ballet dancers cannot achieve this coordination in a spontaneous and natural way. It is because the esthetics and strict rules of classical ballet impose restrictions on movement of limbs and the technique of landing [1]. Very important, then, is the use of feet as biological shock absorbers [19]. Female took advantage of greater ROM in lower limb joints what gives her possibility to more effectively absorb impact occurring in landing phase [20].

Since there are some noticeable differences in landing technique between professional dancers, in amateurs the level of ability to land safe would vary significantly. The positive influence of dancing on different aspects of life and health has been widely documented [2–7]. Therefore in order to improve safety and efficiency of dance therapy future studies should investigate the kinematic and dynamic aspects of dance technique in a group of amateurs.

5 Conclusion

The results of this study may be used in the development and teaching of dancing techniques. The methodology presented in this paper enables one to simultaneously determine the kinematics of movements executed by dancers as well as vGRF values. This is crucial because not only vGRF values indicate safety level of dance movement performance but also profound analysis of kinematics of lower limb joints especially ankle joint and foot [20]. According to directives on technique and esthetics required in classical ballet by Vaganova [1] female performance of landing phase after 'grand pas de chat' was done safer. It was confirmed by choreographer present during the tests. Training sessions carried out in the biomechanical laboratory, with an active participation of dancing teachers would provide feedback on certain ballet element kinematics and vGRF and therefore could improve safety. It concerns professional dancers however even when incorporating dance rehabilitation, dance as a remedy or activity for free time for amateurs it would be reasonable to instruct them on safe landing preserving sufficient amortisation. It will help in injury prevention. In some cases though, having regard to vGRF values occurring in landing after the jump, this expressive elements should be excepted from dance activity.

References

1. Vaganova, A.: Basic Principles of Classical Ballet: Russian Ballet Technique. Dover Publications, New York (2015)
2. Burkhardt, J., Brennan, C.: The effects of recreational dance interventions on the health and well-being of children and young people: a systematic review. Arts Health **4**, 148–161 (2012)
3. Kluge, M.A., Tang, A., Glick, L., LeCompte, M., Willis, B.: Let's keep moving: a dance movement class for older women recently relocated to a continuing care retirement community (CCRC). Arts Health **4**, 4–15 (2012)
4. Lopez-Ortiz, C., Gladden, K., Deon, L., Schmidt, J., Girolami, G., Gaebler-Spira, D.: Dance program for physical rehabilitation and participation in children with celebral palsy. Arts Health **4**, 39–54 (2012)
5. Wilbur, S., Meyer, H.B., Baker, M.R., Smiarowski, K., Suarez, C.A., Ames, D., Rubin, R.T.: Dance for veterans: a complementary health program for veterans with serious mental illness. Arts Health **7**, 96–108 (2015)
6. Houston, S., McGill, A.: A mixed-methods study into ballet for people living with Parkinson's. Arts Health **5**, 103–119 (2013)

7. McRae, C., Leventhal, D., Westheimer, O., Mastin, T., Utley, J., Russell, D.: Long-term effects of Dance for PD on self-efficacy among persons with Parkinson's disease. Arts Health **10**, 1–12 (2018)
8. Nicholas, J., Grossman, R., Hershman, E.: The importance of a simplified classification of motion in sports in relation to performance. Orthop. Clin. North Am. **8**, 499–532 (1977)
9. Dworak, L.B., Gorwa, J., Kmiecik, K., Mączyński, J.: A study characterizing dynamic overloads of professional dancers. Biomechanical approach. Acta Bioeng. Biomech. **7**, 77–84 (2005)
10. Winiarski, S., Rutkowska-Kucharska, A.: Estimated ground reaction force in normal and pathological gait. Acta Bioeng. Biomech. **11**, 53–60 (2009)
11. Dziewiecki, K., Mazur, Z., Blajer, W.: Assessment of external and internal loads in the triple jump via inverse dynamics simulation. Biol. Sport **30**, 103–109 (2013)
12. Gorwa, J., Dworak, L.B., Michnik, R., Jurkojć, J.: Kinematic analysis of modern dance movement "stag jump" within the context of impact loads, injury to the locomotor system and its prevention. Med. Sci. Monit. **20**, 1082–1089 (2014)
13. Kulig, K., Fietzer, A.L., Popovich, J.M.: Ground reaction forces and knee mechanics in the weight acceptance phase of a dance leap take-off and landing. J. Sports Sci. **29**, 125–131 (2011)
14. Forczek, W., Baena-Chicón, I., Vargas-Macias, A.: Movement concepts approach in studies on flamenco dancing: a systematic review. Eur. J. Sport Sci. **17**, 1161–1176 (2017)
15. Fietzer, A.L., Chang, Y., Kulig, K.: Dancers with patellar tendinopathy exhibit higher vertical and braking ground reaction forces during landing. J. Sports Sci. **30**, 1157–1163 (2012)
16. Luke, A., Kinney, S., Dhemecourt, P., Baum, J., Owen, M., Micheli, L.: Determinants of injuries in young dancers. Med. Probl. Perform. Art. **8**, 105–112 (2000)
17. Fryzowicz, A., Murawa, M., Kabaciński, J., Rzepnicka, A., Dworak, L.B.: Reference values of spatiotemporal parameters, joints angles, ground reaction forces, and plantar pressure distribution during normal gait in young women. Acta Bioeng. Biomech. **20**, 49–57 (2018)
18. Wanke, E.M., Koch, F., Leslie-Spinks, J., Groneberg, D.A.: Traumatic injuries in professional dance - past and present: ballet injuries in Berlin, 1994/95 and 2011/12. Med. Probl. Perform. Art. **29**, 168–173 (2014)
19. McNitt-Gray, J.L., Koff, S.R., Hall, B.L.: The influence of dance training and foot position on landing mechanics. Med. Probl. Perform. Art. **7**, 87–91 (1992)
20. Jarvis, D.N., Kulig, K.: Pointing the foot without sickling: an examination of ankle movement during jumping. Med. Probl. Perform. Art. **30**, 61–65 (2015)

Numerical and Experimental Tests on Explosive Material Detonation Effect on the Military Vehicle and Its Occupants

Grzegorz Sławiński(✉), Piotr Malesa, and Marek Świerczewski

Military University of Technology,
Gen. Witolda Urbanowicza 2, 00-908 Warsaw, Poland
grzegorz.slawinski@wat.edu.pl

Abstract. In the paper, an explosive material effect on behaviour of vehicle structure and its occupant through determining the maximum values of parameters deciding about injuries, i.e., forces acting on the abdomen, forces and bending moments in the neck, and other. The obtained results may contribute to select the best structural solutions of the vehicle as well as modifications influencing reduction of blast wave effects on the passenger. From the military vehicles designers point of view, it is vital to know a phenomena of an effect of a blast wave pressure impulse directly on the vehicle structure and indirectly on the soldiers inside the vehicle, which causes injuries posing a threat their lives and health. Numerical and experimental coupling tests are the best method to get knowledge on them. To obtain reliable results of the tests, validated and verified numerical models of the structure, human and physical processes characterizing the explosion and its effect on the structure and soldiers are necessary.

Keywords: Shock wave · Numerical simulations · Criterion of injuries · Spine

1 Introduction

Armed conflicts with the use of military equipment, ongoing all over the world for several decades, as well as resulting from them health and life hazards participating in them, impose a necessity of continuous searching for newer and newer structural solutions for armoured vehicles. Analysis of the events occurred on the battle field indicates that IEDs are presently a dominant means of destruction which at the same time pose the greatest threat. Hence, taking into consideration the threats coming from them is the main stream in the tests on development of effective protective system improving the safety of soldiers in vehicles.

As the statistics obtained based on analyses of documentation of the events occurred in Iraq and Afghanistan show, detonation of IEDs occurs the most frequently through direct running onto them with a wheel or, in the case of big

E. Tkacz et al. (Eds.): IBE 2018, AISC 925, pp. 241–249, 2019.
https://doi.org/10.1007/978-3-030-15472-1_26

charges located near the road, through their remote deployment. In the first case, the part most exposed to damage is a chassis and its subsystems, whereas in the other one, it is the vehicle side structure. In both cases, several biomechanical factors causing different kinds of injuries to a soldier in the vehicle are dealt with.

A basis for immediate injuries is frequently an effect of a blast wave coming from explosion of an explosive charge on the body of a soldier in the military vehicle. HFM-148 Protocol [1] shows a risk of soldier's body injuries resulting from specific loading mechanisms. The document determines criteria for assessment of injuries of soldier's individual parts of body and a methodology for tests for assessment of vehicles (and protection systems) affected by IED explosion [1].

The detonation of an explosive and its effects on a military vehicle [4–6] and its occupants [7–10] have been described in the scientific literature in several different ways for years.

1.1 Pathomorphology of Injuries Resulting from a Shock Wave

Injuries resulting from explosion on a soldier are of a complex character. Several types of injuries acts together posing a serious threat to soldier's life and health. Generally, injuries resulting from explosion may be dived into four categories:

- 1st category: original injuries: caused only by direct interaction of pressure on tissues;
- 2nd category: secondary injuries: caused by shrapnels;
- 3rd category: resulting from explosion of high energy resulting in throwing the body into air (threat of hitting something);
- 4th category: all other injuries (e.g. those related to structure degradation).

From a physical point of view, explosion acts on human tissues with a series of negative factors, including: flash, interaction of explosion gases, acoustic and shock wave and different kinds of radiations. However, up to 30–40% of explosion energy come to the surrounding environment in the form of a shock wave [2]. A shock wave generated as a result of explosive material detonation is characterized by a rapid increase in pressure in the forehead area. This pressure increases dynamically and remains for approximately 5 ms. This chase, called an overpressure (coupling) phase precedes an expansion phase lasting approximately 30 ms. Overpressure causes death if its value exceeds 240 kPa [3].

1.2 Effect of Pressure Wave Coming from Explosive Material Detonation on the Vehicle

During realization of numerical and experimental tests, an effect of selected factors on a risk of injuries in cervical spine and abdomen in dynamic loading conditions, with taking into consideration nonlinear material models of individual elements of a numerical model, was analyzed.

In order to determine the loads in anthropomorphic points of dummies, there were developed model assumptions allowing for development, in LS-Dyna system, models of vehicles and dummies used in field tests. The tests were carried out for various sizes of explosive material and for two variants of its location in respect to the research object, i.e.:

- under the vehicle (central explosion)—Fig. 1;
- next to the vehicle body—Fig. 2.

The developed numerical models of en effect of a blast wave on the vehicle and the crew were validated and verified.

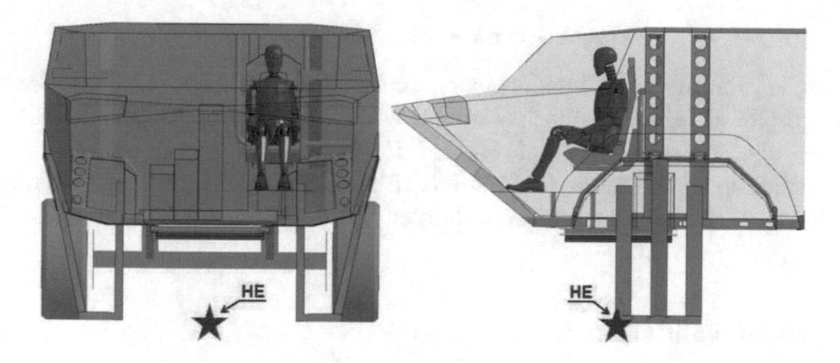

Fig. 1. A view of a numerical model of a vehicle with a visible location of Hybrid III dummy and with location of explosive material detonation under the vehicle.

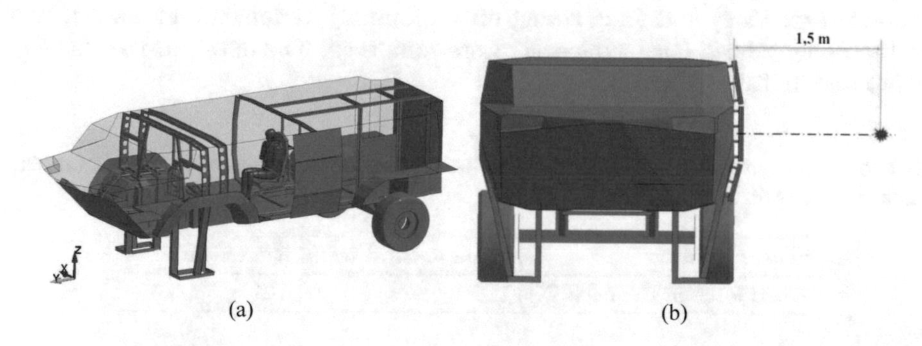

Fig. 2. A view of a numerical model of a vehicle with a visible location of ES-2re dummy and with location of explosive material (vehicle side).

2 Numerical Model

A numerical model of the analyzed object was prepared in such a manner that the most precise mapping of field tests conditions carried out on the real object was possible. Numerical analysis was divided into two stages. The first one lasted 100 ms and included stabilization of the model and placement of the dummy on the vehicle seat with taking into consideration the gravity. In the second stage, loading of the vehicle body with a blast wave was introduced. Conwep algorithm was used for simulation of TNT charge explosion. In this method, the value of shock wave pressure p incident on the surface at angle θ is calculated based on the following Eq. 1:

$$p = p_i(1 + \cos\theta - 2\cos^2\theta) + p_r \cdot \cos^2\theta \tag{1}$$

where p_i is pressure of an incident wave and p_r denotes reflected wave pressure.

A completed vehicle numerical model with a dummy placed inside was composed of, depending on variant Hybrid III or ES-2re, approximately 600,000 finite elements. In the numerical model, 16 material models from LS-Dyna base was used and 34 contact cards were defined. A friction coefficient for wheels and front supports was 0.6.

2.1 Model Validation

The developed numerical models were subjected to a validation process in which the values of the figures registered with sensors located on the dummies were compared. For this purpose, the results of experimental tests for a charge of 1.5 kg (central explosion) and a 2 kg charge of Semtex 1A (equivalent of 2.7 kg of TNT) during a side explosion were used. The values of axial force in the neck (central explosion) and force acting on the dummy abdomen and acceleration of the dummy head (side explosion) were compared. The obtained results were compared in Tables 1 and 2.

Table 1. Comparison of the maximum value of shear force in the upper section of the neck during central explosion.

Sensor location	Experimental tests	Numerical tests
Axial force in the neck [N]	114	310

The obtained considerable convergence of the results allows for statement that the model was properly prepared and conducting further numerical analyses for larger charges of explosive material will be close to reality.

Table 2. Comparison of the maximum values of forces and acceleration during side explosion.

Sensor location		Experimental tests	Numerical tests
Force in the abdomen [N]		25.42	26.14
Acceleration of the head [g]		3.26	3.88
Axial force in the neck [N]	F_{OZ+}	120.35	134.12
	F_{OZ-}	94.37	115.65

3 Results

The numerical analyses conducted in two variants allowed for knowledge of an explosion effect on the soldier in the military vehicle. Injuries in the upper cervical spine (for central explosion) and injuries of the head and abdomen (for side explosion) were analyzed. The results of numerical analyses, in the form of runs of axial forces and bending moments in the spine and forces acting on the abdomen area are presented in Figs. 3, 4, 5 and 6 and in Tables 3, 4, 5 and 6.

Table 3. Maximum values of shear force in the upper cervical spine—central explosion.

Mass of charge [kg]	X Force [kN]
1.5	0.339
6	2.061
8	3.062

Table 4. Maximum values of the bending moment in the upper cervical spine—central explosion.

Mass of charge [kg]	Mg [Nm]
1.5	25.3
6	210
8	255

In the case of a side explosion, a value of force in the dummy abdomen was analyzed. The obtained results in the form of a profile for the case of a 2 kg charge subjected to validation and a larger charge of 10 and 40 kg mass are presented in Fig. 5 and Table 5.

The read values of axial forces and profiles of changes in these forces over time are presented in Fig. 6.

As it was earlier expected, in all the analyzed parts of body, there was an increase in the maximum values of parameters related to traumatism criteria of

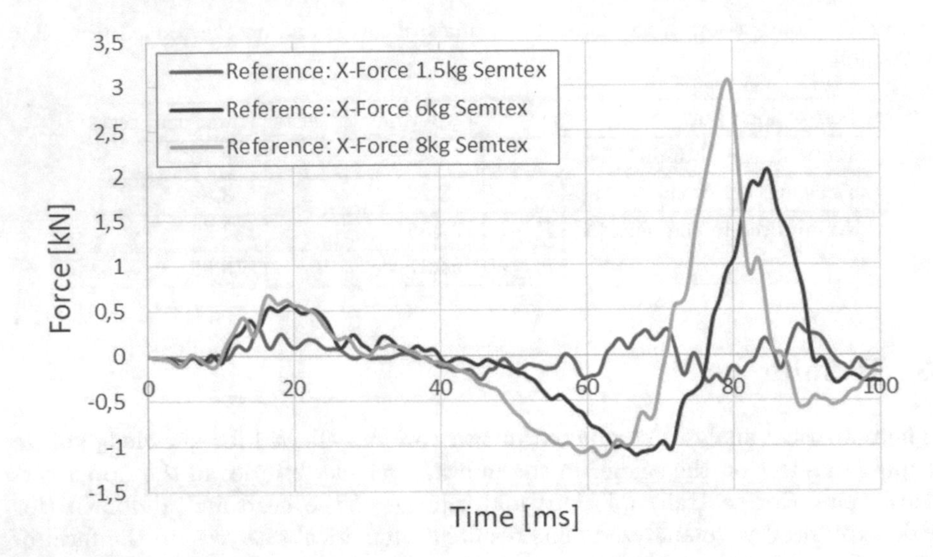

Fig. 3. A profile of changes in a shear force values in the upper cervical spine for various charges of explosive material—comparison of runs for central explosion.

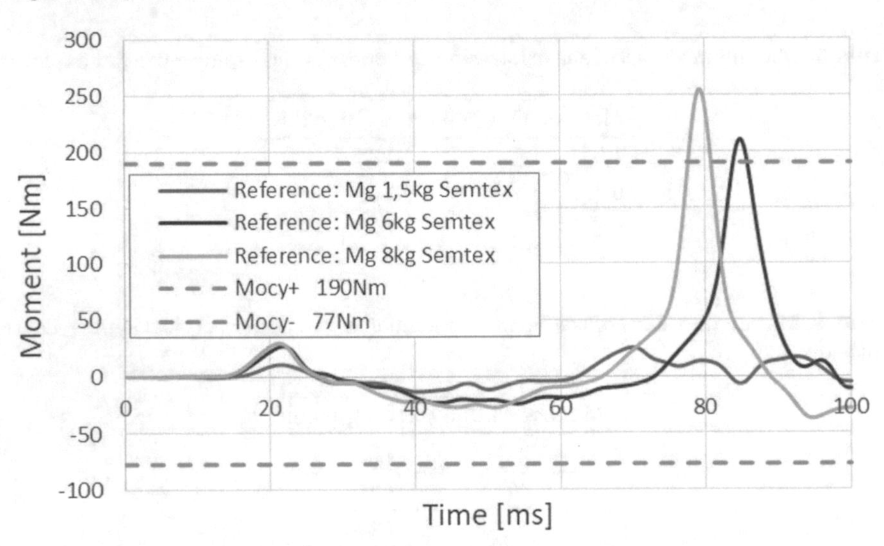

Fig. 4. A profile of changes in a bending moment in the upper cervical spine for various charges of explosive material—comparison of runs for central explosion.

Table 5. Maximum values of the force in the abdomen—side explosion.

Mass of charge [kg]	F [kN]
2	26.14
10	51.76
40	134.97

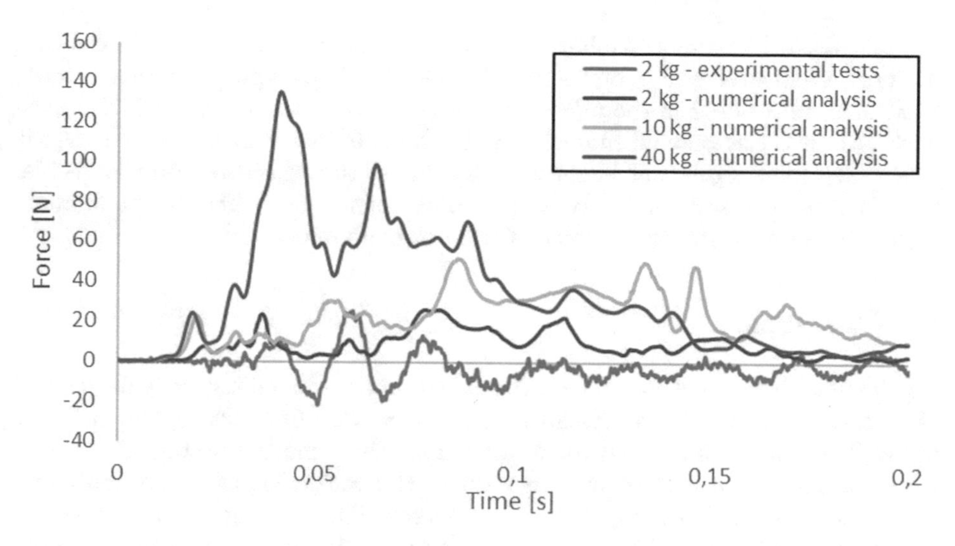

Fig. 5. A profile of changes in force in the dummy abdomen for various charges of explosive material—comparison of runs (side explosion).

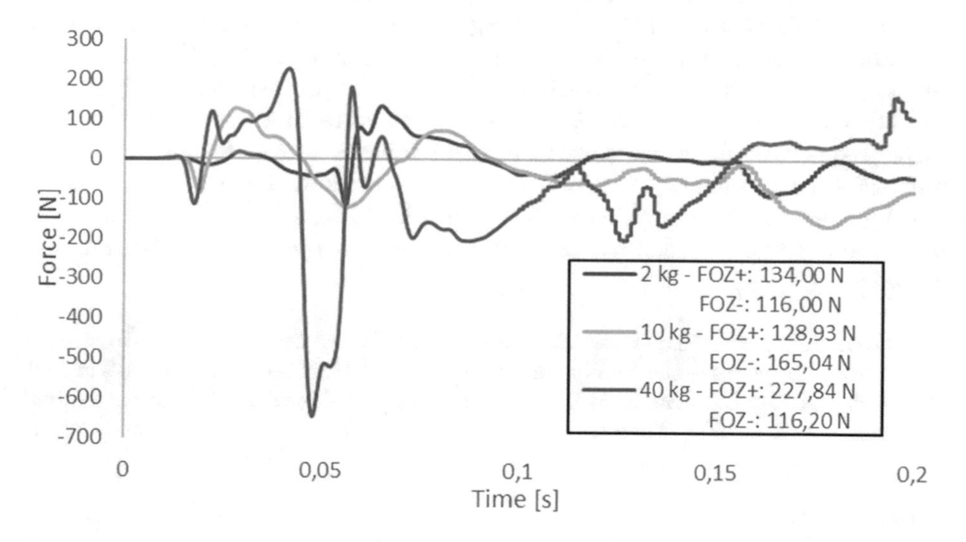

Fig. 6. A profile of changes in force in the dummy abdomen for various charges of explosive material—comparison of runs (side explosion).

Table 6. Values of the axial force in the neck–side explosion.

Mass of charge [kg]	F_{OZ+} [N]	F_{OZ-} [N]
2	134.00	116.00
10	128.93	165.04
40	227.84	116.20

the military vehicle crew member. This trend is observed in the case of explosion under the vehicle as well as on the vehicle side. Applying a pressure wave coming from charges resulted in the maximum force, at the level of 650–3000 N, in the neck and in the maximum moment, at the level of 255 Nm, in the neck. As it was expected, a significant increase in a value of the registered force is visible. Such behaviour results probably from an effect of the pelvis belt on the dummy, which pressure causes generation of force in the abdomen.

4 Conclusion

As a result of explosive material on the side or under the military vehicle detonation, a risk of crew's life and health hazard increases. Considering this incident for soldiers' safety is reduced to complex analysis of mutual interactions of the soldier's body, vehicle seat and elements of the vehicle structure. As an effect of a shock wave interaction, there occur interactions causing vibrations resulting from vibrations of the structure and acceleration of the passenger's body. Determination of a mechanism of soldiers' injuries due to an explosive charge explosion under the vehicle in which they travel is not easy. There is a series of factors contributing to injuries, including vehicle type, initial place of explosion, location of passenger-soldier's seat, inside geometry of the vehicle, unit movement, phase of vehicle body loading. All the mentioned aspects are significant in determination what mechanism, if any, contribute to soldier's injuries. The most frequent mechanisms of cervical spine injuries caused by propagation of a shock wave on passengers in military vehicles are axial compression, bending and hyperextension (connected with compression) and side bending. In the case of the above tests, a compression and a compression-bending mechanism was recorded.

Acknowledgements. The research was done within project no. DOBR-BIO4/022/13149/2013 'Improving the Safety and Protection of Soldiers on Missions Through Research and Development in Military Medical and Technical Areas' supported and co-financed by NCR&D, Poland.

References

1. RTO-TR-HFM-148: Test Methodology for Protection of Vehicle Occupants Against Anti-Vehicular Landmine and/or IED Effects, Technical Report, NATO Unclassified (2012)
2. Przekwas, A.: Model of lung injury and personnel protection from blast overpressures in confined areas, DARPA/DSO SBIR Phase I Final Report (2004)
3. Wildegger-Gaissmaier, A.: Aspects of thermobaric weaponry, ADF Health (2003)
4. Sławiński, G., Świerczewski, M.: Modelling and numerical analysis of explosion under the wheel of light armoured military vehicle. Eng. Trans. **65**(4), 587–599 (2017)

5. Tabatadaei, Z., Volz, J.: A comparison between three different blast method in LS-Dyna: LBE, MM-ALE, coupling of LBE and MMALE. In: 12th International LS-DYNA Users Conference, Detroit (2012). 3. LS-Dyna keyword manual, Vol. 2: Material models. Livermore, 2014
6. Baranowski, P., Malachowski, J., Janiszewski, J., Wekezer, J.: Detailed tyre FE modelling with multistage validation for dynamic analysis. Mater. Des. **96**, 68–79 (2016)
7. Sławiński, G., Świerczewski, M., Malesa, P.: Risk assessment regarding the injuries of the lower limbs of the driver of a military vehicle in the case of an explosion under the vehicle. In: Arkusz, K., Będziński, R., Klekiel, T., Piszczatowski, S. (eds.) Biomechanics in Medicine and Biology. BIOMECHANICS: Advances in Intelligent Systems and Computing, vol. 831, p. 2019. Springer, Cham (2018)
8. Sławiński, G., Malesa, P., Świerczewski, M.: Numerical analysis of the biomechanical factors of a soldier inside a vehicle with the pulse load resulting from a side explosion. In: Arkusz, K., Będziński, R., Klekiel, T., Piszczatowski, S. (eds.) Biomechanics in Medicine and Biology. BIOMECHANICS: Advances in Intelligent Systems and Computing, vol. 831, p. 2019. Springer, Cham (2018)
9. Klekiel, T., Sławiński, G., Będziński, R.: Analysis of the lower limb model response under impact load. In: Arkusz, K., Będziński, R., Klekiel, T., Piszczatowski, S. (eds.) Biomechanics in Medicine and Biology. BIOMECHANICS: Advances in Intelligent Systems and Computing, vol. 831, p. 2019. Springer, Cham (2018)
10. Mackiewicz, A., Będziński, R., Sławiński, G., Niezgoda, T.: Numerical analysis of the risk of neck injuries caused by IED explosion under the vehicle in military environments. Acta Mech. et Automatica **10**(4), 258–264 (2016)

Engineering of Biomaterials

The Influence of Stretch Range on the Hyperelastic Material Model Parameters for Pig's Skin with Consideration of Specimen Taken Direction

Sylwia Łagan(✉) and Aneta Liber-Kneć

Cracow University of Technology, Warszawska 24, 31-155 Cracow, Poland
slagan@mech.pk.edu.pl
http://pk.edu.pl

Abstract. The aim of this work was an analysis of hyperelastic material models to predict the behavior of skin tissue. The most popular Mooney-Rivlin, Humprey, Veronda-Westmann, Yeoh and Ogden models were analized. The parallel and perpendicular to the pig's spinal directions of specimens taken were consider in the tests. The input data to the simulation were defined for parallel direction as 5, 15, 25%, for perpendicular as 10, 20, 30, 40, 50% and also as total range of engineering stretch. The results were used to prediction of mechanical behavior and comparison with experimental and literature data. The strong influence of input data range on the values of model parameters was observed.

Keywords: Skin tissue · Non-linear mechanical behavior · Correlation of fitting material models · Tensile tests

1 Introduction

The interest in simulation of mechanical behavior of soft tissue much increased in the last years. Many different aspects becomes topics of the investigations e.g.: kind of soft tissue (skin, tendon, muscle, liver, aorta), direction of samples taken (longitudinal, transversal), type of mechanical loads (tensile, shear, static, dynamic), kind of animal model (swine, bovine) and range of input data [12,17]. To identify the mechanical response occurring in biological materials during loading, experiments often are realized with the use of a pig's tissue as the model of human one [5,8,15,16]. The study of the mechanical properties of skin (ultimate tensile stress, Young modulus, extensibility or stress relaxation) as well as accurate predictions of its mechanical behavior are crucial in surgical processes simulation [1,4,6,13]. The mathematical description behavior of soft tissue proposed the formulas used for like–rubber material. The aim of this study was estimation of influence of input data range (stress–stretch ratio) on the values of parameters for five hyperelastic material models used to the description of

E. Tkacz et al. (Eds.): IBE 2018, AISC 925, pp. 253–260, 2019.
https://doi.org/10.1007/978-3-030-15472-1_27

pig's skin tissue mechanical behavior. The Mooney–Rivlin, Humprey, Veronda–Westmann, Yeoh and Ogden hyperelastic multiparameters material models were used in this work. The localization (middle of back) and two directions to the long body axis of taken specimens (parallel and perpendicular) and different ranges of stretch were considered.

2 Material and Methodology

2.1 Test Data

To fit constitutive models to experimental data, the test data from the uniaxial tensile test were used. The sample preparation and procedure of mechanical test as well as the procedure of the strength parameters evaluation were the object of earlier researches, presented in [9]. In the uniaxial tensile test, five specimens (n = 5) for every direction were used. The force–elongation curves were obtained and the stress–stretch curves were determined. Next, the average stress–stretch curve was calculated. In the non–linear simulations the experimental average stress–strain curves for parallel and perpendicular direction were used. The input data to the calculation of hyperelastic material behavior were defined for parallel and perpendicular direction of specimens taken respectively as 5, 15, 25% and 10, 20, 30, 40, 50% and also the total range of stretch. The results were used to the prediction of mechanical behavior and compared with experimental and literature data.

2.2 Hyperelastic Material Model Characteristic

A hyperelastic material model is type of constitutive model in which the stress–strain relationship is based on a strain energy density function. Depending on the class (polymers, metals or natural materials) and type of material (iso–, aniso– or orthotropic), strain energy function can be expressed in different ways. In this study, in order to describe the non–linear tensile stress–strain behavior of skin tissue the Mooney–Rivlin, Humprey, Veronda–Westmann, Yeoh and Ogden material models were used. To simplify modeling, biological tissue material can be regarded as isotropic, incompressible and also often modeled via hyperelastic idealization [10,12]. The hyperelastic material models can be classified as phenomenological description of observed behavior or as mechanistic models deriving from arguments about underlying structure of the material or as hybrids of phenomenological and mechanistic models. The strain–energy deformation (W_{iso}), assuming isotropy, can be described as depended on the strain invariants of the deformation tensor of Cauchy–Green I_1, I_2, I_3 defined as (1):

$$W_{iso} = W(I_1, I_2, I_3) \tag{1}$$

Assuming the incompressible material and $\sigma_1 = \sigma_2$ for uniaxial tensile the strain invariants can be simplified and describe as given in formula (2):

$$I_1 = \lambda^2 + 2\frac{1}{\lambda},\ I_2 = \lambda^2 + 2\frac{1}{\lambda^2},\ I_3 = 1 \tag{2}$$

The mathematical formulas of hyperelastic material models used in this work were presented in Table 1. In order to obtain approximation of experimental data with mathematical record, the fitting procedures with the use of software OriginPro7.5 were made. To determine the values of C_i, algorithm of Levenberga–Marquardta was used. Fitting characteristics, the coefficient of determination R^2 (3) and the model parameters C_i were recorded for all defined ranges of stretch, and in the next step were used in the prediction of hyperelastic behavior via extrapolation procedure.

$$R^2 = \frac{\sum_{i=1}^{n} (\hat{y}_i - \bar{y}_i)^2}{\sum_{i=1}^{n} (y_i - \bar{y}_i)^2} \tag{3}$$

where y_i is the real value of the experimental stress, $\hat{y}_i$ is the theoretical value of the stress on the basis on models, $\bar{y}_i$ is the arithmetic mean value of the experimental stress.

Table 1. Utilized stress-stretch ratio formulas

Material model	Incompressible strain energy function
Mooney-Rivlin	$\sigma = 2(\lambda^2 - \frac{1}{\lambda})(C_1 - C_2)$
Humphrey	$\sigma = 2(\lambda^2 - \frac{1}{\lambda})C_1 C_2 \exp(C_2(\lambda^2 + \frac{2}{\lambda} - 3))$
Veronda-Westmann	$\sigma = 2(\lambda^2 - \frac{1}{\lambda})C_1 C_2(\exp(C_2(\lambda^2 + \frac{2}{\lambda} - 3)) - \frac{1}{2\lambda})$
Yeoh	$\sigma = 2(\lambda^2 - \frac{1}{\lambda})(C_1 + 2C_2(\lambda^2 + \frac{2}{\lambda} - 3) + 3C_3(\lambda^2 + \frac{2}{\lambda} - 3)^2)$
Ogden	$\sigma = 2\frac{\mu}{\alpha}(\lambda^{\alpha-1} - \lambda^{-\frac{\alpha}{2}})$

3 Results and Discussion

All results obtained in modelling procedures, as well as the values of the coefficient of determination R^2 for parallel direction of specimens taken were shown in Table 2 and respectively for perpendicular direction in Table 3. The results of the fitting were presented for parallel/perpendicular direction and different ranges of input stretch data. The results showed that for considered directions of specimens taken, different values of range of stretch ratio used as input data in calculation hyperelastic material models are optimal. The parallel oriented specimens, in comparison to perpendicular ones, have lower range of stretch ratio with good fitting correlation to the experimental uniaxial tensile curves. This knowledge is useful for FEM analysis in estimation of skin tissue strength, also for simulation of surgical procedures of skin, with the use of hyperelastic material models (e.g. Mooney–Rivlin, Yeoh or Ogden which are implemented in ANSYS). The Mooney–Rivlin, Humprey, Ogden and Veronda–Westmann (perpendicular) model parameters showed similar trend. The C_1 parameter values increased and C_2 decreased, at the optimal range of stretch ratio. The values of parameters of the Yeoh model showed the reverse trend, the C_1 and C_3 decreased, the C_2 parameter increased, in selected stretch ranges for both analyzed directions of specimens taken. The optimal stretch ranges for applicated hyperelastic material models are 1.25 for parallel and 1.4 for perpendicular direction of specimens taken. On the other hand the correlation of fitting models for total range of experimental data showed uncompatibility as we can see in Fig. 1d and f, but high scope of value of R^2, 0.9703–0.99938 (parallel) and 0.94509–0.99395 (perpendicular) (Tables 2 and 3).

Table 2. Parameters of hyperelastic material models and R^2 for parallel direction

Material model		Stretch ratio ranges			
		1.05	1.15	1.25	Total
Mooney-Rivlin	C_1	4.02	23.32	28.30	22.83
	C_2	−3.75	−24.15	−29.64	−23.21
	R^2	0.85095	0.99069	0.99849	0.99439
Humprey	C_1	0.01	0.03	0.21	0.91
	C_2	48.24	23.34	7.29	2.80
	R^2	0.85433	0.98987	0.97991	0.9703
Veronda-Westmann	C_1	0.47	0.06	0.58	1.80
	C_2	1.76	17.21	5.14	1.93
	R^2	0.82857	0.99305	0.98286	0.96945
Yeoh	C_1	0.35	0.18	0.09	0.40
	C_2	−0.54	19.56	22.61	18.73
	C_3	812.95	−10.97	−36.97	−24.89
	R^2	0.85525	0.99882	0.99988	0.99938
Ogden	μ	0.60	0.86	2.29	4.27
	α	27.38	26.70	15.53	10.08
	R^2	0.85316	0.99479	0.98689	0.97682

According to Flynn et al. [3] the Ogden model parameters were: $\mu = 0.0096$–0.0398 and $\alpha = 33.45$–35.99 for human skin from posterior upper arm and anterior upper forearm. Remache et al. [15] for specimen of pig's skin taken from side of animal in parallel direction to the long axis of the body and 1.25 stretch ratio, reported Ogden model parameters as $\mu = 0.69$ and $\alpha = 23.27$ (in this work: $\mu = 2.29$ and $\alpha = 15.52$ for parallel direction and $\mu = 0.31$ and $\alpha = 23.41$ for perpendicular direction and $1 < \lambda < 1.2$) for porcine back skin, and model Yeoh parameters as $C_1 = 0.26$, $C_2 = 15.5$ and $C_3 = 1.75$ (in this work: $C_1 = 0.09$, $C_2 = 22.61$ and $C_3 = -36.97$ for parallel direction and $C_1 = 0.09$, $C_2 = 3.90$ and $C_3 = 9.68$ for perpendicular direction and $1 < \lambda < 1.2$). Chanda et al. [2] referred results for the human skin surrogate and the Yeoh model and showed the values of $C_1 = 0.95$, $C_2 = 4.94$, $C_3 = 0.01$. Ottenio et al. [14] observed

Table 3. Parameters of hyperelastic material models and R^2 for perpendicular direction

Material model		Stretch ranges					
		1.1	1.2	1.3	1.4	1.5	Total
Mooney-Rivlin	C_1	2.79	9.48	14.20	13.07	9.86	7.36
	C_2	−2.76	−10.04	−15.42	−14.04	−10.00	−6.71
	R^2	0.8472	0.97943	0.99226	0.9957	0.98622	0.97758
Humprey	C_1	0.004	0.01	0.09	0.40	1.39	3.60
	C_2	32.73	15.84	6.65	2.76	1.18	0.51
	R^2	0.87029	0.99291	0.98179	0.96724	0.95832	0.95675
Veronda-Westmann	C_1	0.01	0.03	0.22	1.99	4.30	9.92
	C_2	20.96	12.18	4.95	1.94	0.86	0.49
	R^2	0.86902	0.99552	0.98591	0.96851	0.95279	0.94509
Yeoh	C_1	0.17	0.09	−0.10	−0.18	0.12	0.58
	C_2	−1.42	3.90	7.10	8.03	6.31	4.43
	C_3	101.54	9.68	−4.47	−7.12	−4.65	−2.27
	R^2	0.87506	0.99841	0.99912	0.9996	0.99785	0.99395
Ogden	μ	0.21	0.31	0.78	1.72	2.85	3.78
	α	27.77	23.41	15.86	10.53	7.45	5.84
	R^2	0.86622	0.99678	0.98939	0.97586	0.96486	0.96022

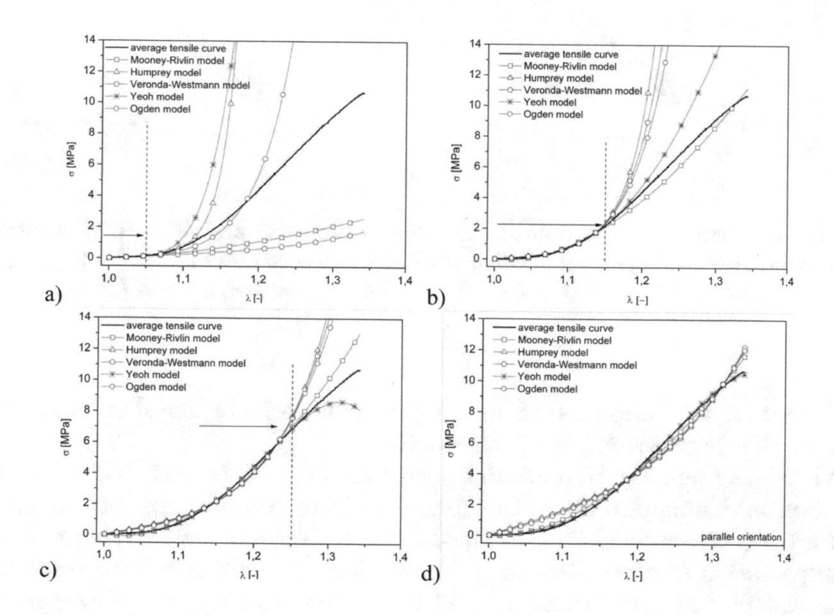

Fig. 1. The average experimental stress–stretch ratio curve and fitting curves extrapolated to total range of λ, for parallel direction of specimens taken for different input data: (a) $1 < \lambda < 1.05$, (b) $1 < \lambda < 1.15$, (c) $1 < \lambda < 1.25$ and (d) total stretch ratio

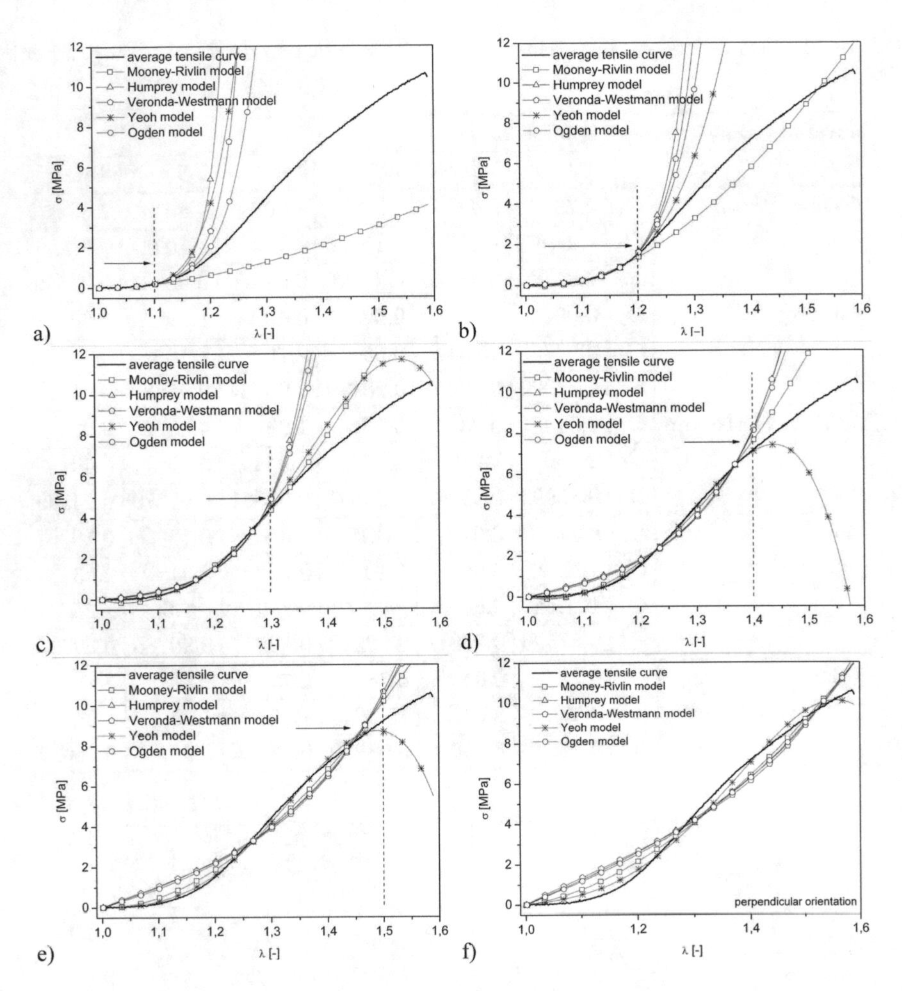

Fig. 2. The average experimental stress–stretch ratio curve and fitting curves extrapolated to total range of λ, for perpendicular direction of specimens taken: (a) $1 < \lambda < 1.1$, (b) $1 < \lambda < 1.2$, (c) $1 < \lambda < 1.3$, (d) $1 < \lambda < 1.4$, (e) $1 < \lambda < 1.5$ and (f) total stretch range

the higher stretch ratio at UTS for perpendicular orientation of specimen than parallel, that is consistent with our results.

There are a number of limitations associated with this work. The analysis is based on an assumption about the direction of animal body axis, but commonly known Langer lines should be identified for each animal model used in researches. An explanation in clear discussion of the effect of strain rates and orientation of the samples at low strains, as well as the influence the stretch ranges and individual factors of animal model, and skin anisotropy is not really possible due to the choice of the load sensor, not optimal for the force levels UTS [9–11]. The separation of the effect of the direction and the effect of the strain ranges in this loading region is difficult. In our studies, we observed a shortening of this toe–region for parallel direction of samples. The next limitation is that the sample size is low, but by eliminating the effects of inter and intra subject variability the data obtained can be used effectively [7]. In this study the average tensile curve was used for the calculation of hyperelastic material model parameters, in the next approach the hyperelastic material model parameters will be calculate for every specimen and than the average value (containing data from many tests) will be computed. Finally, more experimental data obtained from different methods of characterization, such as tensile, compression, shear, torsion or suction tests will be considered in the next studies. As well as the preconditioning procedure and pre–loading are very important to the mechanical parameters of soft tissues. Also dynamical loading and stress relaxation tests should be taken into account, to allow the proper identification of mechanical properties.

4 Conclusions

The research was focused towards comparison of the selected constitutive material models to sufficiently model behavior of skin tissue. An analysis of the stress–stretch curves was made under quasi–static uniaxial tensile tension for the middle back localization and perpendicular/parallel direction of skin specimen taken. The different stretch ranges used as input data showed strong influence on non-linear, hyperelastic nature of pig's skin. It is very important hint for application in the future analysis of skin tissue under different load conditions with the use of finite element method. This type of methods to predict and to understand the mechanical behavior of skin tissues are useful tools. The mathematical validation of the hyperelastic constitutive model was completed by means of comparisons between results from simulating experiment and experimental data. The comparison showed reasonably good agreement for some of the models and stretch range values. Future work will focus on development and validation of material model for the skin, implemented in FEM software.

Acknowledgements. The work was realized due to statutory activities M-1/12/2018/DS.

References

1. Benitez, J.M., Montans, F.J.: The mechanical behavior of skin: structures and models for the finite element analysis. Comput. Struct. **190**, 75–107 (2017)
2. Chanda, A., Graeter, R., Unnikrishnan, V.: Effect of blasts on subject-specific computational models of skin and bone sections at various locations on the human body. AIMS Mater. Sci. **2**(4), 425–447 (2015)
3. Flynn, C., Taberner, A., Nielsen, P.: Mechanical characterisation of in vivo human skin using a 3D force-sensitive micro-robot and finite element analysis. Biomech. Model. Mechanobiol. **10**(1), 27–38 (2011)

4. Groves, R., Coulman, S., Birchall, J., Evans, S.: An anisotropic, hyperelastic model for skin: experimental measurements, finite element modelling and identification of parameters for human and murine skin. J. Mech. Behav. Biomed. Mater. **18**, 167–180 (2013)
5. Joodaki, H., Panzer, M.B.: Skin mechanical properties and modeling: a review. Proc. Inst. Mech. Eng. Part H J. Eng. Med. **232**(4), 323–343 (2018)
6. Lapeer, R.J., Gassona, D., Karri, V.: Simulating plastic surgery: from human skin tensile tests, through hyperelastic finite element models to real-time haptics. Prog. Biophys. Mol. Biol. **103**(2–3), 208–216 (2010)
7. Liber-Kneć, A., Łagan, S.: Factors influencing on mechanical properties of porcine skin obtained in tensile test-preliminary studies. In: Advances in Intelligent Systems and Computing, vol. 623, pp. 255–262 (2018)
8. Lim, J., Hong, J., Chen, W.W., Weerasooriya, T.: Mechanical response of pig skin under dynamic tensile loading. Int. J. Impact Eng. **38**, 130–135 (2011)
9. Łagan, S., Liber-Kneć, A.: Application of the Ogden model to the tensile stress-strain behavior of the pig's. In: Advances in Intelligent Systems and Computing, vol. 526, pp. 145–152 (2017)
10. Łagan, S., Liber-Kneć, A.: Experimental testing and constitutive modeling of the mechanical properties of the swine skin tissue. Acta Bioeng. Biomech. **19**(2), 93–102 (2017)
11. Łagan, S., Liber-Kneć, A.: Influence of strain rates on the hyperelastic material models parameters of pig skin tissue. In: Advances in Intelligent Systems and Computing, vol. 623, pp. 279–287 (2018)
12. Martins, P., Jorge, R.N., Ferreira, A.: A comparative study of several material models for prediction of hyperelastic properties: application to silicone-rubber and soft tissues. Strain **42**, 135–147 (2006)
13. Ní Annaidh, A., Destrade, M., Gilchrist, M.D., Murphy, J.G.: Deficiencies in numerical models of anisotropic nonlinearly elastic materials. Biomech. Model. Mechanobiol. **12**, 781–791 (2013)
14. Ottenio, M., Tran, D., Ní Annaidh, A., Gilchrist, M.D., Bruyère, K.: Strain rate and anisotropy effects on the tensile failure characteristics of human skin. J. Mech. Behav. Biomed. Mater. **41**, 241–250 (2015)
15. Remache, D., Caliez, M., Gratton, M., Dos Santos, S.: The effects of cyclic tensile and stress-relaxation tests on porcine skin. J. Mech. Behav. Biomed. Mater. **77**, 242–249 (2018)
16. Shergold, O.A., Fleck, N.A., Radford, D.: The uniaxial stress versus strain response of pig skin and silicone rubber at low and high strain rates. Int. J. Impact Eng. **32**, 1384–1402 (2006)
17. Wex, C., Arndt, S., Stoll, A., Bruns, C., Kupriyanova, Y.: Isotropic incompressible hyperelastic models for modeling the mechanical behaviour of biological tissues: a review. BioMed Eng/Biomedizinische Technik **60**(6), 577–592 (2015)

Comparison of Mechanical Hysteresis for Chosen Soft Tissues Taken from a Domestic Pig

Aneta Liber-Kneć(✉) and Sylwia Łagan

Cracow University of Technology, Warszawska 24, 31-155 Cracow, Poland
aliber@pk.edu.pl
http://pk.edu.pl

Abstract. The aim of the study was to analyze and compare mechanical hysteresis and ability to energy accumulation by a porcine skin, tendon and aorta sample. The uniaxial tensile tests were carried out on tissues samples taken from a domestic pigs to compare mechanical behavior of tested tissues. Then, the mechanical hysteresis was tested in 5 cycles of loading–unloading. The values of energy dissipation, energy during loading and unloading, hysteresis and residual strain were calculated. The study revealed the differences in ability of energy accumulation by different tissues. Mechanical hysteresis experiments give the possibility to better understand soft tissues biomechanics and provide data for modeling and predicting tissues behavior under the cyclic loading conditions.

Keywords: Porcine model · Skin · Tendon · Aorta · Hysteresis · Energy dissipation · Tensile test

1 Introduction

The mechanical properties of soft tissues are very important in understanding their function and clinical evaluation of aging or choosing appropriate therapies for specific disease. Understanding mechanical response of soft tissue under different loads is an essential assessment for evolution of treatment. Most biological soft tissues exhibit non–linear response under loading and behave like viscoelastic or viscoplastic materials. When soft tissue is subjected to repetitive cyclic loading and unloading, the obtained stress–strain curves form hysteresis loops. These loops continually decrease over time of loading and then stabilize. Mechanical hysteresis is also associated with energy loss [2,11]. This phenomenon was referred by Fung as the preconditioning effect. Despite many studies concerning the preconditioning effect, changes that occur during it in soft tissues, e.g. tendons [9,10] and skin [5,7], are not well understood. An ability of soft tissues to store and efficiently return energy plays important role e.g. in tendons during locomotion. An analysis of energy changes during loading allows to asses tendon thermal damage and the amount of metabolic energy that can be saved during locomotion [1,6]. This study aims to capture the differences in soft tissues

E. Tkacz et al. (Eds.): IBE 2018, AISC 925, pp. 261–268, 2019.
https://doi.org/10.1007/978-3-030-15472-1_28

mechanical behavior under several loading–unloading cycles and the ability to energy accumulation of a porcine skin, tendon and aorta sample.

2 Material and Methodology

Soft tissue samples were taken from several 8–10–month old domestic pigs, weighting about 100 kg. Three types of soft tissues were taken. Skin samples were taken from the back of pig, parallel to the backbone. All samples had the same length of 100 mm and the width of 10 mm, however, these were of different thickness. The average thickness was 2.6 ± 0.2 mm. Tendon samples (tendo calcaneus communis) were extracted from legs of pigs. Their substitute cross–section has been adopted in the shape of a rectangle. The average cross–section area was 48.4 ± 8.1 mm^2. Samples of aorta were cut in the parallel direction. The average width of samples was 11.4 ± 1.2 mm and the thickness was 2.16 $\pm$ 0.23 mm. Samples were stored in the saline solution (0.9%) at the temperature of 4 °C no longer than 12 h (fresh) before the test. To characterize research material, the uniaxial static tensile test was determined with the use of the MTS Insight 50 testing machine. The samples were mounted using scissor action grips with self–tightening and they were extended at a speed of 5 mm/min at a room temperature of 22 $\pm$ 1 °C. The initial gauge length was 50 mm. Registered force elongation curves (force (F) – elongation (Δl)) were recalculated into stress (σ) – strain (ε) curves. The cyclic uniaxial extension test was carried out with the use of the MTS Insight 50 tensile machine, the conditions were the same as described above. Hysteresis was tested in 5 cycles of loading–unloading. The maximum load of each loading–unloading cycle was fixed at 5 N to compare different types of soft tissues under the same conditions. The choice of such force value provided a similiar peak stress value in each load cycle for tested tissues. The recovery time between cycles was 5 s. The initial gauge length was 50 mm for a skin and aorta samples and 30 mm for a tendon samples. Registered hysteresis loops were used to calculate the value of energy dissipation (the area of loop), total work performed on the specimen during stretching (energy loading – the surface area under loading curve), energy unloading (the surface area under unloading curve) and mechanical hysteresis (dissipated energy/total energy) in each loading cycle. Additionally, the residual strain R_s accumulation was calculated according to the following formula [7]:

$$R_s = \frac{R_d(c_n) - R_d(c_2)}{L_0} \tag{1}$$

where $R_d(c_n)$ is the residual deformation accumulated in any n cycle; $R_d(c_2)$ is the residual deformation accumulated in the second cycle, L_0 is the initial gauge length. For each test and tissue at least 5 samples were taken for results analysis. The determined values of parameters were shown as the average values with a standard deviation (X $\pm$ SD).

3 Results and Discussion

The exemplary stress–strain curves for pig's soft tissues samples were shown in Fig. 1. These curves showed typical for soft tissue non–linear characteristic with three phases [2,3]. In the initial loading phase, great deformations of the tissues occurred at a relatively low applied load. In the second phase of stretching, load was higher and the tissue became stiff at higher stresses. In the third phase, the ultimate tensile strength was reached and the fibres began to break leading to failure. Comparison of skin, tendon and aorta samples showed similar mechanical behavior under tension but different ability to load carrying, higher for tendon and skin, as well as different extensibility, the highest for aorta. The exemplary hysteresis loops were shown in Fig. 2 for skin, Fig. 3 for tendon and Fig. 4 for aorta. For all tested tissues, the first hysteresis loop had clearly higher surface area compared to the following loops. The hysteresis loops shifted right and became increasingly narrow as a function of the number of cycles. Between third and fifth cycle of loading and unloading hysteresis stabilized and the progressive superimposition of the curves of loading and unloading can be seen. According to Fung [2], by repeated cycling, the internal structure of the tissue changes and eventually a steady state is reached. Unless the upper or lower limits of the loading are not changed, no further change will occur.

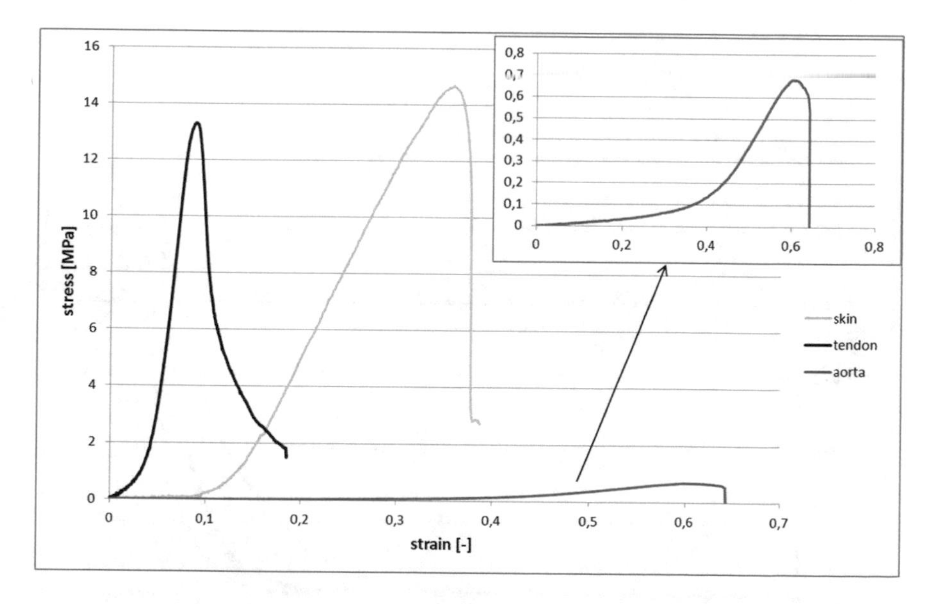

Fig. 1. The exemplary stress-strain curves for pig soft tissues

Selected value of peak load provided similiar values of peak stress in each cycle of loading. The average peak stress value was 0.2 MPa for skin, 0.1 MPa for tendon and 0.2 MPa for aorta. For all analysed tissues, during cyclic loading

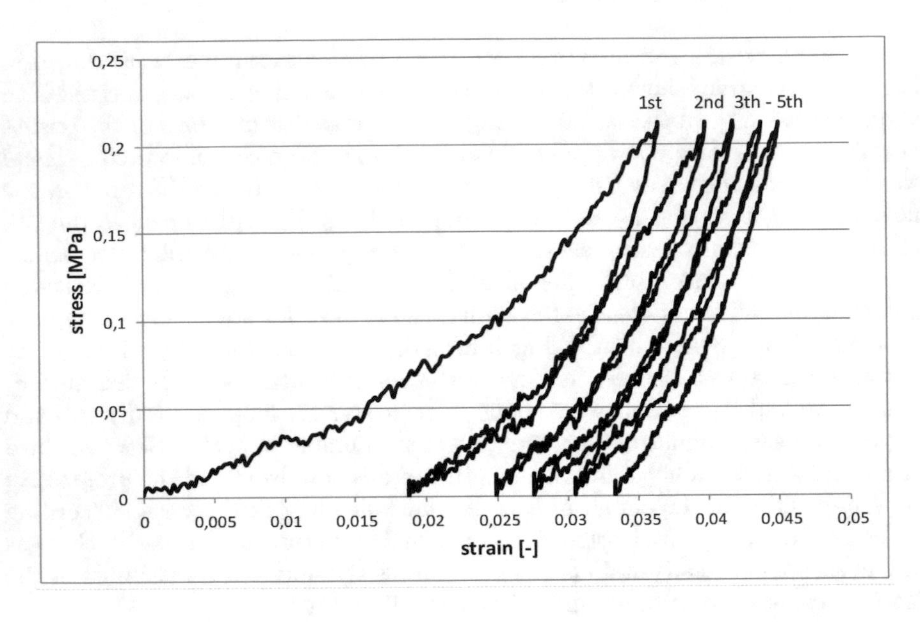

Fig. 2. The exemplary hysteresis loops for a porcine skin sample (load at 5 N)

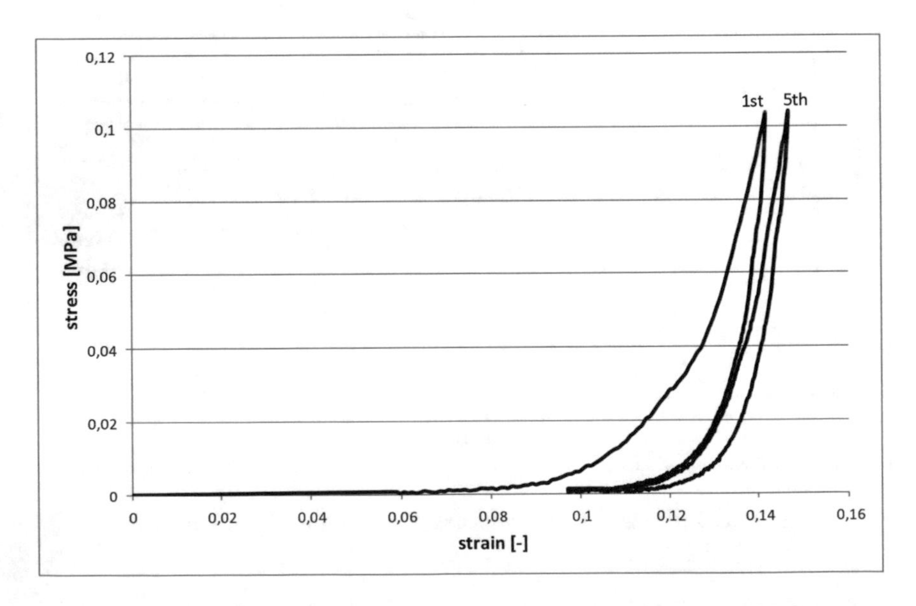

Fig. 3. The exemplary hysteresis loops for a porcine tendon sample (load at 5 N)

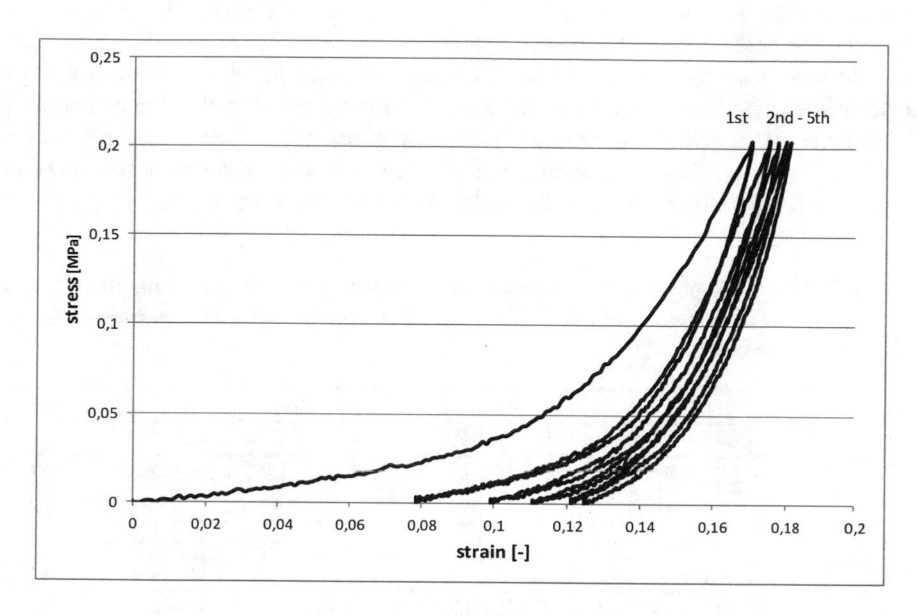

Fig. 4. The exemplary hysteresis loops for a porcine aorta sample (load at 5 N)

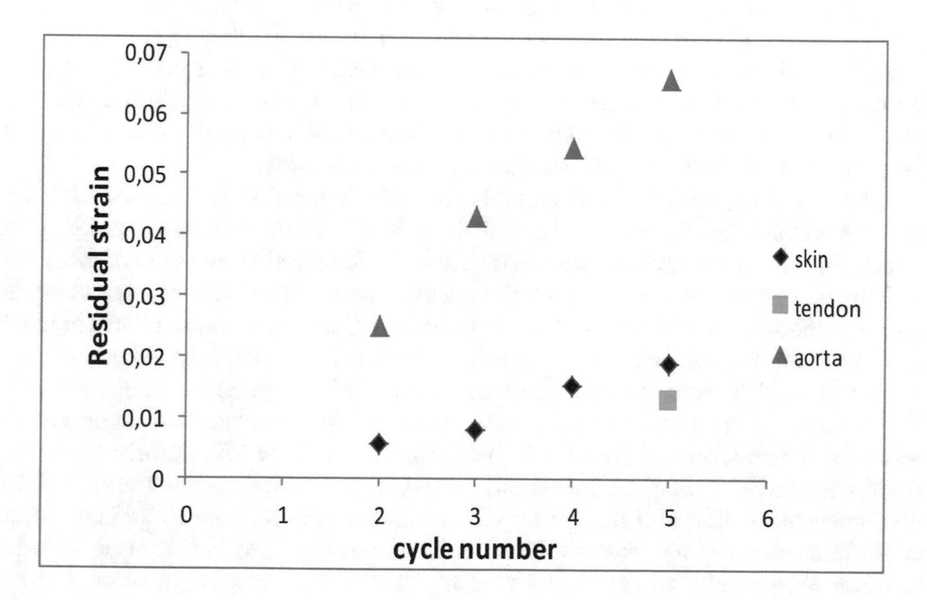

Fig. 5. The average residual strain accumulated by a porcine skin, tendon and aorta samples as a function of cycle number

and unloading hysteresis loops shifted to the right, revealing the presence of plastic component. In Fig. 5, the average residual strain accumulated by a porcine skin, tendon and aorta samples as a function of cycle number was shown. The accumulated residual strain was observed from the second cycle. The amount of nonrecoverable deformation was progressively greater with every loading cycles. The highest value of residual strain was observed for aorta sample in fifth loading cycle and it was about six time lower for skin and tendon sample.

Table 1. The average values of mechanical hysteresis for a porcine skin, tendon and aorta samples (*C. no–cycle number*, *EL–Energy loading* [J], *EU–Energy unloading* [J], *ED–Energy dissipation* [J])

C. no	Skin			Tendon			Aorta		
	EL	EU	ED	EL	EU	ED	EL	EU	ED
1	8.2 ± 0.6	2.8 ± 0.3	5.4 ± 0.8	2.3 ± 0.4	0.9 ± 0.2	1.4 ± 0.4	11.8 ± 1.4	6.5 ± 0.8	5.4 ± 0.7
2	4.3 ± 0.2	2.7 ± 0.1	1.6 ± 0.1	-	-	-	7.8 ± 0.8	5.7 ± 0.5	2.1 ± 0.4
3	3.8 ± 0.1	2.5 ± 0.1	1.3 ± 0.2	-	-	-	6.8 ± 0.5	5.1 ± 0.3	1.7 ± 0.2
4	3.7 ± 0.1	2.3 ± 0.1	1.4 ± 0.1	-	-	-	6.1 ± 0.4	4.7 ± 0.3	1.4 ± 0.1
5	3.2 ± 0.2	2.2 ± 0.1	0.9 ± 0.3	1.3 ± 0.2	0.8 ± 0.3	0.5 ± 0.1	5.6 ± 0.4	4.4 ± 0.4	1.3 ± 0.2

The values of the surface area of each registered loop and surface area under loading and unloading curves were calculated and shown as energies in Table 1. For all tested tissues, an increase of the number of loading cycles caused a decrease of the value of dissipated energy and mechanical hysteresis. The highest percentage reduction of the energy dissipation value between first and fifth cycle of loading was observed for skin samples (app. 83%), then for aorta samples (app. 75%) and the lowest for tendon samples (app. 64%).

The value of mechanical hysteresis was 28.9% for skin samples, 36.1% for tendon samples and 22.6 for aorta in fifth cycle of loading, in which stabilization of the dissipated energy was observed (Table 2). Relatively low value of mechanical hysteresis means that most of the elastic energy stored during stretching is returned once after the tensile load is removed. Calculated value of mechanical hysteresis for tendon was in the range of hysteresis reported by Maganaris et al. for isolated human tendons (between 3 and 38% in tensile testing in vitro). Shadwick et al. conducted the investigation on pig's tendons and showed the value of the mechanical hysteresis between 17.5 and 24.5%. Another investigation showed the value of hysteresis for pig's tendon between 22 and 32% in fifth cycle of loading and unloading but for higher values of load [4]. The differences in mechanical hysteresis values between ones reported in this study and in the literature can be observed due to many factors. The most significant factors are inter study methodological differences (tissue gripping, cross–sectional area), anatomical site, ageing, disuse or physical activity [6,8].

Table 2. The average values of mechanical hysteresis for a porcine skin, tendon and aorta samples

Cycle number	Mechanical hysteresis [%]		
	Skin	Tendon	Aorta
1	65.8 ± 6.0	61.0 ± 9.5	45.3±1.9
2	36.9 ± 0.8	-	26.7 ± 2.8
3	33.3 ± 3.8	-	24.9 ± 2.1
4	37.5 ± 1.3	-	22.5 ± 1.8
5	28.9 ± 8.3	36.1 ± 3.7	22.6 ± 3.9

4 Conclusions

The conducted tests have the character of preliminary assessment and comparision of mechanical hysteresis for selected soft tissues. All tested tissues showed substantial mechanical hysteresis under several loading and unloading cycles and a tendency to stabilize after some cycles if the upper load limit was kept constant. Along with the loading–unloading cycles, remarkably large value of energy was accumulated by a porcine skin and aorta samples and the lowest for tendon samples. The strain–stress curves continuously shifted to the right with the increase in the number of cycles, the amount of hysteresis was reduced and the material accumulated progressively inelastic strain over time during loading. The investigation of hysteresis showed the differences in energy loss between soft tissues under small load. For skin and tendon the peak load was about 1% of the maximum force, for aorta it was 15% but considering the area of cross–section the peak stress was similiar for all tissues. The future study will focus on higher forces matched to the physiological conditions for a given tissue because this property is especially useful to absorb the energies from high forces that may be experienced during catching or accidental impacts. Also fatigue tests should be considered but they are difficult and time–consuming to perform. Uniaxial fatigue tests are easier to perform but may not represent physiological loading conditions. However, the incorporation of constitutive models give the possibility to better understand soft tissues biomechanics and predict tissues behavior under the cyclic loading conditions.

Acknowledgements. The work was realized due to statutory activities M-1/12/2018/DS.

References

1. Finni, T., Peltonen, J., Stenroth, L., Cronin, N.J.: Viewpoint: on the hysteresis in the human Achilles tendon. J. Appl. Physiol. **114**, 515–517 (2013)
2. Fung, Y.C.: Biomechanics. Mechanical Properties of Living Tissues. Springer, New York (1993)

3. Groves, R.B., Coulman, S.A., Birchall, J.C., Evans, S.L.: An anisotropic, hyperelastic model for skin: experimental measurements, finite element modeling and identification of parameters for human and murine skin. J. Mech. Behav. Biomed. Mater. **18**, 167–180 (2013)
4. Liber-Kneć, A., Łagan, S.: Mechanical hysteresis tests for porcine tendons. Eng. Biomater. **20**(139), 26–30 (2017)
5. Liu, Z., Yeung, K.: The preconditioning and stress relaxation of skin tissue. J. Biomed. Pharm. Eng. **2**(1), 22–28 (2008)
6. Maganaris, C.N., Paul, J.P.: Hysteresis measurements in intact human tendon. J. Biomech. **33**, 1723–1727 (2000)
7. Remache, D., Calies, M., Gratton, M., Dos Santos, S.: The effects of cyclic tensile and stress-relaxation tests on porcine skin. J. Mech. Behav. Biomed. Mater. **77**, 242–249 (2008)
8. Shadwick, R.E.: Elastic energy storage in tendons: mechanical differences related to function and age. J. Appl. Physiol. **68**(3), 1033–1040 (1990)
9. Sopakayang, R.: A new viscoelastic model for preconditioning in ligaments and tendons. In: Proceedings of the World Congress on Engineering, WCE 2013, 3–5 July 2013, London, UK, vol. III (2013)
10. Teramoto, A., Luo, Z.P.: Temporary tendon strengthening by preconditioning. Clin. Biomech. **23**, 619–622 (2008)
11. Wiśniewska, A., Liber-Kneć, A.: Influence of a skin tissue anisotropy on mechanical hysteresis. Tech. Trans. Mech. **5–M**(15), 125–136 (2016)

Study of Operating Properties of Surgical Tools

Anita Kajzer(✉), Monika Lewczuk, Marcin Kaczmarek, and Wojciech Kajzer

Department of Biomaterials and Medical Devices Engineering,
Faculty of Biomedical Engineering, Silesian University of Technology,
ul. Roosevelta 40, 44-800 Zabrze, Poland
anita.kajzer@polsl.pl

Abstract. The aim of the work was to assess the operational damage of surgical instruments. As part of the research, the surface, structure, mechanical properties and corrosion resistance of the bone gouge forceps and haemorrhoidal forceps withdrawn from use after the intended period of operation were assessed. The test results showed numerous scratches, dents, cracks and chipping caused by operation and storage on the entire surface of the tested tools, which had a direct impact on the increase of the surface roughness. For the bone gouge forceps, non-metallic inclusions were classified as type D: thin globular oxides with an approximate thickness of up to 8 μm and the pattern number of 1.5, and for haemorrhoidal forceps as type D: thick globular oxides with a thickness of 12 μm and the pattern number of 1.5. Microscopic metallographic examinations showed in the working part of the analyzed tools (both the bone gouge forceps and haemorrhoidal forceps) the structure of low-tempered martensite with visible retained austenite, whereas in the gripping part, the martensite structure in the annealed state. The obtained results of the hardness test showed that in the case of the working part of the bone gouge forceps, the gripping and the working part of the hemorrhoid forceps, the values met the recommendations given in the standards. The test of droplet corrosion resistance showed that all surgical instruments selected for the test are corrosion resistant.

Keywords: Surgical tools · Structure · Mechanical properties · Corrosion resistance

1 Introduction

Surgical procedures are associated not only with the appropriate choice of the type of implant with surface properties that ensure biocompatibility, adapted to the addressed implantation time [1–6] but also surgical instruments [7,8]. The right choice of material determines the success of the surgery. In terms of construction and operation, surgical instruments should be characterized by: high reliability, safety of use and ease of use, resistance to corrosion in operating conditions, shape allowing for a specific surgical procedure, dismantlement

E. Tkacz et al. (Eds.): IBE 2018, AISC 925, pp. 269–277, 2019.
https://doi.org/10.1007/978-3-030-15472-1_29

allowing complete sterilization, ergonomics and aesthetics of construction as well as fabrication with a specific set of materials mechanical properties. The surgical instruments used in health care facilities, depending on the type of medical activity performed, are operated with varying intensity and frequency. As part of one surgical operation, some instruments are used throughout its duration, while others are used sporadically. Notwithstanding this, European standards require the use of sterile tools to remove and effectively eliminate microorganisms. A prerequisite for the safe use of reusable medical devices is to correctly complete all steps of the operation: cleaning, disinfection, packaging, sterilization, drying, storage and transport. Working in such different conditions means that the tools are worn or damaged in various areas. Before first use, surgical instruments should be autoclaved. Due to the possibility of damage to the blades, thin tips and other delicate elements of surgical instruments, special attention should be paid during their transport, cleaning, disinfection, maintenance, sterilization, packaging and storage. Contact of tools with the environment of tissues and body fluids, disinfectants and acids is the cause of corrosion changes on their surface. The place of tool storage also influences the possibility of any damage to the equipment. It should be a dry, clean and moisture-free place, and each tool should be stored in individual packaging. Examples of operational damage include: mechanical damage to the surface, discoloration, cracks, scratches, lack of smoothness of the tool, change of sharpness of the cutting edge, etc. All surgical instruments should be used according to their purpose. Any use of tools not in accordance with the manufacturer's instructions may cause damage to it or cause permanent damage to the material structure, which may lead to patient injury or death. Due to the lack of appropriate regulations, subjective assessment of the product is possible. An organoleptic method of quality assessment is acceptable when the use of objective evaluation is unprofitable or impossible [9,10]. All tools found to have any operational damage should be withdrawn [11–13].

2 Materials and Methods

Two surgical tools were selected for the study: a bone gouge forceps belonging to the group of cutting, double-edged, shaped tools and hemorrhoid forceps from the group of gripping tools with special purpose.

2.1 Roughness Measurement and Macroscopic Evaluation of the Surface Using a Stereoscopic Microscope

The surface roughness measurement by means of the contact method was made using a Taylor - Hobson profilometer. The place of roughness measurement on the analyzed surgical instruments was selected on the basis of observations with an unaided eye. The result of the test was the arithmetic mean of the roughness profiles Ra [μm] [14,15]. Five measurements were taken within the determined areas of the surface selected for testing. Due to the lack of normative recommendations regarding the surface roughness of tools selected for testing, the data contained in the following standards were used to compare the obtained results: PN–87/Z–5402 [16] and PN–88/Z-54123 [17]. The macroscopic evaluation of the surface was carried out using a stereoscopic microscope - SteREO Discovery.V8 by Zeiss with AxioVision software, with a total magnification of 10x.

2.2 Metallographic Microscopic Examinations

In order to reveal non-metallic inclusions and the actual structure of the material, metallographic specimens from the selected areas of surgical instruments were performed. The ZEISS AxioObserver Z1m light microscope with the AxioVision software was used for the study. The metallographic preparation steps included cutting, mounting, grinding and polishing, and etching in a reagent of 100 cm^3 of ethanol + 3 g $FeCl_3$ + 1.5 cm^3 HCl. The assessment of non-metallic inclusions was carried out using the comparative method with the standards included in the ISO 4967-1979 (E) standard [18]. The grain structure and size were assessed at 100x and 500x magnifications. To determine the size of the former austenite grains, the Jeffries method was used in accordance with the recommendations of PN-EN ISO 643 [19].

2.3 Hardness Measurement

The hardness measurement in particular areas of the selected tools was carried out using the Vickers method on the Struers DuraScan microhardness tester in accordance with the recommendations of the PN-EN ISO 6507–1–2007 standard. For each of the instrumentariums selected for the tests, three measurements were carried out with a load of 9.81 N. The hardness of the heat–treated clamping arms of haemorrhoidal forceps should be in the range 35–56 HRC in accordance with the PN-88 Z-54030 standard [20]. According to the PN–74/Z–54122 standard, the hardness of the heat-treated cutting parts should be in the range 50–54 HRC [21].

2.4 Corrosion Resistance Test

The corrosion resistance test was carried out by a droplet method. It consisted in applying a few drops of a solution composed of 1616 g of distilled water, 57 g of sulfuric acid, 142 g of copper sulfate pentahydrate on the degreased surface of the tested tools in three arbitrarily selected points. After 10 min, the solution was removed. Next, surface observations were carried. Evidence of corrosion or a presence of red tarnish of copper indicate lack of corrosion resistance. After the test, the surface was evaluated macroscopically.

3 Results and Discussion

3.1 Results of Roughness Measurement and Macroscopic Evaluation of the Surface Using a Stereoscopic Microscope

The macroscopic examinations of the surfaces selected for testing revealed a diverse number of scratches, dents and deep scuffs (Figs. 1 and 2). To identify the surface area of the examined bone gouge forceps, the nomenclature of pages A and B was assumed, and for the identification of the surface area of the examined hemorrhoid forceps, the nomenclature of pages C and D was assumed (Table 1, Fig. 3).

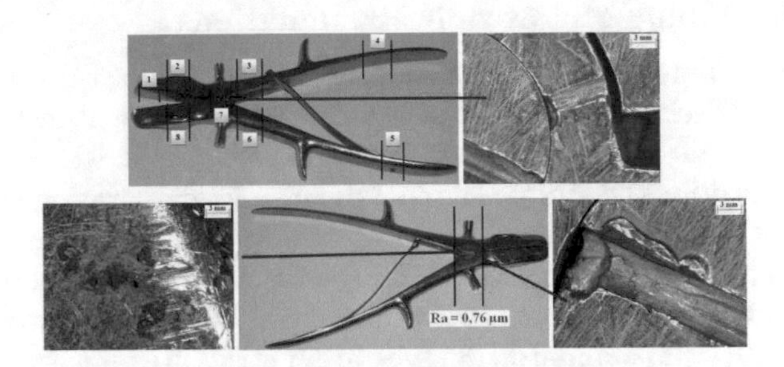

Fig. 1. Exemplary mechanical damage of the surface of the bone gouge forceps near the lock – stereoscopic microscope, magnification 10x

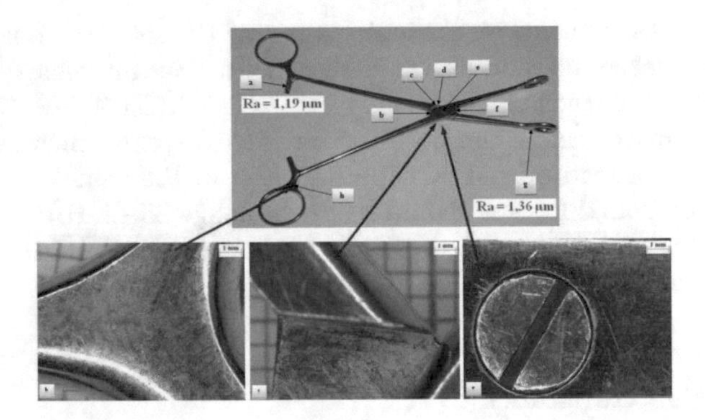

Fig. 2. Haemorrhoidal forceps with exemplary surface damage - stereoscopic microscope: (c–h) magnification 10x

On the basis of the obtained surface roughness results, it can be stated that the Ra parameter in the grip part of the bone gouge forceps exceeds the recommendations given in the standard. The roughness value measured in the working part of haemorrhoidal forceps is Ra = 1.36 µm and exceeds the recommendations of the standard more than twice. The increase in the Ra parameter can be related to mechanical damage to the tool surfaces that arise during their use, sterilization and storage.

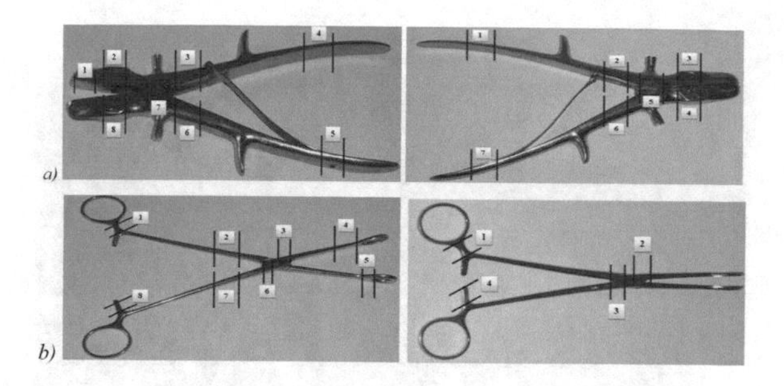

Fig. 3. Areas selected for measuring roughness: (a) bone gouge forceps, (b) haemorrhoidal forceps

Table 1. Results of roughness measurements for: bone gouge forceps - side A and B, haemorrhoidal forceps - side C and D

Material	Ra for 1 [μm]	Ra for 2 [μm]	Ra for 3 [μm]	Ra for 4 [μm]	Ra for 5 [μm]	Ra for 6 [μm]	Ra for 7 [μm]
Bone gouge forceps side A	1.01 ±0.09	0.57 ±0.07	1.56 ±0.17	0.70 ±0.03	0.43 ±0.07	1.14 ±0.09	0.59 ±0.13
Bone gouge forceps side B	1.34 ±0.11	1.40 ±0.16	0.48 ±0.04	0.65 ±0.07	0.76 ±0.13	1.17 ±0.14	1.22 ±0.11
Haemorrhoidal forceps side C	1.19 ±0.13	0.24 ±0.07	0.20 ±0.04	0.31 ±0.04	1.36 ±0.15	0.16 ±0.01	0.40 ±0.09
Haemorrhoidal forceps side D	1.19 ±0.13	0.13 ±0.04	0.18 ±0.03	0.34 ±0.08	-	-	-

3.2 Results of Roughness Measurement and Macroscopic Evaluation of the Surface Using a Stereoscopic Microscope

For the bone gouge forceps, non-metallic inclusions were classified as type D: thin globular oxides with a thickness of up to 8 μm and a standard pattern of 1.5. For the haemorrhoidal forceps, the presence of non-metallic inclusions was classified as type D: thick globular oxides with a thickness of up to 12 μm and a standard pattern of 1.5. Structure of both tools was low tempered martensite with visible retained austenite (Figs. 4 and 5).

3.3 Results of Hardness Measurement

The results of hardness measurements are presented in Table 2. The mean hardness values for the bone gouge forceps are equal to: 224 HV1 for the gripping part and 227 HV1 in the area of the connector (between gripping and working part). However, in the area of the working part associated with higher loads and greater abrasion resistance, the hardness was equal to 578 HV1, which is roughly 53.9 HRC on the Rockwell scale. The value of hardness in the grip area of the haemorrhoidal forceps was 441 HV1, which in comparison to Rockwell hardness is 44.9 HRC. In the working part, the hardness was equal to 51.4 HRC.

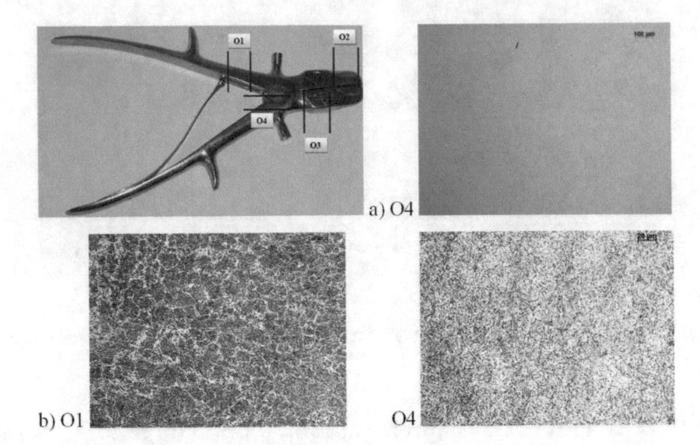

Fig. 4. Bone gouge forceps: (a) non–metallic inclusions from selected areas of the surface, magnification 100x (b) O1 – martensite structure in the annealed state, O2 – low-tempered martensite structure, magnification 500x

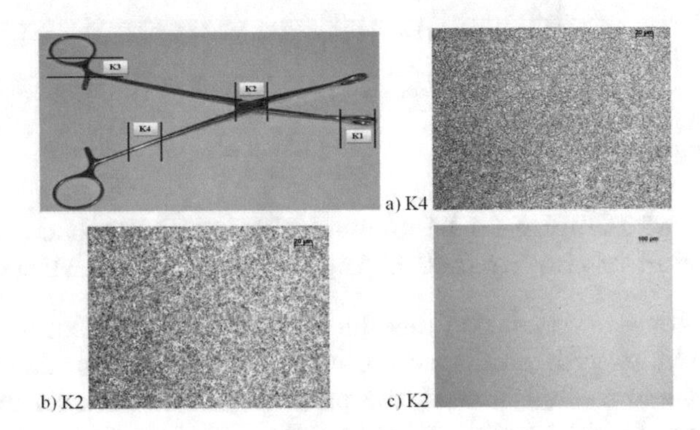

Fig. 5. Haemorrhoidal forceps: (a, b) structure of low-tempered martensite, magnification 500x, (c) non-metallic inclusions from selected areas of the surface, magnification 100x

Table 2. Hardness test results for the bone gouge forceps and haemorrhoidal forceps

Side of the tested area		$HV1_{av}$	HRC_{av}
Bone gouge forceps	Gripping part	227 ± 4	-
	Working part	528 ± 7	53.9
Haemorrhoidal forceps	Gripping part	441 ± 5	44.9
	Working part	534 ± 9	51.4

3.4 Results of Corrosion Resistance Test

On the basis of the corrosion tests, it was found that the both bone gouge forceps and haemorrhoidal forceps are resistant to corrosion. No pits, discolorations and other changes indicating initiation of the corrosion process were found on the analyzed surfaces.

4 Conclusion

Conditions of operating, cleaning, sterilization and storage cause that tools are used with varying intensity and frequency. Increasing the surface roughness, exceeding the acceptable values of the Ra parameter, is associated with numerous mechanical damage to the tool surfaces in the form of scratches, dents, cracks and chipping, both in the working and gripping parts. On the basis of normative requirements, the roughness of the tool (Ra) should not exceed: for external shiny surfaces −0.2 μm, external matte surfaces −0.63 μm, and the surface of the lock, grooves, notches on jaws of 2.5 μm. However, the observed damage did not affect the corrosion resistance. After the test, no changes in the form of spots or pits on the surface were found. Based on the obtained test results, it was found that in terms of the number and size of non-metallic inclusions, as well as the structure and size of the grain, the material from which the tools were produced complied with the recommendations of the manufacturers. For the bone gouge forceps, non-metallic inclusions were classified as type D: thin globular oxides with a thickness of up to 8 μm and a standard pattern of 1.5. For haemorrhoidal forceps they were classified as type D: thick globular oxides with a thickness of up to 12 μm and a standard pattern of 1.5. Observations of metallographic specimens revealed the presence of a low-tempered martensite structure with visible retained austenite with grain size reference number G = 10 for the bone gouge forceps and G = 9 for haemorrhoidal forceps (Figs. 3 and 4). It was also observed that in the gripping part and the area of the lock, the hardness was 224 HV1 and 227HV1, which indicates the use of martensitic steel in the annealed state. However, in the working part associated with higher loads and higher abrasion resistance, the hardness was 578 HV1, which is equal to 53.9 HRC on the Rockwell scale. The obtained value indicates that quenching and tempering were used in the working part of this tool to obtain a martensitic structure exhibiting greater hardness. According to the PN–74/Z–54122 standard, the hardness of heat treated cutting parts should be in the range 50–54 HRC [21]. On this basis, it was found that the hardness was within the limits required by the standard. In turn, in the gripping part of the haemorrhoidal forceps, the hardness was 44.9 HRC, and for the working part it was equal to 51.4 HRC. According to the PN–88 Z–54030 standard, the hardness of clamping arms after heat treatment should be in the range 35–56 HRC [20]. It can therefore be concluded that the results obtained fall within the given range.

Acknowledgements. The work has been financed from research project no $BK - 210/RIB2/201807/020/BK\_18/0028$.

References

1. Kajzer, A., Kajzer, W., Gołombek, K., Knol, M., Dzielicki, J., Walke, W.: Corrosion resistance, eis and wettability of the implants made of 316 LVM steel used in chest deformation treatment. Arch. Metall. Mater. **61**(2a), 767–770 (2016)
2. Szewczenko, J., Marciniak, J., Kajzer, W., Kajzer, A.: Evaluation of corrosion resistance of titanium alloys used for medical implants. Arch. Metall. Mater. **61**(2), 695–699 (2016)
3. Bączkowski, B., Ziębowicz, A., Ziębowicz, B., Wojtyńska, E.: Laboratory evaluation of the fit of anti-rotational elements at the hybrid implant abutment used in prosthetic dentistry. In: Innovations in Biomedical Engineering. Advances in Intelligent System and Computing, vol. 526, pp. 11–16. Springer, Heidelberg (2017)
4. Mondrzejewska, A., Ziębowicz, A., Bączkowski, B.: Periimplantitis as the cause of separation the prosthetic bridge based on implants. In: Innovations in Biomedical Engineering. Advances in Intelligent System and Computing, vol. 526, pp. 57–66. Springer, Heidelberg (2017)
5. Kajzer, W., Kajzer, A., Grygiel-Pradelok, M., Ziebowicz, A., Ziebowicz, B.: Evaluation of physicochemical properties of TiO_2 layer on AISI 316LVM steel intended for urology. In: Information Technologies in Medicine. Advances in Intelligent Systems and Computing, vol. 472, pp. 385–398 (2016)
6. Kaczmarek, M., Kurtyka, P.: Corrosion resistance of surface treated NiTi alloy tested in artificial plasma. In: Innovations in Biomedical Engineering. Advances in Intelligent System and Computing, vol. 526, pp. 17–24. Springer, Heidelberg (2017)
7. Basiaga, M., Paszenda, Z., Walke, W.: Study of electrochemical properties of carbon coatings used in medical devices. Przegląd Elektrotechniczny **87**(12B), 12–15 (2011)
8. Basiaga, M., Paszenda, Z., Szewczenko, J.: Biomechanical behaviour of surgical drills in simulated conditions of drilling in a bone. In: Information Technologies in Biomedicine. Advances in Intelligent and Soft Computing, vol. 69, pp. 473–481 (2010)
9. Hendzel, M.: Narzędzia i urządzenia medyczne. Wydawnictwo Szkolne i Pedagogiczne, Warszawa (1982). (In Polish)
10. Paszenda, Z., Tyrlik-Held, J.: Instrumentarium chirurgiczne. Wydawnictwo Politechniki Śląskiej, Gliwice (2003)
11. Katalog firmy Chifa – Chirurgia
12. BN-74/5909-02: Narzędzia medyczne i weterynaryjne. Pakowanie, przechowywanie i transport. Wspólne wymagania i badania
13. http://www.weron.eu
14. PN-EN ISO 4284: Specyfikacje geometrii wyrobów. Struktura geometryczna powierzchni: metoda profilowa. Terminy, definicje i parametry struktury geometrycznej powierzchni. (In Polish)
15. PN-87 M-04251: Struktura geometryczna powierzchni. Chropowatość powierzchni. Wartości liczbowe parametrów. (In Polish)
16. PN-87/Z-5402: Narzędzia medyczne. Rękojeści i brzeszczoty skalpeli z wymiennymi brzeszczotami. Wymiary przyłączeniowe

17. PN-88/Z-54123: Narzędzia medyczne. Nożyczki chirurgiczne. Wspólne wymagania i badania. (In Polish)
18. ISO 4967: Skala wzorca ASTM
19. PN-EN ISO 643: Stal. Mikrograficzne określanie wielkości ziarna. (In Polish)
20. PN-88 Z-54030: Narzędzia medyczne. Narzędzia zaciskające z zapadkami. Ogólne wymagania i metody badań
21. PN-74/Z-54122: Narzędzia medyczne. Odgryzacze kostne. Wspólne wymagania i badania

Electrochemical Corrosion of Magnesium Alloy AZ31 in NaCl Solutions After Rolling

Joanna Przondziono[1(✉)], Eugeniusz Hadasik[1(✉)], Witold Walke[2(✉)], Janusz Szala[1(✉)], Jakub Wieczorek[1(✉)], and Marcin Basiaga[2]

[1] Faculty of Materials Engineering and Metallurgy, Silesian University of Technology, 8 Krasińskiego Street, 40-019 Katowice, Poland
{joanna.przondziono,eugeniusz.hadasik,janusz.szala, jakub.wieczorek}@polsl.pl

[2] Faculty of Biomedical Engineering, Silesian University of Technology, Roosevelta 40, 41-800 Zabrze, Poland
witold.walke@polsl.pl

Abstract. The purpose of this study was to evaluate the electrochemical corrosion resistance of magnesium alloy AZ31 after rolling. Corrosion tests were conducted in NaCl solutions containing various concentrations of chloride ions (0.01–2 M NaCl). Potentiodynamic tests were conducted to obtain anodic polarisation curves. Immersion tests were conducted over periods of 1–5 days. The microstructure of AZ31 was examined by scanning electron microscopy (SEM) after the immersion tests. Electrochemical impedance spectroscopy (EIS) was applied to evaluate the electrochemical phenomena occurring at the surface of the tested alloy. Geometrical features of the AZ31 alloy surface were also measured after the corrosion tests. The results of all of the tests carried out demonstrate a clear deterioration in the corrosion properties of magnesium alloy AZ31 with the increase in the molar concentration of the NaCl solution. Irrespective of the molar concentration of the NaCl solution, pitting corrosion on the surface of the tested alloy was observed.

Keywords: Magnesium Alloy AZ31 · Rolling · Electrochemical corrosion · SEM · EIS

1 Introduction

Magnesium and its alloys are garnering increasing interest from the medical industry. These materials combine low density with strong mechanical and physical properties in a way that makes them perfect for use in lightweight construction (stent, plate) [1,2]. The development of deformable magnesium alloys and methods for their plastic working has been extremely limited to date. Magnesium alloys that undergo plastic working are used sporadically, which results from technological problems (during plastic working) and high costs of production [3].

E. Tkacz et al. (Eds.): IBE 2018, AISC 925, pp. 278–292, 2019.
https://doi.org/10.1007/978-3-030-15472-1_30

The favourable properties of magnesium alloys are obtained by heat treatment and plastic working. As it has been proved, in magnesium alloys, intensive dynamic recrystallisation takes place during plastic working, which enables size reduction and improvement in mechanical characteristics [4]. Depending on the content of the alloy components, plastic working of magnesium and its alloys may be carried out only over a limited range of temperatures. Through heat treatment and plastic working, it is possible to obtain grains with average diameters below 10 μm (in magnesium alloys). Magnesium alloys, depending on their chemical composition and formability, can undergo hot forming by means of the following methods [5–18]: rolling, open die forging and stamping, extrusion forging and sheet press forming in heated matrices after rolling.

Rolling of ingots composed of magnesium alloy is extremely costly and time-consuming due to the need for intermediate annealing; thus, the resulting goods processed in this manner are expensive. Currently, not only conventional methods but also novel technologies for magnesium sheet production are still being developed. Magnesium sheets are rarely produced from ingots. Stock material typically comes in the form of strand cast slabs with a minimum thickness of 120 mm. Processing is carried out over the temperature range of 200–450 °C. Technologies for casting between rollers are currently in the pre-production phase of testing. High expectations are held for these technologies in terms of reducing cost and improving quality [2]. However, the application of magnesium alloys is limited due to their insufficient resistance to corrosion [19–30]. As a highly electronegative element, magnesium is extremely susceptible to dissolving in electrolyte solutions. Magnesium is extremely prone to electrochemical and chemical corrosion, particularly in environments containing chloride ions, which substantially limits the range of its alloys' application. The reason for the low corrosion resistance of magnesium is the inadequate protective properties of the oxide layer that is formed on the surface of magnesium in an oxidising atmosphere or the hydroxide layer formed in aqueous solutions. The literature contains reports on the properties (e.g., corrosion resistance, among others) of cast AZ31 alloy and other magnesium alloys obtained by the casting method.

The purpose of this study was to evaluate the electrochemical corrosion resistance of magnesium alloy AZ31 after rolling. Corrosion tests were conducted in NaCl solutions containing various concentrations of chloride ions (0.01–2 M NaCl). Potentiodynamic tests were conducted to obtain anodic polarisation curves. The Stern method was applied to determine the corrosion resistance parameters of the alloy. Immersion tests were carried out over periods of $1 \div 5$ days. Qualitative and quantitative analyses of the chemical composition in microscopic areas of the alloy were conducted using an S-4200 Hitachi field-emission scanning electron microscope (FE SEM) in conjunction with an EDS Voyager 3500 spectrometer from Noran Instruments. Electrochemical impedance spectroscopy (EIS) was performed to evaluate the electrochemical phenomena occurring at the surface of the tested alloy. Measurements were performed by using an AutoLab PGSTAT 302N measurement system equipped with an FRA2 (Frequency Response Analyser) module. Impedance spectra were obtained from the measurement system, and the data obtained data were matched to an equivalent circuit. Geometrical features of the AZ31 alloy surface after the corrosion tests were also measured in this study.

2 Materials and Methods

Samples composed of magnesium alloy AZ31 - ASTM designation (MCMgAl3Zn1) after hot rolling - were used as the initial testing material. The chemical composition of the alloy is presented in Table 1. After hot rolling, AZ31 alloy was annealed at 50 °C for 1 h and then cooled in air.

Table 1. Chemical composition of magnesium alloy AZ31, % of mass.

Al	Zn	Mn	Cu	Mg
2.83	0.80	0.37	0.02	Balance

The corrosion tests were carried out in NaCl solutions containing various concentrations of chloride ions. Measurements were performed in 0.01, 0.2, 0.6, 1 and 2 M NaCl. The solution temperature during testing was $21 \pm 1\,°C$.

Resistance to electrochemical corrosion was determined based on the registered anodic polarisation curves. To perform potentiodynamic tests, a VoltaLab®PGP201 system by Radiometer was used. A saturated calomel electrode (SCE) of the KP-113 type served as the reference electrode, whereas a platinum electrode of the PtP-201 type was used as the auxiliary electrode. The tests were begun by determining the open-circuit potential EOCP. Later, anodic polarisation curves were obtained, beginning with the measurement of the potential according to the relation $E = E_{OCP} - 100\,\mathrm{mV}$. The potential was varied in the anodic direction at a rate of 1 mV/s. When the anodic current reached a density of $10\,\mathrm{mA/cm^2}$, the polarisation direction was reversed. Thus, return curves were also measured. The open-circuit potential E_{OCP} of the tested samples was stabilised after 30 min.

Based on the registered anodic polarisation curves, typical parameters describing resistance to electrochemical corrosion were determined: corrosion potential, corrosion current density and corrosion rate. The Stern method was applied to determine the polarisation resistance. Immersion tests were carried out at ambient temperature in 0.01–2 M NaCl solutions over periods of 1–5 days. After special preparation of the surface of the samples, the samples were weighed and their mass m_0 was determined. After immersion of the alloy in NaCl solutions for 1–5 days, the samples were taken out and corrosion products were removed using a reagent containing $200g/lCrO_3$ and $10g/lAgNO_3$. Next, the samples were flushed with distilled water, degreased with acetone, dried and weighed again; their final mass was denoted m1. These tests allowed for the determination of the corrosion rate. In this study, scanning electron microscopy was also carried out to examine the microstructure of AZ31 alloy after the immersion tests. Using an S-4200 Hitachi field-emission scanning electron microscope (FE SEM) equipped with an EDS Voyager 3500 spectrometer from Noran Instruments, qualitative and quantitative analyses of the chemical composition in microscopic areas of the samples were conducted.

To obtain information about the physical and chemical characteristics of the surface of the AZ31 alloy samples, electrochemical impedance spectroscopy (EIS) was employed. Measurements were performed by using an AutoLab PGSTAT 302N measurement system equipped with an FRA2 (Frequency Response Analyser) module. The applied measurement system allowed for tests to be performed over the frequency range 103–104 Hz. The impedance spectra of the system were thus determined, and data obtained from the measurement were matched to an equivalent circuit. Using this circuit, numerical values for the resistance R and capacity C of the analysed systems were determined. The impedance spectra of the tested system are presented as Nyquist diagrams for various frequencies and as Bode diagrams. The obtained EIS spectra were interpreted after matching the data to an equivalent circuit by the smallest squares method. This method allowed for the characterisation of the impedance of the AZ31 alloy surface layer NaCl solution phase boundary by approximating the impedance data via an equivalent electric circuit model [31,32].

Geometrical characteristics of the surface of the alloy AZ31 samples after the corrosion tests were also measured. These tests were carried out using an optical profile measurement gauge Microprof. (CWL3000) made by FRT. The conditions of the surface geometry measurements were as follows: measurement field: 4×2 mm, resolution 2000×1000 pixels.

3 Results and Discussion

3.1 Potentiodynamic and Immersion Tests

Potentiodynamic tests carried out in NaCl solutions of various molar concentrations allowed for the corrosion properties of hot-rolled magnesium alloy AZ31 to be determined. The corrosion resistance test results (mean values of measurements) are compared in Table 2. Anodic polarisation curves of the selected samples are shown in Fig. 1.

It was determined that as the chloride ion concentration increases, there is a decrease in the corrosion potential and polarisation resistance and an increase in the corrosion current density and corrosion rate. The corrosion rate in the immersion tests was determined according to the following formula 1:

$$V = \frac{m_0 - m_1}{St} \quad (1)$$

Table 2. Results of electrochemical corrosion resistance tests of magnesium alloy AZ31 (mean measurement values).

Molar concentration NaCl, M	Ecorr mV	Icorr A/cm^2	Rp2 Ocm^2	Corr mm/year
0.01	−1490	0.007	2600	0.260
0.2	−1488	0.023	1100	0.527
0.6	−1538	0.107	281	2.460
1	−1537	0.128	210	2.940
2	−1566	0.127	208	3.440

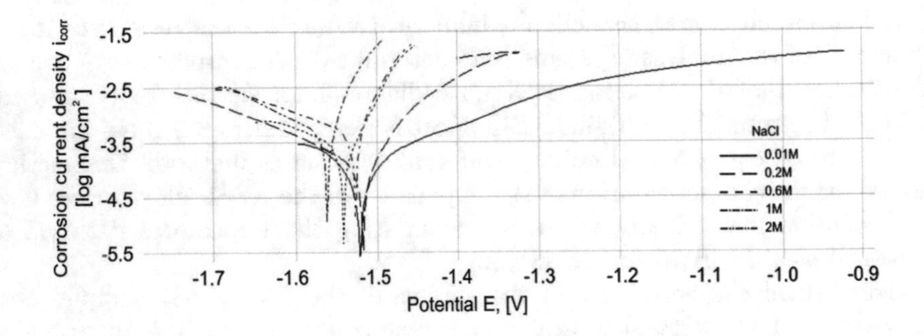

Fig. 1. Anodic polarisation curves

where V—corrosion rate, mg/(cm^{2}day), m_0—initial sample mass, m_1—sample mass after immersion test and removal of corrosion products, S—sample surface area, t—exposure time (day).

Table 3 shows the results of the immersion test.

Table 3. Results of the immersion test

NaCl concentration, M	Corrosion rate, mg/(cm^{2}day)	NaCl concentration, M	Corrosion rate, mg/(cm^{2}day)
	1 day	3 day	5 day
0.01	0.007133	0.033287	0.064197
0.2	0.009516	0.052309	0.106995
0.6	0.021399	0.052631	0.149793
1	0.038043	0.095107	0.256788
2	0.066575	0.156926	0.506443

4 SEM and Optical Profile

Scanning electron microscopy revealed that the alloy structure consists of a solid solution of αMg and individual inclusions of Al-Mn and MgSi. After 1 day of testing in 0.01 M NaCl, corrosion showed a selective and non-uniform character. Pits could be observed on the surface of the alloy (Fig. 2). Undissolved intermetallic phases of M_2Si were also observed (Fig. 2). Figure 3 shows the results of the qualitative and quantitative analyses of the chemical composition of the Mg_2Si phase performed using the aforementioned FE SEM system.

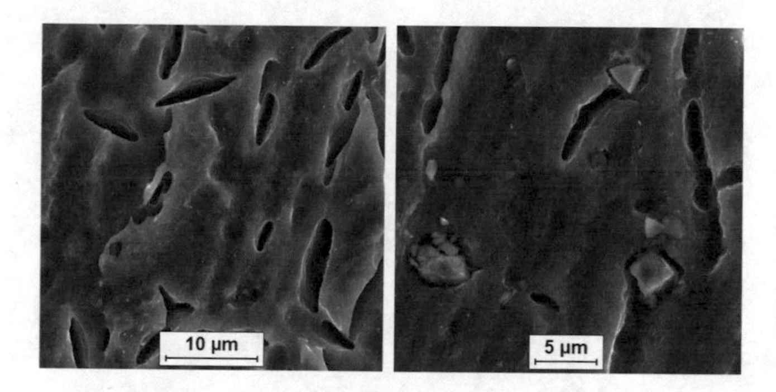

Fig. 2. On the left: Pits on the surface of AZ31 (0.01 M NaCl; 1 day). and on the right: Phases of Mg2Si type (0.01 M NaCl; 1 day).

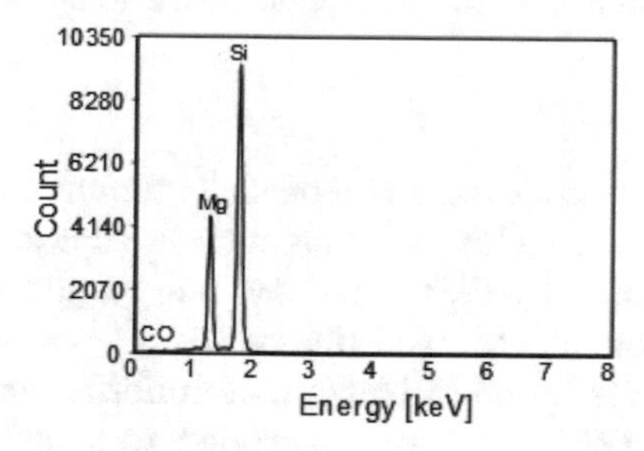

Fig. 3. X-ray spectrum and the results of quantitative analysis of the Mg2Si phase

Element	Wt [%]
Si	71.09
Mg	28.91

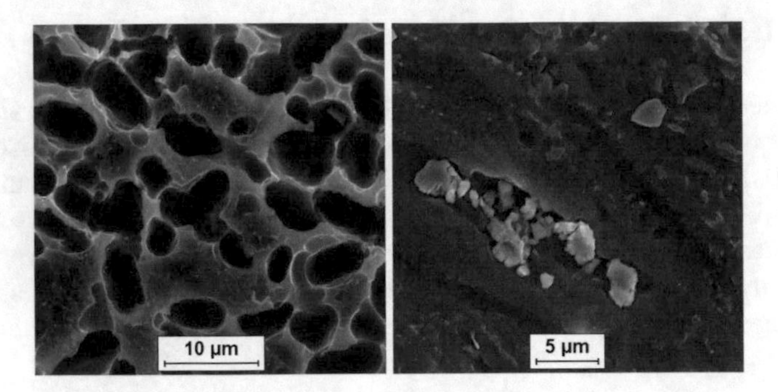

Fig. 4. On the left: Pits on the surface of AZ31 (0.01 M NaCl; 3 days). and On the right: Phases of Mg-Si and Al-Mn (0.01 M NaCl; 5 days).

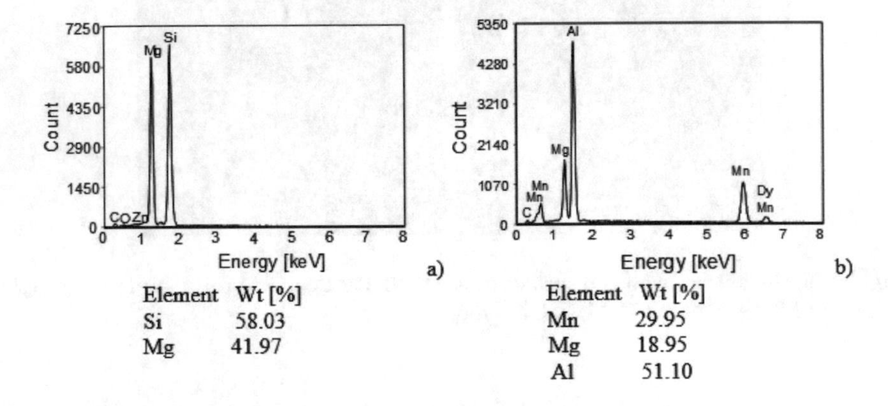

Fig. 5. X-ray spectrum and the results of quantitative analysis of the phases Mg-Si (a) and Al-Mn (b).

After 3 and 5 days of exposure, an increasing number of pits could be observed on the surface of the alloy (Fig. 4). The surface showed non-corroded phases of Mg-Si and Al-Mn (Fig. 4). Figure 5 shows the results of the qualitative and quantitative analyses of the observed phases.

After immersion tests in 0.6 M NaCl, non-uniform corrosion could be observed. Test samples of AZ31 alloy were corroded to a different extent (Fig. 6 a, b). Moreover, there was a part of each surface that was not corroded. On the corroded surface, the pits exhibited larger depths and diameters (Fig. 7).

After immersion in 2 M NaCl, only few non-corroded regions were observed on the samples. Corrosion showed a more uniform character (Fig. 8a). A large number of pits could be observed on the surface of the alloy samples (Fig. 8b).

At higher magnification, areas with oriented microstructure, associated with plastic strain, could be observed on the samples immersed in 0.6 M NaCl (Fig. 9) and in 2 M NaCl (Fig. 10). Due to the anisotropic properties of the rolled alloy, it is expected that there is crystallographic direction along which corrosion will proceed with greater intensity.

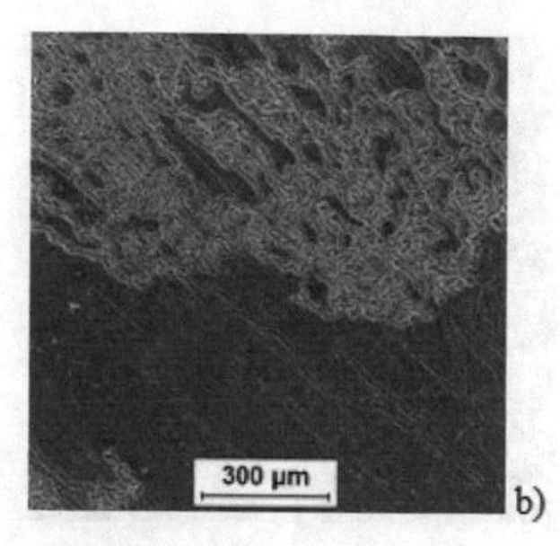

Fig. 6. Surface of alloy sample after exposure to 0.6 M NaCl for 3 days (a) and 5 days (b).

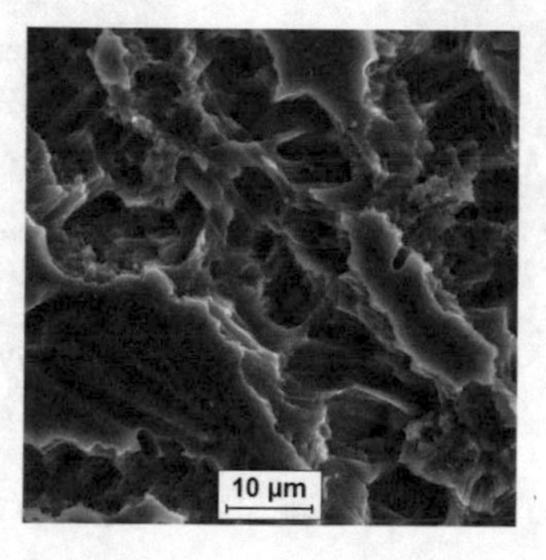

Fig. 7. Pits on sample alloy surface after immersion in 0.6 M NaCl for 3 days.

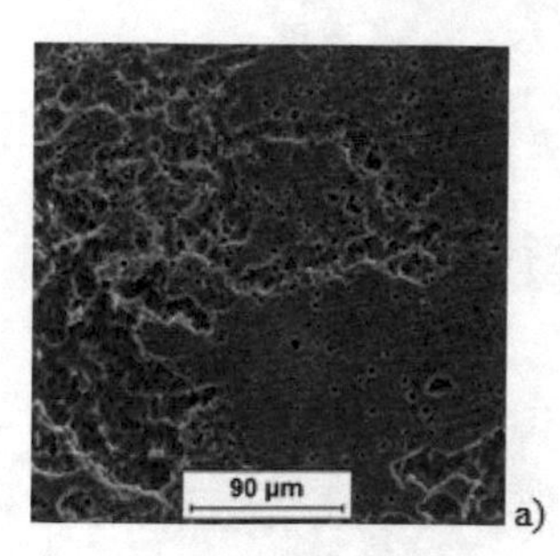

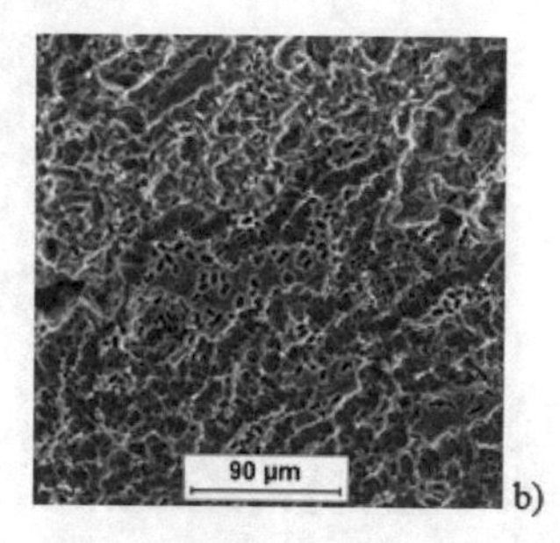

Fig. 8. Surface of alloy sample after exposure to 2 M NaCl for 3 days (a) and 5 days (b).

Tests carried out using an optical profile measurement gauge confirmed that the corrosion of AZ31 alloy after 5 days of exposure was uniform. After the test in 0.01 M NaCl, the depths of pits did not exceed 4 μm (Fig. 11), whereas after the test in 2 M NaCl, the pits reached depths exceeding 10 μm (Fig. 12).

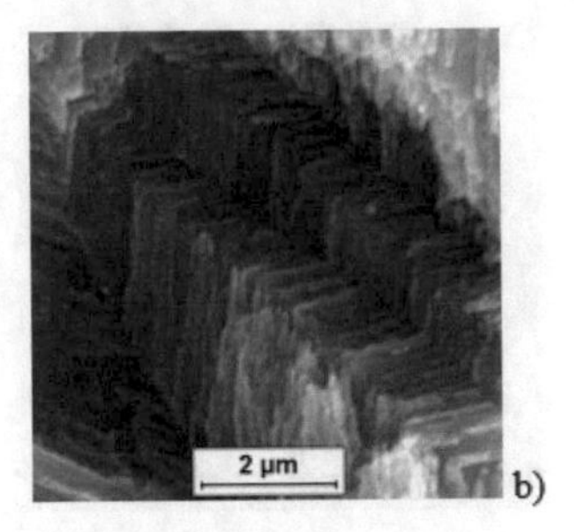

Fig. 9. Areas with directed microstructure:

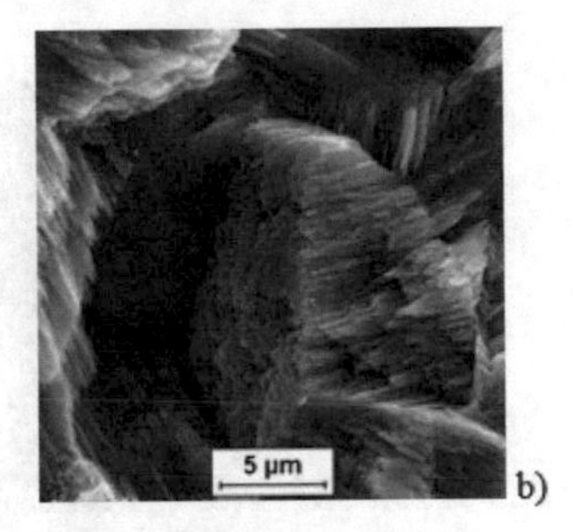

Fig. 10. Areas with directed microstructure: and (a) 2 M NaCl (3 days) and (b) 2 M NaCl (5 days).

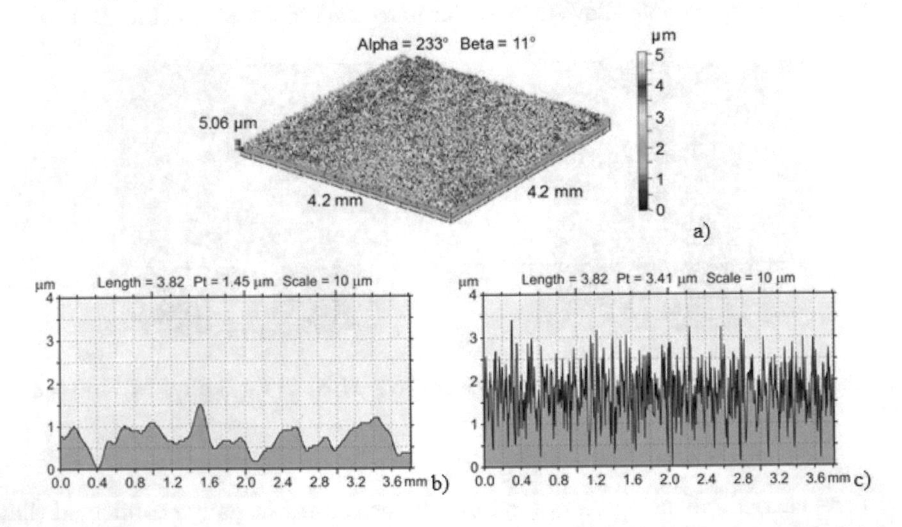

Fig. 11. 3D image of the surface (a), corrugation of the selected cross-section (b) and roughness of the selected cross-section (c) of AZ31 alloy after exposure for 5 days in 0.01 M NaCl solution.

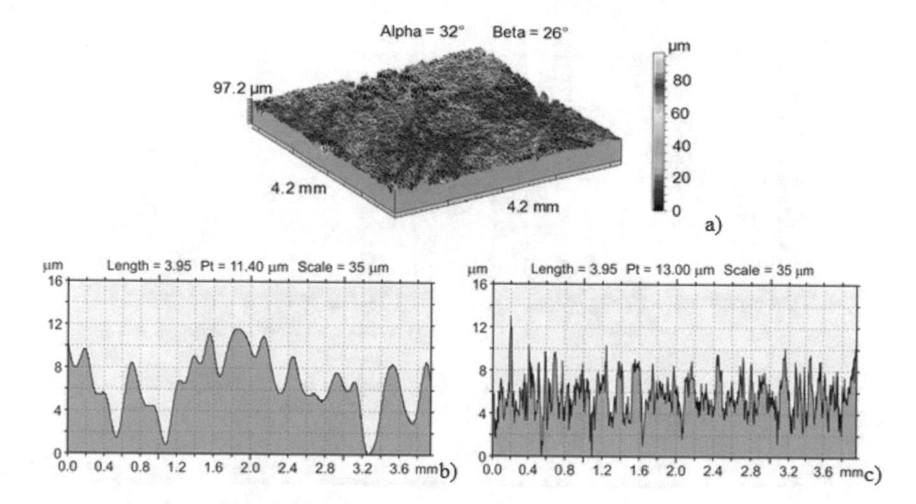

Fig. 12. 3D image of the surface (a), corrugation of the selected cross-section (b) and roughness of the selected cross-section (c) of AZ31 alloy after exposure for 5 days in 2 M NaCl solution.

5 EIS

Impedance spectra of rolled AZ31 alloy as a function of NaCl concentration are presented in Figs. 13, 14, 15 and 16. Table 4 presents the results of the EIS analysis.

Based on the obtained diagrams, the data were matched to equivalent circuit models, which are physical models describing phenomena that occur in given systems. It was determined that the best matching of the experimental impedance spectra was achieved by applying the following electrical equivalent circuit:

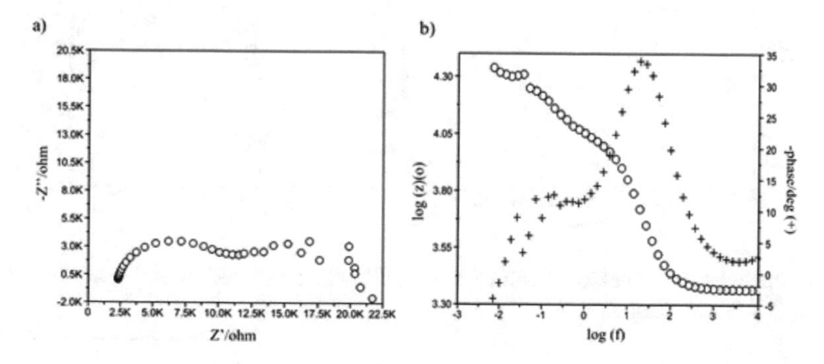

Fig. 13. Impedance spectra obtained for AZ31 alloy in 0.01 M NaCl: and (a) Nyquist diagram, (b) Bode diagram.

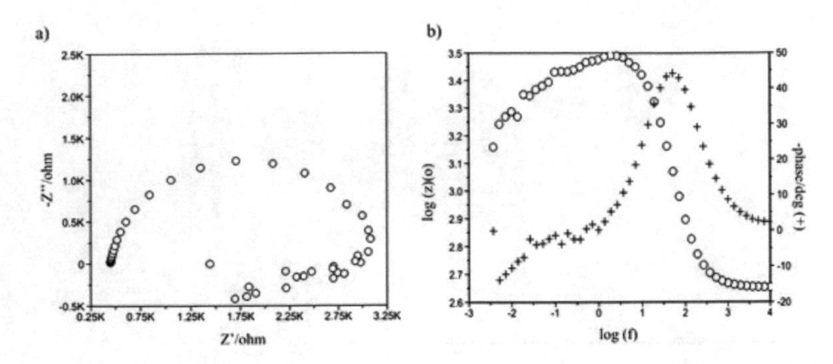

Fig. 14. Impedance spectra obtained for AZ31 alloy in 0.2 M NaCl: and (a) Nyquist diagram, (b) Bode diagram.

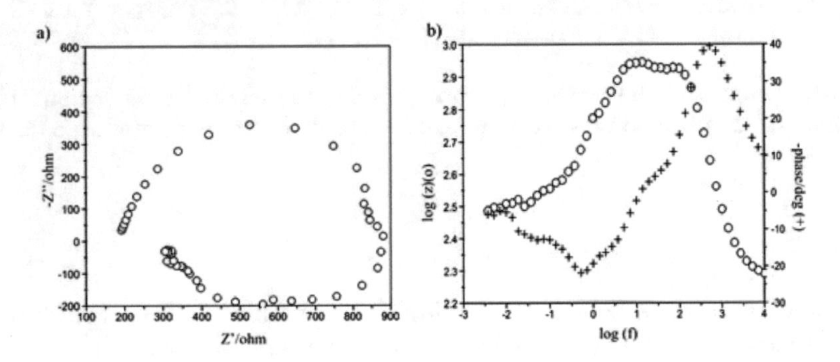

Fig. 15. Impedance spectra obtained for AZ31 alloy in 0.6 M NaCl: and (a) Nyquist diagram, (b) Bode diagram.

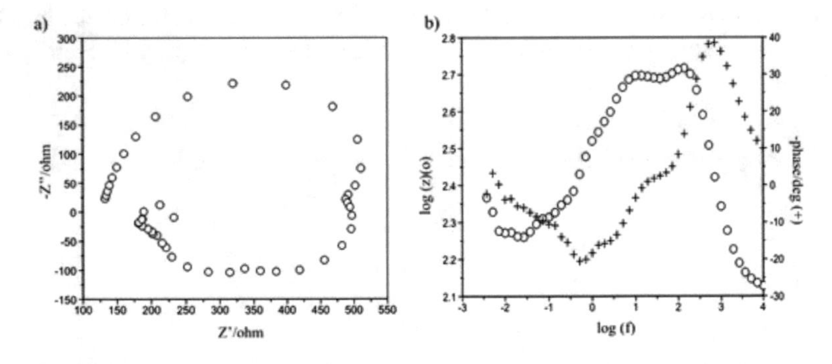

Fig. 16. Impedance spectra obtained for AZ31 alloy in 1 M NaCl: and (a) Nyquist diagram, (b) Bode diagram.

- For the sample exposed to a 0.01 M NaCl solution, the circuit consisted of two consecutive parallel systems of CPE connected to a transition resistance and a resistance at high frequencies that can be attributed to the ohm resistance of the electrolyte.

In Fig. 18, Rct represents the resistance to charge transfer and $C\neg PEdl$ represents the capacity of the double (porous) layer, whereas Rp and CPEp indicate the resistance and capacity of the thin passive layer.

- For samples exposed to solutions of 0.2–2 M NaCl, at low frequencies, the equivalent circuit consisted of a parallel capacity system connected to an ion transition resistance Rct associated with the metal solution phase boundary and a resistance RL associated with a coil L (electric conductor with electromagnetic induction) representing corrosion and resistance processes at high frequencies, which can be attributed to electrolyte resistance Rs (Table 4). In Fig. 19, Rct represents the resistance to charge transfer and $CPEdl$ the capacity of the double (porous) layer, whereas RL and L compose the induction loop representing the initialisation and development of pitting corrosion.

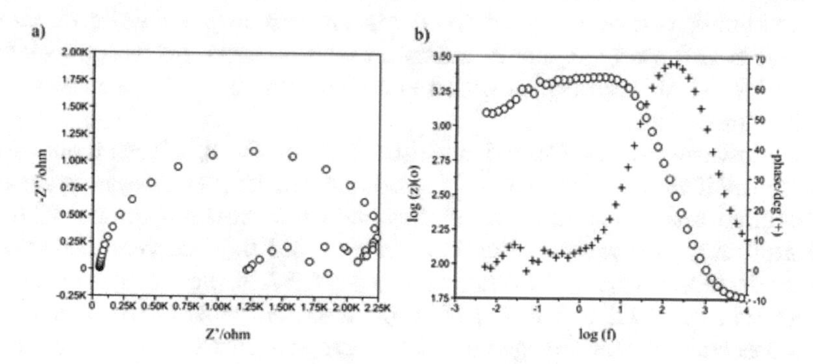

Fig. 17. Impedance spectra obtained for AZ31 alloy in 2 M NaCl: and (a) Nyquist diagram, (b) Bode diagram.

Table 4. EIS analysis results

NaCl, M	Rs, kΩcm²	Rct, kΩcm²	Cdl, μFcm⁻²	L, Hcm⁻²	CPEdl		Rp kΩcm²	CPEp		RL, kΩcm²
					Y01, Ω⁻¹ cm?²s?n	n1		Y01,Ω⁻¹ cm?²s?n	n2	
0.01	2.31	8.45	-	-	0.3671e-5	0.87	9.62	0.1305e-3	0.83	-
0.2	0.46	1.22	0.17	22.9	-	-	-	-	-	0.01
0.6	0.19	0.65	0.83	117	-	-	-	-	-	0.17
1	0.13	0.37	1.05	542.9	-	-	-	-	-	0.09
2	0.06	2.06	2.83	1000.2	-	-	-	-	-	6.41

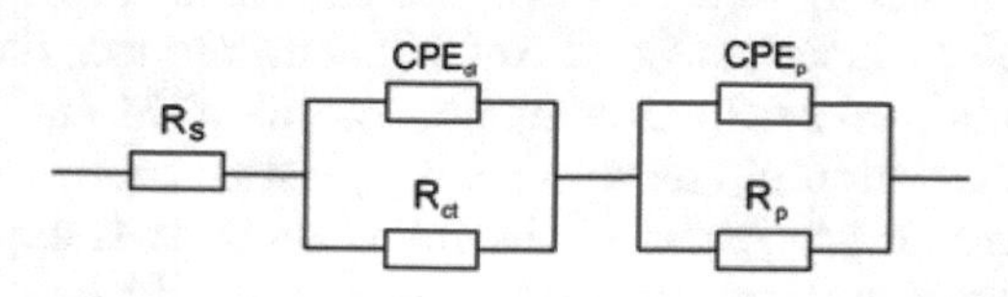

Fig. 18. Physical model of electrical equivalent circuit for the metal-passive layer-solution configuration.

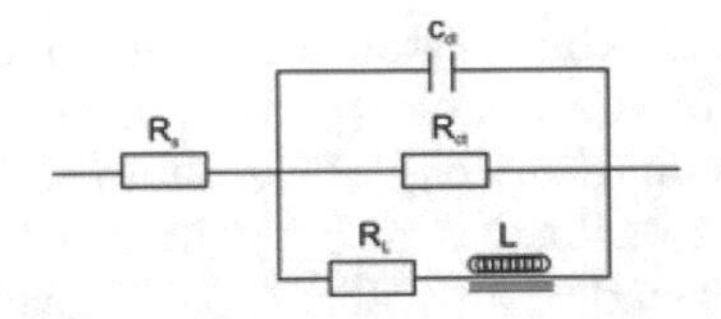

Fig. 19. Physical model of equivalent electrical circuit for the metal-solution corrosion system.

6 Conclusions

The applications of magnesium alloy AZ31 after plastic working are largely dependent on the alloys' resistance to electrochemical corrosion, especially in chloride solutions. Hence, it was advisable to carry out corrosion tests in solutions containing a wide range of NaCl concentration. The results of the tests performed in this study explicitly indicate that the corrosion resistance of magnesium alloy AZ31 deteriorates with the increase in the molar concentration of NaCl in solution.

Potentiostatic tests performed in solutions with 0.01–2 M NaCl demonstrated that with the increase in the chloride ion concentration, a decrease in the corrosion potential and polarisation resistance and an increase in the corrosion current density and corrosion rate of AZ31 alloy could be observed. The corrosion potential decreased from −1490 mV to 1566 mV. With the increase in the molar concentration of NaCl in solution, the polarisation resistance decreased from 2600 Omegacm^2 to 208 Omegacm^2. The corrosion current density of the cast alloys increased from 0.007 A/cm^2 to 0.127 A/cm^2. The corrosion rate increased from 0.26 mm/year to 3.44 mm/year.

The results of the immersion tests also indicate the deterioration of the corrosion properties of magnesium alloy AZ31 with the increase in the molar concentration of NaCl in solution. The corrosion rate of the sample immersed in 0.01 M NaCl increased from 0.007133 mg/(cm^2*day*) after 1 day to 0.064197 mg/(cm^2*day*) after 5 days. Tests carried out over 1 to 5 days in 2 M NaCl demonstrated that the corrosion rate increased substantially, from 0.066575 mg/(cm^2*day*) up to 0.506443 mg/(cm^2*day*).

The alloy structure revealed the formation of micro-cells between solid solution ?Mg and an intermetallic phase. At low NaCl concentrations, corrosion was selective. When the test solution was more aggressive, the corrosion process was more intense. Anodic solubilisation of the alloy surface occurred, and a large number of pits could be observed. The surface of the AZ31 alloy sample submitted to corrosion testing in 2 M NaCl was the most corroded one. With the increase in the molar concentration of NaCl in solution, the diameter and depth of the corrosion pits increased. After immersion in 0.6 M and 2 M NaCl, areas with directed microstructure, associated with plastic strain, could be observed. Due to the anisotropic properties of the rolled alloy, it is expected that there is a crystallographic direction along which corrosion will be more intense. The equivalent electrical circuit developed using EIS data suggests the presence of an upper zone in the tested systems, which features high roughness and a great extent of surface corrosion. EIS spectra show that the corrosion process occurs through reagents = diffusion. To summarise, it should be highlighted that irrespective of the molar concentration of NaCl in solution, pitting corrosion was observed in the tested alloy. This result indicates the lack of corrosion resistance of magnesium alloy AZ31 after hot rolling. The test results further prove that a protective coating should be deposited on elements composed of AZ31 alloy.

Acknowledgements. The financial support of Structural Funds in the Operational Programme Innovative Economy (IE OP) financed by the European Regional Development Fund—Project "Modern material technologies in aerospace industry", No. POIG.0101.02-00-015/08 is gratefully acknowledged.

References

1. Hadasik, E.: Tests of metal plasticity. Monograph, Printing House of the Silesian University of Technology, Gliwice (2008, in Polish)
2. Kawalla, R.: Magnesium and magnesium alloys. Monograph: Metal processing. Plasticity and structure, Printing House of the Silesian University of Technology, Gliwice (2006, in Polish)
3. Kiełbus, A., Kuc, D., Rzychoń, T.: Magnesium alloys' microstructure, properties and application. Monograph. Modern metallic materials - presence and future, Department of Materials Engineering and Metallurgy, Katowice (2009, in Polish)
4. Brooks, C.R.: Heat treatment, structure and properties of nonferrous alloys. ASM Internationale, Metals Park, Ohio (1984)
5. Xu, Z., Tang, G., Tian, S., Ding, F., Tian, H.: Research of electroplastic rolling of AZ31 Mg alloy strip. J. Mater. Process. Technol. **182**(1–3), 128–133 (2007)
6. Zhang, H., Yan, Q., Li, L.: Microstructures and tensile properties of AZ31 magnesium alloy by continuous extrusion forming process. Mater. Sci. Eng. A **486**(1–2), 295–299 (2008)
7. Qudong, W., Yinhonga, W., Chinob, Y., Mabuch, M.: High strain rate superplasticity of rolled A291 magnesium alloy. Rare Metals **27**(1), 46–49 (2008)
8. Kawalla, R., Lehmann, G., Ullmann, M., Voght, H.P.: Magnesium semi-finished products for vehicle construction. Arch. Civil Mech. Eng. **8**(2), 93–101 (2008)
9. Xu-yue, Y., Miura, H., Sakai, T.: Recrystallization behaviour of fine-grained magnesium alloy after hot deformation. Trans. Nonferrous Met. Soc. China **17**(6), 1139–1142 (2007)
10. Čížek, L., Greger, M., Dobrzański, L.A., Juřička, I., Kocich, R., Pawlica, L.: Structure and properties of alloys of the Mg-Al-Zn system. J. Achievements Mater. Manuf. Eng. **32**(2), 179–187 (2009)
11. Xing, J., Yang, X., Miura, H., Sakai, T.: Superplasticity of magnesium alloy AZ31 processed by severe plastic deformation. Mater Trans. **48**(6), 1406–1411 (2007)
12. Bryła, K., Dutkiewicz, J., Lityńska-Dobrzyńska, L., Rokhlin, L.L., Kurtyka, P.: Influence of number of ECAP passes on microstructure and mechanical properties of AZ31 magnesium alloy. Arch. Metall. Mater. **57**(3), 711–717 (2012)
13. Gontarz, A., Dziubińska, A., Okoń, Ł.: Determination of friction coefficients at elevated temperatures for some Al, Mg and Ti alloy. Arch. Metall. Mater. **56**(2), 379–384 (2011)
14. Walke, W., Hadasik, E., Przondziono, J., Kuc, D., Bednarczyk, I., Niewielski, G.: Plasticity and corrosion resistance of magnesium alloy WE43. Arch. Mater. Sci. Eng. **51**(1), 16–24 (2011)

15. Tomczak, J., Pater, Z., Bulzak, T.: Thermo-mechanical analysis of a lever preform forming from magnesium alloy AZ31. Arch. Metall. Mater. **57**(4), 1211–1218 (2012)
16. Pater, Z., Tomczak, J.: Experimental tests for cross wedge rolling of forgings made from non-ferrous metal alloys. Arch. Metall. Mater. **57**(4), 919–928 (2012)
17. Gontarz, A., Pater, Z., Drozdowski, K.: Hammer forging process of lever drop forging from AZ31 magnesium alloy. Metalurgija **52**(3), 359–362 (2013)
18. Cyganek, Z., Tkocz, M.: The effect of AZ31 alloy flow stress description on the accuracy of forward extrusion FE simulation results. Arch. Metall. Mater. **57**(1), 199–204 (2012)
19. Maker, G.L., Kruger, J.: Corrosion of magnesium. Int. Mater. Rev. **38**(3), 138–153 (1993)
20. Song, G., Trens, A., Wu, X., Zhang, B.: Corrosion behaviour of AZ21, AZ501 and AZ91 in sodium chloride. Corros. Sci. **40**(10), 1769–1791 (1998)
21. Amira, S., Dubé, D., Tremblay, R., Ghali, E.: Influence of the microstructure on the corrosion behavior of AXJ530 magnesium alloy in 3.5% NaCl solution. Mater. Charact. **59**(10), 1508–1517 (2008)
22. Dobrzańska-Danikiewicz, A.D., Tański, T., Domagała-Dubiel, J.: Unique properties, development perspectives and expected applications of laser treated casting magnesium alloys. Arch. Civil Mech. Eng. **12**(3), 318–326 (2012)
23. Przondziono, J., Walke, W., Szala, J., Hadasik, E., Wieczorek, J.: Evaluation of corrosion resistance of casting magnesium alloy AZ31 in NaCl solutions. In: IOP Conference Series: Materials Science and Engineering, vol. 22, no. 012-017, pp. 1–12 (2011)
24. Rzychoń, T., Michalska, J., Kiełbus, A.: Effect of heat treatment on corrosion resistance of WE54 alloy. J. Achievements Mater. Manuf. Eng. **20**(1–2), 191–194 (2007)
25. Przondziono, J., Walke, W., Hadasik, E., Jasiñski, B.: Electrochemical corrosion of magnesium alloy AZ31 in NaCl solutions. Acta Metallurgica Slovaca **16**(4), 254–260 (2010)
26. Ambat, R., Aung, N., Zhou, W.: Evaluation of microstructural efects on corrosion behaviour of AZ91D magnesium alloy. Corros. Sci. **42**(8), 1433–1455 (2000)
27. Altun, H., Sen, S.: Studies on the influence of chloride ion concentration and pH on the corrosion and electrochemical behaviour of AZ63 magnesium alloy. Mater. Des. **25**(7), 637–643 (2004)
28. Anik, M., Celikten, G.: Analysis of the electrochemical reaction behavior of alloy AZ91 by EIS technique in H3PO4/KOH buffered K2SO4 solutions. Corros. Sci. **49**(4), 1878–1894 (2007)
29. Przondziono, J., Walke, W., Hadasik, E.: Galvanic corrosion test of magnesium alloys after plastic forming. Light Metals Their alloys II Solid State Phenomena **191**, 169–176 (2012)
30. Przondziono, J., Walke, W., Hadasik, E., Szala, J., Wieczorek, J.: Corrosion resistance tests of magnesium alloy WE43 after extrusion. Metalurgija **52**(2), 242–246 (2013)
31. Basiaga, M., Paszenda, Z., Walke, W.: Study of electrochemical properties of carbon coatings used in medical devices. Electr. Rev. **87**(12B), 12–15 (2012)
32. Walke, W., Paszenda, Z., Pustelny, T., Opilski, Z., Drewniak, S., Kocielniak-Ziemniak, M., Basiaga, M.: Evaluation of physicochemical properties of SiO_2 coated stainless steel after sterilization. Mater. Sci. Eng. C - Mater. Biol. Appl. **63**, 155–163 (2016)

Application of Novel Polymeric Materials Supporting 3D Printing Technology in the Development of Anatomical Models and Regenerative Medicine

Andrzej Szymon Swinarew[1](✉), Jarosław Paluch[2], Klaudia Kubik[1], Beata Dorzak[3], Anna Kwaśniewska[4], Tomasz Flak[1], Jadwiga Gabor[1], Marta Łężniak[1], Hubert Okła[1], Grzegorz Bajor[3], Damian Kusz[5], Robert Wilk[5], Hanna Sikora[5], Krzysztof Aniołek[1], and Adrian Barylski[1]

[1] Faculty of Computer Science and Materials Science, Institute of Materials Science, University of Silesia in Katowice, Katowice, Poland
andrzej.swinarew@us.edu.pl

[2] Department of Laryngology, School of Medicine in Katowice, Medical University of Silesia, Katowice, Poland

[3] Department of Anatomy, School of Medicine in Katowice, Medical University of Silesia in Katowice, Katowice, Poland

[4] Department of Radiology, Medical University of Silesia in Katowice, Hospital SPSK M, Katowice, Poland

[5] Department of Orthopedics and Traumatology, School of Medicine in Katowice, Medical University of Silesia in Katowice, Katowice, Poland

Abstract. The article is focused on the new polymer material which could be used for hip endoprosthesis. Nowadays the most common materials which are used are metal alloys, ceramics and polyethylene. The aim of this study was to present the properties of modified polycarbonate (PC) with the particular consideration of tribological and mechanical properties and proving that this material could be successfully used beside existing and currently chosen materials. For mentioned materials conducted a few tests. Among of them were hardness measurements, static stretching tests and study of abrasive wear resistance. Obtained results allowed to conclude about using the material as not only a part of hip endoprosthesis but a whole implant. The novel obtained polymeric materials based on the PC modified with nanosilica presents instead of bacteriostatic properties improved resistance to volumetric wear. The nanoparticles of silica do not negative affect on the friction coefficient. The obtained results clearly presents that the material possess better coefficient of friction combined with lower volumetric wear.

Keywords: Regenerative medicine · Polymeric materials · 3D printing · Biomaterials

E. Tkacz et al. (Eds.): IBE 2018, AISC 925, pp. 293–300, 2019.
https://doi.org/10.1007/978-3-030-15472-1_31

1 Introduction

Hip replacement surgery involves the insertion of foreign elements into the human body to create artificial replacements of a damaged joint to restore the anatomical structure of lost functions. The hip endoprosthesis is used to treat degenerative changes, femur neck injuries and trochanteric fractures [1,2]. The hip joint is classified as a ball-and-socket joint connecting a lower limb to the pelvis, in which the rounded head of the femoral bone fits into the acetabulum (socket of the Ilium). Both articular surfaces are covered with a hyaline cartilage. In case of the cartilage degradation, called coxarthrosis, the weight bearing area of the hip joint is decreased leading to an increase of the contact hip stress and degenerative changes with a decrease of the lower limb function as a support device. Nowadays the most popular method of coxarthrosis treatment is total hip arthroplasty (THA) [3,4].

Currently, the different types of prosthesis are available: with short or long stem, cemented or cementless etc. The choose of prosthesis depends on a bone stock, patients age and primary disease (coxarthrosis, fracture or a tumour) as well. In a similar way, we choose a type of articular surface. Nowadays liner could be ceramic or different types of polyethylene. The femoral head could be metal or ceramic. There is also metal on metal bearings. Every combination of acetabular insert and the femoral head has some advantages and disadvantages [5,6] Main disadvantages of differently used bearing surfaces are high liner wear rate, component wear debris-induced osteolysis, and component-related noise. For example, Amanatullah et al. reported that the mean liner wear rate in the CoC (ceramic-on-ceramic) group was 30.5 ± 7.0 µm/year, however, in the CoP (ceramic-on-polyethylene)group it was 218.2 ± 13.7 µm/year. On the other hand, the rates of ceramic implant fracture (2.6%) and audible component-related noise (3.1%) were statistically higher in the CoC group than in the CoP group ($P < 0.05$). Patients with ceramic surfaces mainly experience clicking and creaking noises, while patients with MoM (metal-on-metal) and MoP (metal-on-polyethylene) bearings mainly experience clicking, grating, and creaking noises [6,7].

The aim of our study was to presented new material which could be used for the prosthesis and showed how modification affected to properties allowing to control and create the materials with desired characteristics.

2 Materials and Methods

For research used PC modified silicon oxide and PC modified rhodamine to show that modification has a vital impact on properties of the material. The modified materials were produced at Silesian University according to a procedure—for 700 g of PC (Makrolon 2600®) used 4.2 g of silicon oxide and 11.2 of polyol synthesized earlier. Similarly was in case PC doped with derivative of rhodamine. The rhodamine derivative was obtained follow: p-Ph2CC6H4CHO was condensed with 1,3-indendione in $EtOH/H_2SO_4$ reflux to prepare rhodamine derivative used then as a bacteriostatic modifier. The modified polymer showed antibacterial activity against *Escherichia coli.* for 700 g of PC (Makrolon 2600®) used 1.75 mg of previously obtained rhodamine derivative. All components were combined and after that, a strings were made. Next step was to prepare samples—to mechanical and tribological tests. All sapless where prepared by the use of 3D printer, the test specimens shape and dimensions were prepared according to standards. The 3D printer used in this project works in FDM (Fused Deposition Modeling) technology that means that, objects are created layer by layer from melted polymeric material, in this case from PC with additives. Temperatures of 3D printer were set to: extruder 275 °C; bed 130 °C. Unfortunately, it is not possible to print from pure commercially available PC.

The hardness was measured according to the Vickers method using a Wolpert Wilson Instruments hardness gauge. The method described in PN-EN ISO 6507-1:2007 standard [8]. The indenter pressure was 980 mN. The samples were layered what was the result of manufacturing (3D printing). In order not to lead to delamination of the material and to conceal the measurement result, the hardness measurement was made perpendicular to the material layers. The hardness of the material was calculated according to the formula 1:

$$HV = 0.1891\frac{F}{d^2} \tag{1}$$

where: F—force [N], d—arithmetic mean of two diagonals of imprint [mm].

Determination of mechanical properties in static stretching of plastics was performed using the norms: PN-EN ISO 527: 2012 [9,10]. The values measured during the stretching test were deformation and deformation forces. The Young modulus was calculated based on that results. The tensile test was performed using the Instron 5982 strength machine.

The initial length of the samples was $L_0 = 80$ mm, the cross-section of the stretched part was $A_0 = 40$ mm^2. The maximum stress for the material was calculated from the formula 2:

$$\sigma_M = \frac{F_M}{A_0}[MPa] \tag{2}$$

where: F_M – maximum force obtained during stretching [N], A_0—initial cross-section of the sample [mm^2].

Young's modulus was calculated from the formula 3:

$$E = \frac{\sigma_M}{\epsilon_M}[MPa] \tag{3}$$

where: σ_M—maximum stress, ϵ_M—maximum relative elongation.

Abrasive wear resistance tests were carried out at the Anton Paar TRB tribology site in the sphere-disk layout. Shields with a diameter of 30 mm were made of modified PC. As a countertop, a ceramic ball of diameter 10 mm was used. Tribological studies were performed in rotary motion under technically dry friction conditions. The research was carried out with the following parameters: load 10 N, sliding speed – 0.1 m/s, friction distance - 1000 m, friction distance diameter: 16 mm. Within the tests, volumetric consumption and friction coefficient of cooperating friction pair were determined. Volume consumption of the disc was calculated according to the formula 4:

$$V_v = \frac{V}{F_n \cdot s} \tag{4}$$

where: V_v—volumetric wear $\left[\frac{mm^3}{N \cdot m}\right]$, F_n—normal force [N], s—friction distance [m], V- average volume of material removed during the tribological wear, calculated as: $V = P \times 2\pi r[mm^3]$, P—average surface area of the removed material [mm^2], r—radius of the friction distance: 8 mm.

The average surface of the wear trace - P was determined using the Mitutoyo Surftest SJ-500 profilograph.

3 Results

The amount of rhodamine and its derivatives less then 0.1% due to the chemical structure and its chemical compatibility to PC doesn't effect on hardness [14].

Due to rhodamine doesn't affect on mechanical properties, the hardness measurements were conducted only for PC modified with silicon oxide and compared to pure one. It was noticed that hardness of modified PC (14.7 HV) is higher than non-modified (11.52 HV) [11].

The results form mechanical tests are presented on Fig. 1. It was noticed that silicon oxide PC is characterized by worse mechanical properties than rhodamine PC e.g. the average of tension σM for PC modified with rhodamine was 69.94 MPa and for PC modified with silicon oxide was 52.2 MPa. In case of the hip prosthesis, one of the most important parameters is proper Young modulus which is vary depending on the type of bones. For cortical bone should be 5–22 GPa and for spongy bone 0.1–5 GPa [12]. After mechanical tests noticed that

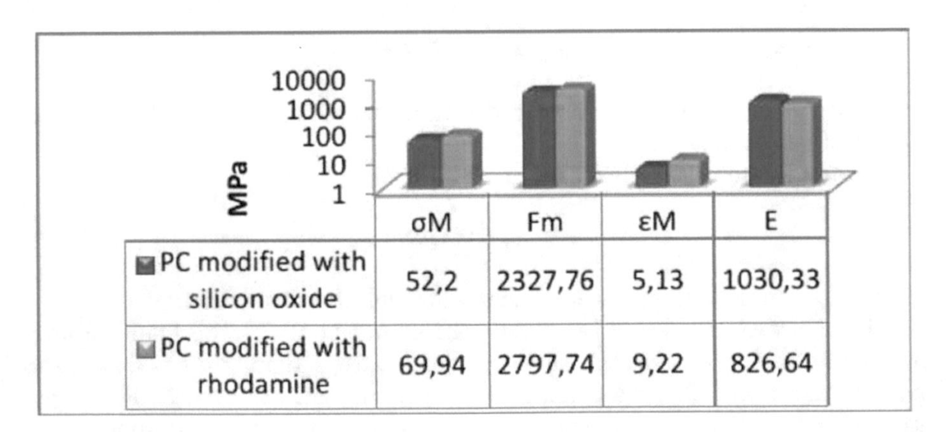

Fig. 1. Average values of selected parameters determined during static tensile test expressed in MPa.

Young modulus for rhodamine PC and silicone oxide PC is too small, for non modified PC Young modulus is in the range 2.1–2.5 GPa [13]. In case of using for hip prosthesis, the material should be reinforcement.

The last conducted research was connected with tribological properties. A significant difference in the tribological properties of the tested materials was found (Fig. 2). Sample with silicone oxide has a significantly lower wear resistance compared to sample with rhodamine. For sample with rhodamine, there was 11 times greater abrasive wear resistance.

Studies show that both materials differ significantly in course and friction coefficient values. In the case of sample with silicone oxide, it was found that after a friction path of about 35 m the coefficient of friction reaches a stabilized value of about 0.57. The value obtained is invariable until the end of the tribological test. At the same time, it was observed that this sample has a wide amplitude

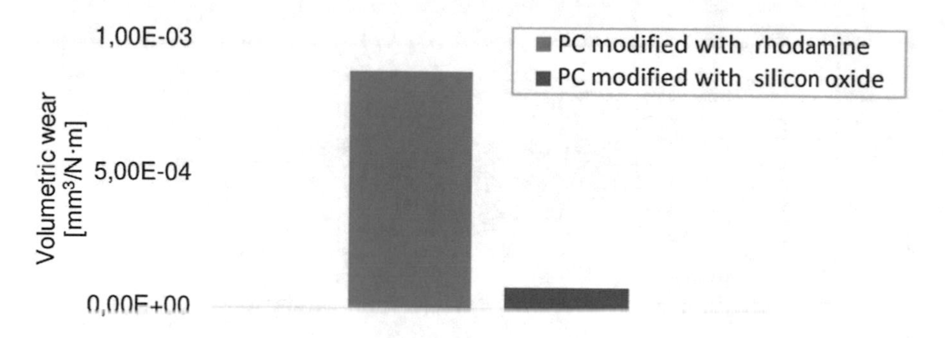

Fig. 2. Volumetric wear of femoral head prosthesis made from PC modified rhodamine and PC modified silicon oxide.

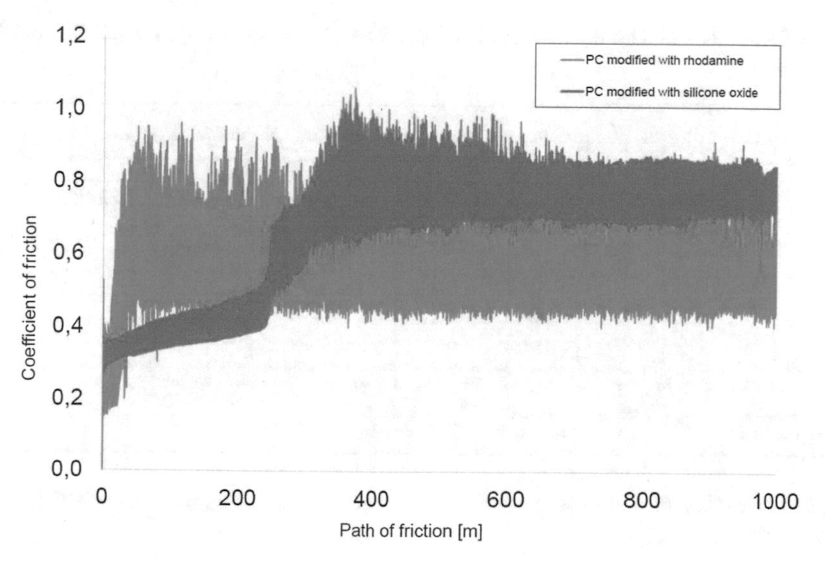

Fig. 3. Graph of changes in friction coefficient depending on the friction path.

of friction coefficient (which may be due to uneven surface). The maximum friction coefficient during the test was 0.97. Sample with rhodamine is characterized by a completely different course of friction coefficient. It has been shown that the course of the friction coefficient and its amplitude changed significantly more in comparison to sample with silicone oxide. In the friction path, up to 230 m, the coefficient of friction increased linearly to 0.45. At the same time, this parameter was characterized by relatively narrow amplitude. After exceeding the friction path of 230 m, the coefficient of friction was increased to approximately 0.82. At the same time, the amplitude of the friction coefficient was increased, especially in the range of friction from 300 to 600 m. The maximum value of friction coefficient was equal to 0.84 (Figs. 3, 4, 5).

The femoral head prosthesis before and after the test doesn't show volumetric were changes.

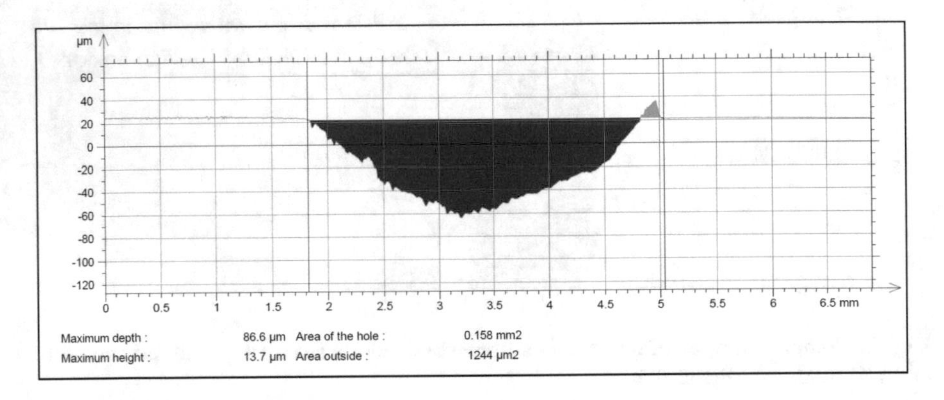

Fig. 4. Geometry of the cross-section of the wear trace for sample wear PC modified rhodamine.

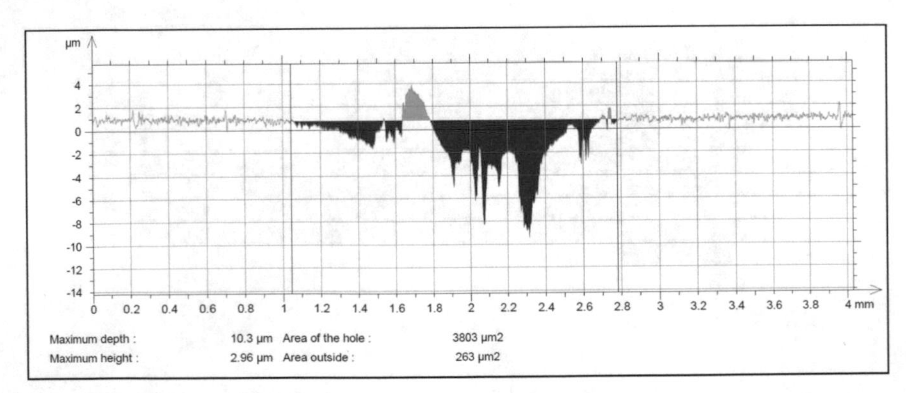

Fig. 5. Geometry of the cross-section of the wear trace for sample wear PC modified siliconoxide.

4 Discussion

There are several reconstruction methods of the joint after resection. In the all of them the material type plays the main role due to the mechanical and biological interactions. Interface material properties and surface geometry are the key areas that can influence the lubrication mechanism of artificial hip joints and, consequently, fluctuation in coefficient of friction and wear rate. In this study, an investigation was conducted on the behavior of future implant material, such as modified PC for hip joint prostheses. The most important finding of this study was the effect of nanosilica particles modification of PC and impact on its prostheses. Significantly lower and steady coefficient of friction profiles and lower volumetric ware were obtained for nanosilica PC modified compared to those of non-doped PC prostheses surface. Due to modification method restricted in the patent pending both materials are printable by the use of the FDM method as well as they can be used in the process of injection molding. Empirical tests and experience during 3D printing let spouse that method of modification affect on the adhesion properties to the table surface during printing. The phenomena opens wide area of potential application in the medical market. This property was not observed for any other printable material presented in the literature.

5 Conclusion

The obtained results clearly show that the presented material can successfully be used as a hip prosthesis and not just as a cartridge placed in the metal shell, because of exhibit very good wear resistance and above average strength properties as well as biocompatibility. The material presented is characterized by low cytotoxicity and biocompatibility what was already published in the chemical industry journal. The resulting polymer-mineral composites can be oriented in the manufacturing process to obtain a porous surface that allows the bone to be independently mounted with bone cement. Low usage for artificial femoral head tests shows the possibility of using this type of acetabulum in combination with not only the ceramic femoral head but also metal. In the future, according to the obtained results, it is worth to make the PC femoral head reinforced with suitable fibers to use the innovative combination polymer to eliminate the need for variable-pressure ceramics.

References

1. Affatato, S.: The history of biomaterials used in total hip arthroplasty (THA). In: Advances in Biomaterials and Their Tribological interactions, pp. 19–36 (2014)
2. Niemczewska-Wójcik, M.: Wear mechanisms and surface topography of artificial hip joint components at the subsequent stages of tribological tests. Meas. J. Int. Meas. Confederation **107**, 89–98 (2017)
3. Houcke, J.V., Khanduja, V., Pattyn, C., Audenaert, E.: The history of biomechanics in total hip arthroplasty. Indian J. Orthop. **51**(4), 421–433 (2017)

4. Hodge, W.A., Fijan, R.S., Carlson, K.L., Burgess, R.G., Harris, W.H., Mann, R.W.: Contact pressures in the human hip joint measured in vivo. Proc. Nat. Acad. Sci. U.S.A. **83**(9), 2879–2883 (1986)
5. Askari, E., Flores, P., Dabirrahmani, D., Appleyard, R.: A review of squeaking in ceramic total hip prostheses. Tribol. Int. **93**, 239–256 (2015)
6. Alvarado, J., Maldonado, R., Marxuach, J., Otero, R.: Biomechanics of hip and knee prostheses. Engineering 1–20 (2003)
7. Si, H.B., Zeng, Y., Cao, F., Pei, F.X., Shen, B.: Is a ceramic-on-ceramic bearing really superior to ceramic-on-polyethylene for primary total hip arthroplasty? A systematic review and meta-analysis of randomised controlled trials. Hip Int. **25**(3), 191–198 (2015)
8. The Polish Committee for Standardization: Pomiar twardości sposobem Vickersa. Część 1: Metoda badań (PN-EN ISO 6507-1:2007) (2007)
9. The Polish Committee for Standardization: Tworzywa sztuczne. Oznaczanie właściwości mechanicznych przy statycznym rozciąganiu. Część 1: Zasady ogólne. (PN-EN ISO 527-1:2012) (2012)
10. The Polish Committee for Standardization: Tworzywa sztuczne. Oznaczanie właściwości mechanicznych przy statycznym rozciąganiu. Część 2: Warunki badań tworzyw sztucznych przeznaczonych do prasowania, wtrysku i wytłaczania (PN-EN ISO 527-2:2012) (2012)
11. Covestro (Bayer). Makroln 2600 PC Datasheet. http://cn.tccves.com/materials/148/makrolon-2600. Accessed 2 Feb 2018
12. Madej, T., Ryniewicz, A.M.: Modelling and strength simulations in a hip joint equipped with an overlay prosthesis as a diagnostic procedure before the hip. Tribologia **2**, 115–128 (2013)
13. https://www.tworzywa.pl/wiedzopedia/baza-tworzyw/81,poliweglan-pc,polimer.html. Accessed 15 July 2018
14. Swinarew, A., Flak, T., Okła, H., Kubik, K., Rozwadowska, B., Gabor, J., Łężniak, M.: Organiczny materiał bakteriostatyczny, nr zgłoszenia: P. 420670. (in Polish)

Corrosion Resistance of Passivated NiTi After Long-Term Exposure to Simulated Body Fluids

Marcin Kaczmarek(✉), Przemysław Kurtyka, and Zbigniew Paszenda

Faculty of Biomedical Engineering,
Department of Biomaterials and Medical Devices Engineering,
Silesian University of Technology, Zabrze, Poland
marcin.kaczmarek@polsl.pl

Abstract. Corrosion resistance of the implant alloy is a very important determinant of its biocompatibility. Both, the environment and the applied surface treatment have a significant influence on corrosion. The main aim of the research was evaluation of corrosion resistance of NiTi alloy after long-term exposure to various simulated body fluids. The evaluation of the electrochemical behavior of NiTi alloy was realized by recording of anodic polarization curves with the use of the potentiodynamic method. The tests were carried out for differently modified surfaces both before and after the 6 months exposure to the simulated body fluids. Surface condition of metal biomaterial determines its corrosion resistance. In the course of the work it was observed that the corrosion resistance of all the samples (with different surface conditions) before long-term exposure is similar. The exposure did not change the corrosion resistance of the NiTi alloy significantly. Only the electropolished samples revealed the decrease of the corrosion resistance. The obtained results could be the basis optimization of physicochemical properties of the NiTi alloy. The future research should be focused on selected specific implants specially with respect to their application features. On the basis of the obtained results it can be stated that the suggested surface treatment can be applicable for NiTi alloys due to the increase of the corrosion resistance. The long-term exposure to the simulated body fluids did not deteriorate the corrosion resistance of the investigated alloy. The proposed surface treatment methods can be applied to implants intended for diverse medical applications, especially in cardiology and urology.

Keywords: NiTi alloy · Biomaterials · Corrosion resistance · Potentiodynamic tests · Simulated body fluids

1 Introduction

Equiatomic NiTi shape memory alloys belong to the group of materials known as 'smart materials'. The outstanding properties exhibited by them are the thermal

E. Tkacz et al. (Eds.): IBE 2018, AISC 925, pp. 301–311, 2019.
https://doi.org/10.1007/978-3-030-15472-1_32

shape memory effect and superelastic behavior. Shape memory refers to the ability of the material to 'remember' its shape, even after significant deformations. Once deformed at low temperatures, in their martensitic phase, these materials will spontaneously return to their original predeformated shape when heated. Besides the shape memory effect, this family of alloys exhibits remarkable characteristics such as a pseudoelastic behaviour and a very high damping capacity. Pseudoelasticity is the effect by which a material recovers, upon unloading, a strain produced by a stress-induced phase transformation. The amount of this reversible strain is much greater than the elastic strain produced within a single phase. In contrast to the shape memory effect, here the temperature remains constant.

The basis of the shape memory effect and superelasticity are both based on a thermoelastic, reversible martensitic transformation. The high temperature austenite phase transforms martensitically upon cooling below an alloy specific temperature to a distorted monoclinic martensite structure [1]. Nitinol is characterized by a specific stress-strain relation that is different from the deformation behavior of conventional materials but similar to living tissues. The similarity in the deformation behavior between Nitinol and living tissues contributes to the harmonic performance of dynamic implants under loading-unloading conditions in the body and represents such a feature as biomechanical compatibility of Nitinol. The absence of high recoverable strain similar to that of living tissues may be one of the major reasons for the fracture of fixation plates made of conventional materials in the body where they can be subjected to plastic deformation [2].

When shape memory alloys are considered as candidates to be applied in medical devices, they must be able to fulfill functional requirements related not only to their mechanical reliability but also to their chemical reliability (in vivo degradation, decomposition and dissolution, corrosion, etc.) and their biological reliability (biocompatibility, cytotoxicity, carcinogenicity, anti-thrombogenicity, antigenicity, etc.). NiTi is the most important shape memory alloy used as a biomaterial without any deterioration in its properties and in its specific function after implantation [1–3]. Due to its extraordinary properties, e.i. elastic deployment, thermal deployment, kink resistance, biocompatibility, constant unloading stresses, biomechanical compatibility, dynamic interference, hysteresis, MR compatibility, fatigue resistance and uniform plastic deformation nitinol is considered as the material of first choice in medical applications. Since the very beginning Nitinol was found to be applied in orthopaedics (osteosynthesis plates, intramedullary nails, staples for fastening bones, Harrington rods, jaw plates), heart and vascular surgery (filters and stents, artificial heart muscle, cranial aneurysm clips), dentistry and orthodontics (oral implants, orthodontic archwires) and other medical devices (micropump for an artificial kidney, rehabilitation devices, guidewires, arthroscopic suture needles) [4].

Various studies have shown that the metal components of the alloys used in implantology may be toxic and dissolve in body fluids due to corrosion. Every metal has its own intrinsic toxicity to cells, but the corrosion mostly determines the existing concentration. Thus, the corrosion resistance of the alloy and the toxicity of individual metals in the alloy are the main factors determining its biocompatibility [5–12,17,18]. The corrosion of metals in aqueous solutions takes place via an electrochemical mechanism. Different metals have different intrinsic aptitudes to corrode. The more noble the metal, the lesser is its aptitude to corrode. Reactions taking place on the metal surface and in the specific environment may cause radical changes in this theoretical nobility. After implantation, the metal is surrounded by serum ions, proteins and cells, which may all modify the effect on local corrosion reactions. There are numerous factors which affect metal corrosion. Porosity and rough surfaces increase the reacting surface area of the implant and thus the total amount of corrosion. The loading areas of the implant are more sensitive to corrosion compared to the less loading areas. The structure, composition and thickness of the passive layer are highly dependent on the metal itself and its environment.

Metals contain various elements, such as lattice defects, impurities and contaminants, which may affect the corrosion reaction. The different heat treatments and working processes change the grain size and energy state of the metal and cause surface heterogenecity. All these factors may affect the passive layer. The corrosion resistance of metals and metal alloys is mainly based on a passivation phenomenon. The passivation of a metal is due to the compact coat, the passive layer, which contains hardly any original metal, but forms a metal-oxide layer. The corrosion resistance of the implant alloy is a very important determinant of its biocompatibility. The wide spectrum of application in implantology imposes special requirements on the biocompatibility of Nitinol. The two main factors that determine the biocompatibility of a material are: the host reaction induced by the material and the degradation of the material in the body environment. Much concern exists over both these issues in the case of Nitinol. Dissolution of Ni ions and the possibility of inducing allergic, toxic and carcinogenic [13] effects associated with the biological properties of Ni is the greatest problem that can be faced after Nitinol implantation [2]. The body is a complicated electrochemical system that constitutes an aggressive corrosion environment. Body fluids consist of an aerated solution containing 0.9% NaCl, with minor amounts of other salts and organic compounds. High acidity of certain body fluids is especially hostile for metallic implants. Acidity can increase locally in the area adjacent to an implant due to inflammatory response of surrounding tissues mediating hydrogen peroxide and reactive oxygen and nitrogen species [2]. Many experimental set-ups have been used to evaluate the corrosion resistance of NiTi with the use of two main methods: the potentiodynamic and the potentiostatic. There is also a third approach to corrosion, which may even be applied in vivo. It consists of measuring the concentration of leached ions in a solution by spectroscopic means after exposure of the sample to a given environment.

Numerous attempts have been undertaken to characterize the corrosion behavior of NiTi used in medical devices [14]. However, using comparable testing strategies, the results differ considerably, conferring poor to excellent corrosion resistance to NiTi in physiological fluids. The characteristic values of Ep taken from the literature vary over a wide range, from −200 to 1.300 mV [6,8,15]. But the main reason is differences in the samples' surface quality, which may be affected by various factors. In studies where samples from different providers are tested in the as-received state, a marked difference in corrosion resistance is noted depending on the brand. The difference in corrosion resistance may also be due to sample processing steps (mechanical and electropolishing, passivation, modifications with the use of energy sources and chemical vapors) [4,16].

2 Materials and Methods

The corrosion tests were carried out on equiatomic NiTi alloy. The chemical composition of the alloy (Ni - 55.5%, Ti - balance) met the requirements of the ASTM 2063 standard. The tests were carried out on flat samples ($10 \times 10 \times 1$ mm). In order to evaluate the influence of diverse methods of surface modification on corrosion resistance of the alloy, the following surface treatments were applied:

- electropolishing,
- H_2O chemical passivation,
- H_2SO_4 electrochemical passivation.

Since passivation is often considered as the first choice, surface treatment assuring formation of the dense, stable TiO_2 oxide layer, different methods of passivation were adopted in the study. Both chemical and electrochemical methods were adopted. The applied methods of surface treatment and their parameters were presented in Table 1.

Table 1. Parameters of the applied surface modifications

Surface treatment	Applied baths	Time[min.]	Temp.[°C]	Potential [V]
Electropolishing	HF-based	15	60	50
Chemical passivation	H_2O	60	Boiling	-
Electrochemical passivation	H_2SO_4	3–20	10–30	25

Since NiTi alloy can be used in different fields of medical applications, the following simulated body fluids were selected for the corrosion studies:

- Tyrode's physiological solution,
- artificial urine,
- artificial plasma.

Chemical compositions of the selected simulated body fluids were presented in Table 2.

Table 2. Parameters of the applied surface modifications

Tyrode's solution								
Concentration of components [g/l]								
NaCl	$CaCl_2$	KCl	$NaHCO_3$	$NaHPO_4$	$MgCl_2$			
8.000	0.200	0.220	1.000	0.050	0.200			
Artificial urine								
Concentration of components [g/l]								
$CaCl_2(aq)$	Na_2SO_4	$MgSO_47$(aq)	NH_4Cl	KCl	$NaHPO(aq)$	$NaHPO_4$	$Na_3(aq)$	NaCl
1.765	4.862	1.143	4.643	12.130	6.8	0.869	1.168	13.545
Artificial plasma								
Concentration of components [g/l]								
NaCl	CaCl2	KCl	$NaHCO_3$	NaH_2PO_4	$MgSO_4$	Na_2HPO_4		
6.8	0.2	0.400	2.200	0.026	0.100	0.126		

The electrochemical tests of the investigated alloy were performed with the use of a potentiodynamic method by recording of anodic polarization curves. In the tests the scan rate was equal to 1 mV/s. The PGP 201 (Radiometer) potentiostat with the software for electrochemical tests was applied - Fig. 1. The saturated calomel electrode (SCE) was applied as the reference electrode and the auxiliary electrode was a platinum wire.

All samples were immersed in the artificial plasma for 60 min before the scanning started at a potential of about 100 mV below the recorded open circuit potential (EOCP). The scanning direction was reversed when the anodic current density reached 1000 $\mu A/cm^2$. The tests were carried out at the temperature of $37 \pm 1O$ °C. On the basis of the recorded curves characteristic values describing the resistance to pitting corrosion i.e.: corrosion potential Ecorr (V), breakdown potential Eb (V) or transpassivation potential Etr (V), polarization resistance Rp ($\Omega * cm^2$) and corrosion current density (A/cm^2) were determined. To determine the value of polarization resistance Rp the Stern method was applied.

The aim of the second stage of research was evaluation of corrosion resistance of the NiTi alloy after the 6 months exposure to the simulated body fluids. In order to carry our this stage of research the experimental set-up was designed - Fig. 2. The set-up allowed to simulate flow conditions adequate to physiological. Furthermore, pH value and temperature (37 ± 1 °C) were continuously controlled.

3 Results and Discussion

The electrochemical tests showed diverse resistance of NiTi alloy to pitting corrosion, depending on the applied surface treatment and the applied artificial

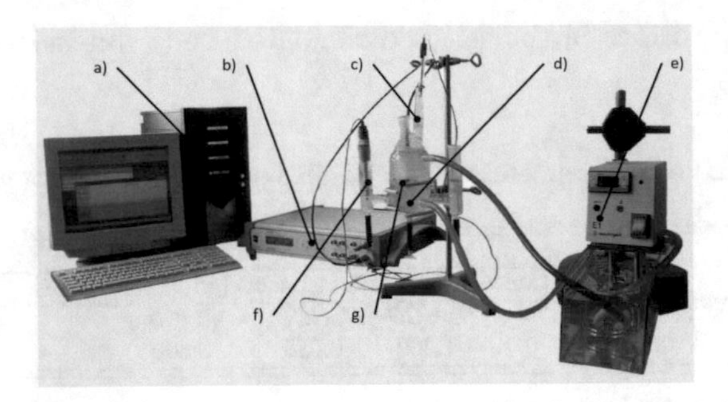

Fig. 1. Corrosion study set-up: (a) computer with software recording anode polarization curve curves, (b) VoltaLab®PGP201 potentiostat, (c) reference electrode - saturated calomel electrode (SCE), (d) cell for electrochemical tests, (e) thermostat, (f) auxiliary electrode - platinum electrode, (g) anode - test sample.

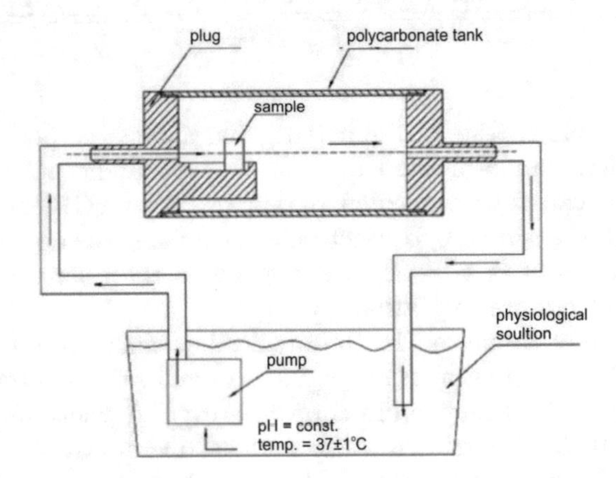

Fig. 2. Set-up designed for tests of 6 month exposure to simulated body fluids, simulating physiological flow conditions.

body fluids. Results of the pitting corrosion tests for the electropolished and passivated reference samples as well as the samples after the 6 months exposure to the given simulated body fluids were presented in Tables 3 and 4 and in Fig. 3. The results presented in the tables are mean values.

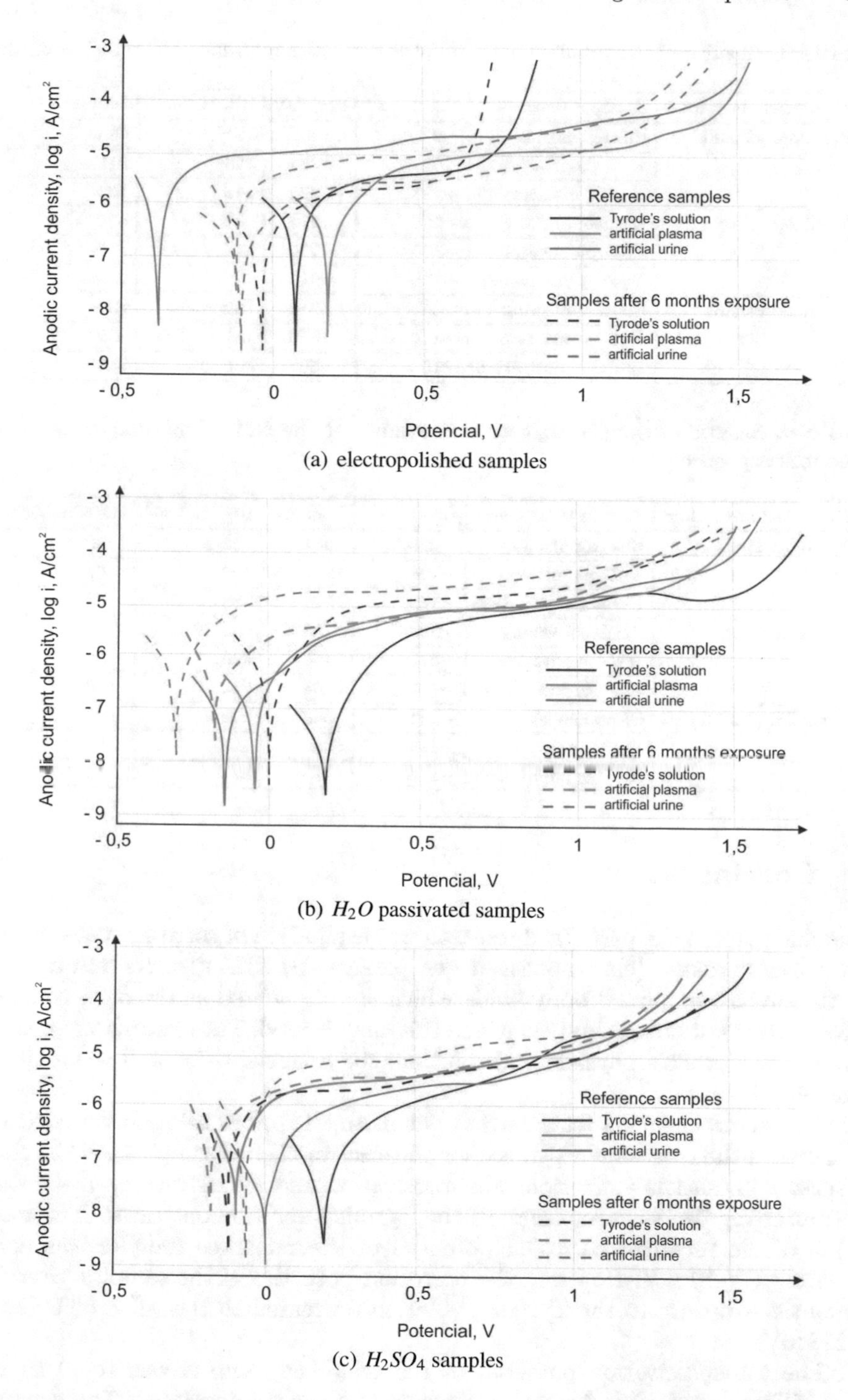

(a) electropolished samples

(b) H_2O passivated samples

(c) H_2SO_4 samples

Fig. 3. Exemplary anodic polarization curves for: (a) the electropolished samples, (b) the H_2O passivated samples, (c) the H_2SO_4 samples, before and after the 6 months exposure to the applied simulated body fluids

Table 3. Results of the pitting corrosion studies of the reference NiTi alloy samples

Simulated body fluid	Surface treatment	Ecorr, mV	Etr, mV	Rp, kω*cm^2	icorr, nA/cm^2
Tyrode's solution	Electropolishing	−81	+1357	37	688
	H_2O passivation	+87	+1372	135	211
	H_2SO_4 passivation	+121	+1395	143	172
Artificial urine	Electropolishing	−200	1222	40	650
	H_2O passivation	−188	1245	120	217
	H_2SO_4 passivation	−136	1316	89	292
Artificial plasma	Electropolishing	−33	1216	113	230
	H_2O passivation	−126	1400	137	190
	H_2SO_4 passivation	−153	1307	106	245

Table 4. Results of the pitting corrosion studies of the NiTi alloy samples after the 6 months exposure

Simulated body fluid	Surface treatment	Ecorr, mV	Etr, mV	Rp, kω*cm^2	icorr, nA/cm^2
Tyrode's solution	Electropolishing	−228	641	23	1130
	H_2O passivation	−108	1307	406	64
	H_2SO_4 passivation	−146	1375	738	35
Artificial urine	Electropolishing	−394	1428	562	46
	H_2O passivation	−377	1454	489	53
	H_2SO_4 passivation	−135	1405	641	40
Artificial plasma	Electropolishing	−141	338	42	604
	H_2O passivation	−193	1339	427	61
	H_2SO_4 passivation	−174	1358	571	45

4 Conclusion

The main aim of the performed research was evaluation of corrosion resistance of the surface treated (electropolished and passivated) NiTi samples and exposed to the diverse simulated body fluids which are considered as the most common environments of human body in which this alloy is used. The research was carried out in the Tyrode's physiological solution, the artificial urine and the artificial plasma.

The electrochemical tests carried out in the Tyrode's solution revealed the decrease of the corrosion resistance for the electropolished samples. All the characteristic potentials (corrosion, transpassivation and repassivation) of the samples after the 6 months exposure to the Tyrode's solution decreased. The mean value of the recorded corrosion potential of the reference samples was equal to Ecorr = −81 mV. However, the corrosion potential of the samples after the 6 months exposure to the Tyrode's solution decreased to the value of Ecorr = −228 mV.

The transpassivation potential of the reference samples was equal to Etr = +1357 mV and after 6 month exposure significantly decreased. The decrease was so considerable that the recorded potential should be referred to as the breakdown potential, since the hysteresis loop was also observed. The obtained mean value of the breakdown potential was equal to Ebr = 641 mV.

For the H_2O passivated samples tested in the Tyrode's solution no decrease of the corrosion resistance was observed. However, the significant change of the corrosion potential was recorded (Ecorr changed from +87 mV to −108 mV). But this change did not influence the transpassivation and repassivation potentials (the mean valued were similar). Moreover, the significant change of the polarization resistance was observed. The obtained results revealed that long term exposure to the Tyrode's solution increased the protective properties of the generated oxide layer. The recorded polarization resistance was three times higher than for the reference samples.

The electrochemical characteristics of the H_2SO_4 passivated samples tested in the Tyrode's solution were similar to the H_2O passivated samples. No deterioration of the corrosion resistance with was observed. The corrosion potential decreased significantly (Ecorr = +121 mV to −146 mV), however the obtained values for the transpassivation potential were similar (Etr = +1395 and +1397 mV). For this method of passivation (H_2SO_4 electrochemical) even greater (seven times) increase of the polarization resistance was observed.

In general, the corrosion resistance of the reference samples and the samples after the 6 months exposure to the artificial urine showed good corrosion resistance. However, the electrochemical tests carried out in the artificial urine showed changes in the corrosion and transpassivation potentials as well as polarization resistance and anodic current density.

The exposure of the electropolished samples to the artificial urine caused significant (almost two times) decrease of the corrosion potential, but at the same time, increase of the transpassivation potential (Etr = from 1222 mV to 1428 mV). The most spectacular increase was observed for the polarization resistance - 1400.

The same electrochemical behavior was observed for the H_2O passivated samples. For these samples significant decrease of the corrosion potential (200%) as well as increase of polarization resistance (400%) was recorded. For the H_2SO_4 samples tested in the artificial urine no significant changes of the corrosion potentials before and after the 6 months exposure were observed (Ecorr = −136 mV and −135 mV, respectively). For the transpassivation potential a slight increase (statistically insignificant) of the mean values was observed, whereas with reference to the polarization resistance, a statistically significant increase (720%) was recorded.

The electrochemical tests carried out in the artificial plasma revealed the decrease of the corrosion resistance for the electropolished samples. All the characteristic potentials of the samples after the 6 months exposure to the artificial plasma decreased. The corrosion potential changed significantly of almost 400% (decrease). The transpassivation potential decrease of the electropolished samples was equal 72%. Both, the decrease of the transpassivation potential (Etr = +1216 mV vs Ebr = +338 mV), the presence of the hysteresis loop and the decrease of the polarization resistance, indicate significant decrease of the corrosion resistance of the electropolished samples tested in the artificial plasma.

For both, the H_2O and H_2SO_4 passivated samples the corrosion and the transpassivation potentials have not changed significantly. However, for both applied passivation methods, a statistically significant increase of the polarization resistance was observed (311% - for the H_2O passivated samples, and 538% - for the H_2SO_4 passivated ones). The obtained results indicate that the applied methods of passivation provide similar electrochemical behavior of the tested NiTi alloy, and thus similar corrosion resistance.

On the basis of the obtained results it can be concluded that surface treatment consisting of electropolishing only, does not provide an appropriate level of corrosion resistance of the NiTi alloy, both in the environment of the Tyrode's solution and artificial plasma. Thus in order to increase corrosion resistance of the alloy, additional subsequent surface treatment is required. The proposed method of surface modification was passivation. It could be summarized that both of the applied passivation methods (chemical - H_2O and electrochemical - H_2SO_4) favorably influence the corrosion resistance of the NiTi alloy tested in all the analyzed simulated body fluids (Tyrode's solution, artificial urine and artificial plasma).

References

1. Jani, J.M., Leary, M., Subic, A., Gibson, M.A.: A review of shape memory alloy research, applications and opportunities. Mater. Des. **56**, 1078–1113 (2014)
2. Morgan, N.B.: Medical shape memory alloy applications - the market and its products. Mater. Sci. Eng. A **378**, 16–23 (2004)
3. Shabalovskaya, S., Van Humbeeck, J.: Biocompatibility of shape memory alloys. In: Yoneyama, T., Miyazaki, S. (eds.): Shape Memory Alloys for Medical Applications. Woodhead Publishing Limited, Cambridge (2008)
4. Kaczmarek, M.: Kształtowanie własności fizykochemicznych warstw powierzchniowych nadsprężystego stopu NiTi przeznaczonego na stenty wieńcowe. Wydawnictwo Politechniki Śląskiej, Gliwice (2015). ISBN 978-83-7880-305-8
5. Kajzer, A., Kajzer, W., Gołombek, K., Knol, M., Dzielicki, J., Walke, W.: Corrosion resistance, EIS and wettability of the implants made of 316 LVM steel used in chest deformation treatment. Arch. Metall. Mater. **61**(2a), 767–770 (2016)
6. Szewczenko, J., Marciniak, J., Kajzer, W., Kajzer, A.: Evaluation of corrosive resistance of titanium alloys used for medical implants. Arch. Metall. Mater. **61**(2a), 695–700 (2016)
7. Basiaga, M., Kajzer, W., Walke, W., Kajzer, A., Kaczmarek, M.: Evaluation of physicochemical properties of surface modified Ti6Al4V and Ti6Al7Nb alloys used for orthopedic implants. Mater. Sci. Eng. C **68**, 851–860 (2016)
8. Kajzer, A., Kajzer, W., Gołombek, K., Knol, M., Dzielicki, J., Walke, W.: Corrosion resistance, EIS and Wettability of the implants made of 316 LVM steel used in chest deformation treatment. Arch. Metall. Mater. **61**(2a), 767–770 (2016). (30 pkt IF = 1,09)
9. Szewczenko, J., Marciniak, J., Kajzer, W., Kajzer, A.: Evaluation of corrosion resistance of titanium alloys used for medical implants. Arch. Metall. Mater. **61**(2a), 695–700 (2016). https://doi.org/10.1515/amm-2016-0118
10. Kaczmarek, M.: Investigation of pitting and crevice corrosion resistance of NiTi alloy by means of electrochemical methods. Prz. Elektrot. **86**(12), 102–105 (2010)

11. Kaczmarek, M., Kurtyka, P.: Corrosion resistance of surface treated NiTi alloy tested in artificial plasma. In: Gzik, M., Tkacz, E., Paszenda, Z., Piętka, E. (eds.) Innovations in Biomedical Engineering. Advances in Intelligent Systems and Computing, vol. 526. Springer, Cham (2017)
12. Kajzer, W., Jaworska, J., Jelonek, K., Szewczenko, J., Kajzer, A., Nowińska, K., Hercog, A., Kaczmarek, M., Kasperczyk, J.: Corrosion resistance of Ti6Al4V alloy coated with caprolactone-based biodegradable polymeric coatings. Eksploat. i Niezawodn. **20**(1), 30–38 (2018)
13. Pulletikurthi, C., Munroe, N., Gill, P., Pandya, S., Persaud, D., Haider, W., Iyer, K., McGoron, A.: Cytotoxicity of Ni from surface-treated porous nitinol (PNT) on osteoblast cells. J. Mater. Eng. Perform. **20**, 824–829 (2011)
14. Tian, H., Schryvers, D., Liu, D., Jiang, Q., Van Humbeeck, J.: Stability of Ni in nitinol oxide surfaces. Acta Biomater. **7**, 892–899 (2011)
15. Shabalovskaya, S., Rondelli, G., Anderegg, J., Xiong, J.P., Wu, M.: Comparative corrosion performance of black oxide, sandblasted, and fine-drawn nitinol wires in potentiodynamic and potentiostatic tests: effects of chemical etching and electropolishing. J. Biomed. Mater. Res. Part B Appl. Biomater. **69B**, 223–231 (2004)
16. Khalil-Allafi, J., Amin-Ahmadi, B., Zare, M.: Bio-compatibility and corrosion behavior of the shape memory NiTi alloy in the physiological environments simulated with body fluids for medical applications. Mater. Sci. Eng. C **30**, 1112–1117 (2010)
17. Basiaga, M., Paszenda, Z., Walke, W.: Study of electrochemical properties of carbon coatings used in medical devices. Przeglad Elektrotechniczny **87**(12B), 12–15 (2011)
18. Walke, W., Paszenda, Z., Pustelny, T., Opilski, Z., Drewniak, S., Kościelniak-Ziemniak, M., Basiaga, M.: Evaluation of physicochemical properties of SiO2-coated stainless steel after sterilization. Mater. Sci. Eng. C-Mater. Biol. Appl. **63**, 155–163 (2016)

The Effect of Surface Modification of the Zirconia Abutment on Its Physicochemical Properties

Klaudia Tokarska[1(✉)], Anna Ziębowicz[1], Przemysław Kurtyka[1], Marcin Kaczmarek[1], and Jarosław Serafińczuk[2]

[1] Faculty of Biomedical Engineering, Department of Biomaterials and Medical Devices Engineering, Silesian University of Technology, Zabrze, Poland
klaudia.tokarska@polsl.pl

[2] Faculty of Microsystem Electronics and Photonics, Department of Metrology, Micro and Nanostructures, Wroclaw University of Technology, Wroclaw, Poland

Abstract. Nowadays aesthetic plays a very important role in prosthetic restorations. Over the past several years the metal restorations are being replaced by full ceramic restorations. Dental crowns built on the foundation of the zirconia are characterized by high aesthetics and high bending strength in comparison to other available materials. Ceramics, as a biocompatible material, are very willingly used in dental prosthetics. Long-term clinical observations have shown that the restorations on the zirconia abutments are often accompanied by destructive incidents such as chipping and cracking of the ceramics or fractures in the abutments. The appropriate connection of the porcelain with abutment can be achieved by modifying the outer surface of the ceramics. The aim of the project was to define the influence of the sandblasting and etching with hydrofluoric acid (HF) on the physicochemical properties of the surface of the zirconia abutment. The study consisted of the wettability angle measurement, Atomic Force Microscopy (AFM), X-ray Diffraction (XRD) and Scanning Electron Microscopy (SEM). The research has shown that the process of sandblasting and HF acid-etching influences the change on physicochemical properties of zirconia surface, its topography and wettability. After conducting the sandblasting and acid-etching manipulation of the zirconia surface, the wettability angle corresponds with poorly wettable materials, which are characterized by hydrophobic properties. The biggest variations on the zirconia surface were obtained after etching with hydrofluoric acid.

Keywords: Surface engineering · Zirconium oxide · Ceramic abutments · All-ceramic restorations · Dental prosthetics

E. Tkacz et al. (Eds.): IBE 2018, AISC 925, pp. 312–320, 2019.
https://doi.org/10.1007/978-3-030-15472-1_33

1 Introduction

Currently, aesthetics is an important aspect of the proposed prosthetic solutions. Awareness of dental possibilities intensifies the demand for a beautiful, perfect smile, and thus healthy and white teeth [3]. In recent years, significant development of dental materials and production technologies, has been observed.

Classic dental crowns on a metal abutment were replaced with all-ceramic restorations [4]. They are characterized by high aesthetics and satisfactory mechanical properties [2,5,6]. The zirconium oxide turned out to be a revolutionary discovery for dental prosthetics, which made it possible to obtain the transparency of a natural tooth and in terms of mechanical properties it is comparable to metals [1,7,8]. All-ceramic restorations on the zirconia abutment can be used not only in the front section of the dental arch but also lateral and posterior. Long-term clinical observations have shown that restorations based on zirconium oxide are often accompanied by delamination incidents related to the durability of the bond strength of layered porcelain fused to zirconia. There are also fracture cases including cracking and crushing of abutments, but the reasons responsible for this phenomena are not yet well known and require further research [9–11]. Works are still in progress to improve and optimize the adhesion of the layer of fused porcelain to ceramic abutments, which depend on the physicochemical properties of the substrate [12–15].

The paper focuses on examining the influence of the method of surface modification of the ceramic abutments on the change of its physicochemical properties and integration of ceramic restoration layers. The following methods of surface modification of the ceramic abutments were analysed: abrasive blasting with aluminium oxide (Al_2O_3) and hydrofluoric acid (HF) etching. The scope of the study included sample preparation, sinterization, surface modification and evaluation of the modification impact on physicochemical properties of zirconium oxide. The characteristic of the obtained surfaces was determined using the following research methods: contact angle investigation, atomic force microscopy (AFM), X-ray diffraction (XRD) and scanning electron microscopy (SEM).

2 Materials and Methods

Twenty flat samples were prepared from presintered zirconium oxide blocks Zr-i Y-TZP Copran stabilized by yttrium oxide and produced by Whitepeaks Dental Solutions GmbH & Co. KG - Tables 1 and 2. The samples were obtained by machining on a Robocam CNC milling machine. After treatment the dimensions of the samples were 22 mm × 6.3 mm × 0.7 mm.

Samples were randomly divided into 4 groups of 5 samples - Fig. 1. The first group consisted of samples in the presintered initial state. The other three were subjected to sintering process, one of which was a control group, while the other two were modified respectively by sandblasting and HF etching.

The sintering process allows to obtain the target mechanical properties, including hardness, and is required for clinical use. The process was performed

Table 1. Chemical composition of zirconium oxide Zr-i Y-TZP Copran

Chemical composition	ZrO_2	Y_2O_3	Al_2O_3	Fe hyd.	Er_30_3	Co_3O_4	Other
[%]	Balance	4.95–5.35	0.15–0.35	0–0.01	0	0	0–0.06

Table 2. Technical data of zirconium oxide Zr-i Y-TZP Copran

Density $[g/cm^3]$	6.05
Flexural strength [MPa]	1400

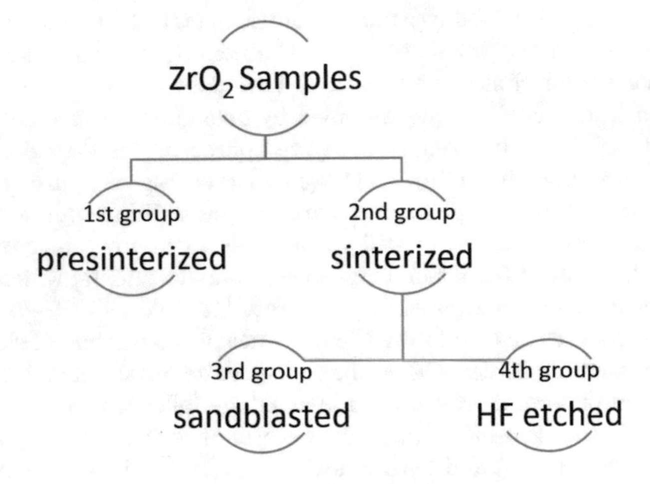

Fig. 1. Diagram of the research group

within a Robocam sintering furnace according to the manufacturer's instructions. Initially the samples were heated at a rate of 10 °C/min up to 950 °C - 95 min, then the heating rate was reduced to 6 °C/min and samples were heated at this rate until the maximum temperature of 1500 °C was reached - 250 min. The samples were maintained at 1500 °C for 90 min. The last step was controlled cooling, initially at a rate of 10°C/min till the temperature of 300 °C was reached, and then at a rate of 6 °C/min until the furnace was fully cooled down. The total time of the sintering process was 10 h 5 min - Fig. 2.

Within the surface modification, one group of 5 samples was subjected to sandblasting with the use of aluminium oxide as an abrasive factor with a grain size of 110 μm. The angle of incidence of the stream was equal 45°. Each sample was sandblasted under the pressure of 2.5 bar in distance of 10 mm for a period of 15 s. After the treatment, the samples were cleaned using a stream of steam. The second group was treated with 9% hydrofluoric acid (HF) to etch the surface of the samples. The process was carried out by immersing the samples in the acid solution for 5 min, then rinsing in distilled water and drying. In order to determine the surface quality of the zirconium oxide and the nature of its

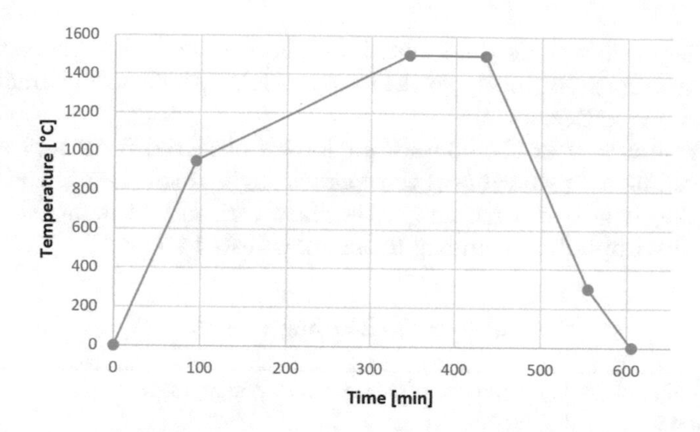

Fig. 2. Diagram of the research group

connection with the fused material, a wettability test was carried out by measuring the contact angle for the control group and modified surfaces. Material wettability affects the absorption as well as the aggregation and depends on its surface energy.

The investigation of the contact angle was carried out on samples at 20 °C on the device produced by Möller-Wedel Optical using goniometric method. The distilled water was applied on the surface of the sample with a volume of the measuring drop equal 1.5 μl. In order to preserve statistics, 5 measurements were taken for each sample. Samples were cleaned and dried from the residual water using compressed air before each measurement. Then the analysis of obtained surfaces was performed using atomic force microscopy (AFM) produced by Park System model XE-100 and scanning electron microscopy (SEM) from Hitachi S-570 at the voltage of 2.00 kV. In the case of AFM, the contact mode was used in which there is a direct contact of the measuring blade (100 μm × 10 μm × 1 μm) with atoms of the sample at a distance less than 0.1 nm. The phase composition of zirconium oxide was analysed by X-ray diffraction (XRD) using an X'Pert diffractometer from Philips. The monochromatic radiation of wavelength 1.518Å was used in the conducted research, which corresponds to the CuKα line and a scintillation counter for detection and recording of the signal. Then the diffractometric data processing was performed using the HighScore software.

3 Results

The results of contact angle test for samples after sintering and after surface modification are presented in Table 3. However, the presinterized samples were characterized by an unmeasurable wetting angle, which indicates the high hydrophilicity of such form of the material. While, the average value of the contact angle for samples after sintering process was 79°. In the case of samples after sandblasting an increase in the contact angle to about 95° was observed in relation to the test performed immediately after the sintering process. The value of the contact angle determined for the samples after the sandblasting process corresponds to poorly wettable materials, which are characterized by hydrophobic properties.

A similar increase in the contact angle was observed for samples after HF acid-etching. The average value of the contact angle in this case is equal 100.78°. Therefore, the surface of these samples is characterized by the lowest wettability and highest hydrophobicity among all samples tested so far.

Table 3. Contact angle results

Type of surface modification	Sample no.	Average wettability angle	
		After sintering	After surface modification
Sandblasting	1	78.25	95.92
	2	77.40	96.24
	3	79.04	101.80
	4	79.62	91.03
	5	82.40	91.94
HF etching	6	78.34	100.72
	7	76.98	99.83
	8	76.18	97.82
	9	79.90	103.41
	10	79.67	102.10

As a result of the surface granularity test within the AFM analysis, histograms of the surface morphology were obtain considering the number of grains and their sizes. The grain size for samples after sintering process is in the range from 100 to 580 nm with simultaneous decrease in their density compared to presinterised samples whose grain size is in the range from 30 to 300 nm. In the AFM image for presinterized sample, the surface is irregular with numerous depressions and protrusions. After the sintering process the sample surface has been smoothed. The sandblasting process and HF acid-etching affected the surface morphology of samples, what is presented in example F and H - Fig. 3.

The phase identification of presinterized samples was obtained using X-ray diffraction. The study showed the presence of monoclinic phase for the following peaks 28.068, 29.980, 31.15, with the grain size of: 28.068 - 277Å, 29.980 - 1713Å, 31.150 - 506 Å- Fig. 4. However, in case of samples subjected to sintering process, the presence of the tetragonal phase was detected - reflex (111) - Fig. 4. As a result of the sintering process, a polycrystalline single-phase ceramic material was formed. X-ray diffraction showed no phase changes after surface modification of the samples by sandblasting or HF acid-etching. In case of samples after sintering and sandblasting or etching the grain size for the 29.993 peak was greater than 1 μm. Therefore, it may be concluded that the sintering of zirconium oxide

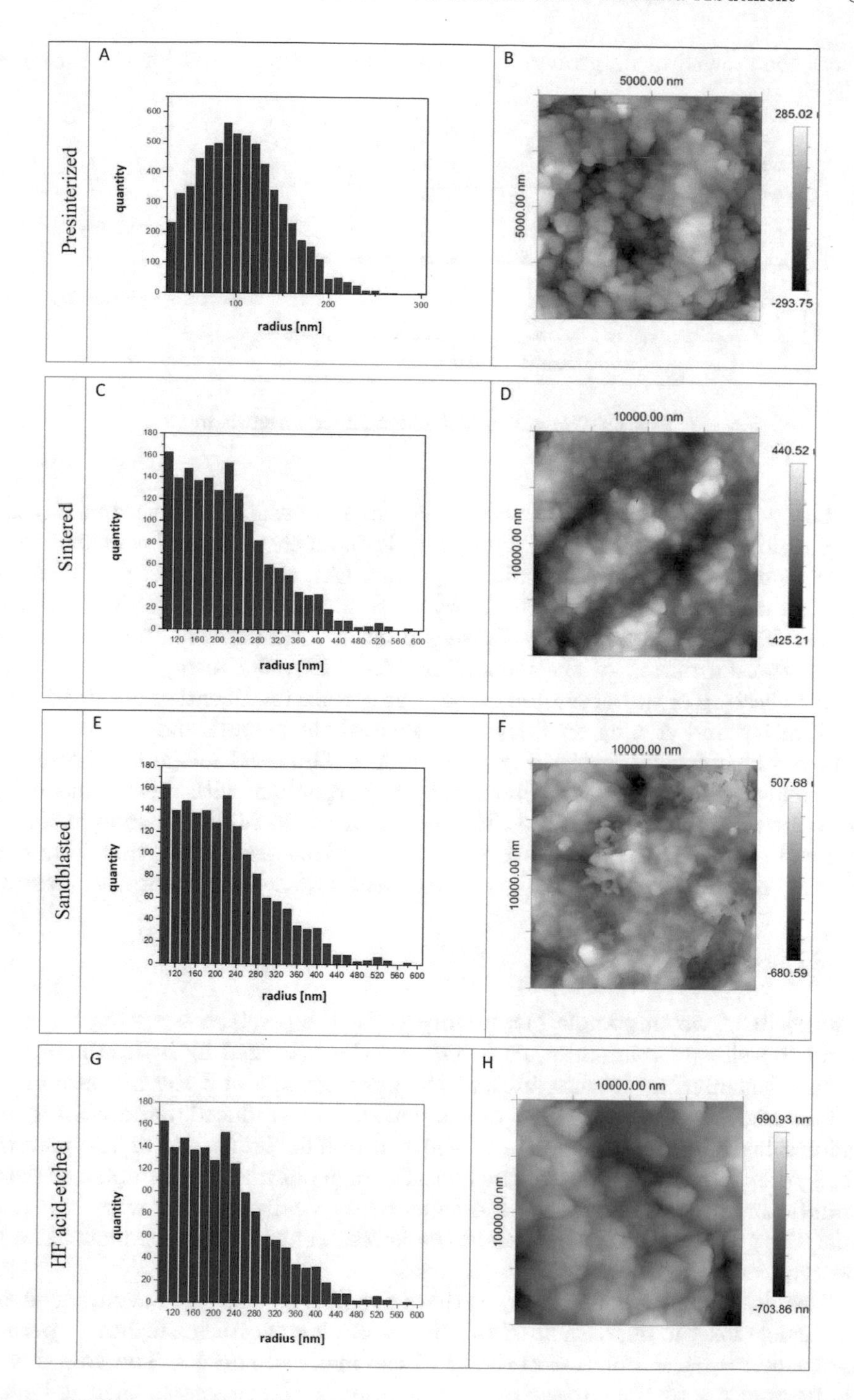

Fig. 3. AFM method - histograms of the amount of grains per radius (nm) and AFM images

affects the growth of its grains. The grain size was calculated for the strongest reflections.

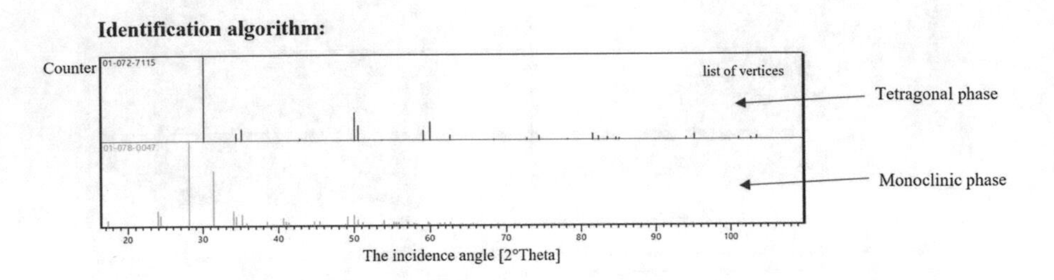

Fig. 4. The vertex list identification algorithm

The results obtained within SEM present the surface structure of zirconia samples for individual research groups - Fig. 5. In the image of a presinterized sample subjected only to the milling process (A), numerous irregularities and gaps between the grains can be observed. However, the sample after sintering process (B) is characterized by densely packed crystals, which may be caused by the transformation of the monoclinic phase into the tetragonal phase as a result of heating of the zirconium oxide. The surface modification obtained using sandblasting and etching with HF acid affected the smoothening of the sample surface, which can be seen in Figure C and D. The SEM image for the sample after acid-etching (D) is characterized by a morphology with greater inequality and convexity compared to the SEM image of the sandblasted sample. In the image of sample after sandblasting, the depressions are visible, that may have occurred as a result of breaking grains from the surface during abrasive blasting.

4 Discussion

The results of wetting angle measurements for the presinterized samples correspond to well-wettable materials, which are characterized by hydrophilic properties. The sintering process affected the grain growth and the increase of the contact angle, so the wettability of the material was reduced. Sandblasting and acid-etching affected the change of the hydrophilic properties of the material into hydrophobic. The value of the contact angle obtained for samples after the sandblasting process and after HF etching corresponds to poorly wettable materials. The greatest contact angle and the lowest wettability of the material were observed for samples etched with HF acid.

The literature presents many articles suggesting that the favouring factor responsible for the deposition of the bacterial plaque on the surface of permanent dentures, is the high wettability of the material surface. The adhesion of microorganisms to the surface of prosthetic materials promotes their colonization, which may result in various infections [16–18]. Therefore, it can be concluded that changing the hydrophilic properties of the zircionium oxide surface

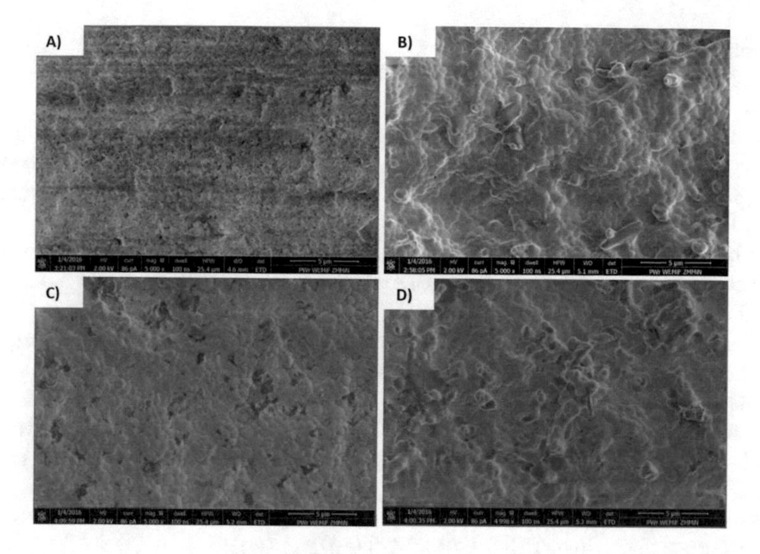

Fig. 5. SEM images for: A - presintered sample, B - sample after sintering, C - sandblasted sample, D - HF acid-etched sample

into a hydrophobic is a favourable and desirable phenomenon for such dental abutments. The material roughness is a factor affecting wettability. The greater roughness results in better wettability, and therefore a lower contact angle, which was confirmed by AFM. The sintering process affected the crystallographic structure of the zirconium oxide in the course of transition of the monoclinic phase into the tetragonal phase with the concomitant growth of grains. This transformation entails changes in mechanical properties, including increase in the hardness, which is desired in clinical use of ceramic abutments. The use of the presented surface modifications affects the inequality of the surface topography of the zirconium oxide, including its concavity and convexity. Both methods, Al_2O_3 sandblasting and HF acid-etching, are responsible for conversion of the hydrophilic properties of the material into hydrophobic. However, either sandblasting and etching does not cause phase changes in the material structure. Whereas, the greatest diversity of the surface of the zirconium oxide was obtained by etching using hydrofluoric acid, which can have a positive effect on the integration of fused porcelain to zirconia.

References

1. White, S.N., Miklus, V.G., McLaren, E.A., Lang, L.A., Caputo, A.A.: Flexural strength of a layered zirconia and porcelain dental all-ceramic system. J. Prosthet. Dent. **94**(2), 125–131 (2005)
2. Denry, I., Kelly, J.R.: State of the art of zirconia for dental applications. Dent. Mater. **24**(3), 299–307 (2008)

3. Okoński, P., Lasek, K., Mierzwińska-Nastalska, E.: Kliniczne zastosowanie wybranych materiałów ceramicznych, Protetyka stomatologiczna, Warszawa, LXII 3, str. 181-189 (2012)
4. Pietruski, J.K., Pietruska, M.D.: Materiały i technologie używane we współczesnej protetyce stałych uzupełnień zębowych - wady i zalety przedstawione na podstawie przeglądu piśmiennicwa i doświadczeń własnych, Stomatologia Estetyczna, tom 9, nr 3, lipiec-wrzesień, str. 89–99 (2013)
5. Stendera, P., Grochowski, P., Łomżyński, Ł.: Zastosowanie tlenku cyrkonu w protetyce stomato-logicznej, Protetyka stomatologiczna, Warszawa, LXII 2, str. 115–120 (2012)
6. Lasek, K., Okoński, P., Mierzwińska-Nastalska, E.: Tlenek cyrkonu - właściwości fizyczne i zastosowanie kliniczne, Protetyka stomatologiczna, Warszawa, LIX 6, str. 415–422 (2009)
7. Powers, J.M., Wataha, J.C., Kobierska-Brzoza, J.: Materiały stomatologiczne, Wyd. 1 pol./red. Urszula Kaczmarek, Elsevier Urban and Partner, Wrocław, ISBN 978-0-323-07836-8 (2013)
8. Powers, J.M., Sakaguchi, R.L.: Craig. Materiały stomatologiczne, Elsevier Urban and Partner, Wrocław, ISBN 978-83-7609-072-6 (2013)
9. Lemcke, H., Nakonieczny, D., Paszenda, Z.: Tlenek cyrkonu dla brukserów. Przegląd piśmien-nictwa, Twój przegląd stomatologiczny, str. 36–41, December 2013
10. Majchrzak, K., Mierzwińska-Nastalska, E., Bączkowski, B., Szczyrek, P.: Kliniczna ocena uzupełnień ceramicznych na podbudowie z tlenku cyrkonu, Protetyka stomatologiczna, Warszawa, LXIII 6, str. 431–440 (2013)
11. Subotowicz, K.: Ceramika dla każdego, Wydawnictwo Elamed, Katowice, ISBN 978-83-61190-05-9 (2009)
12. Nakonieczny, D., Ziębowicz, A., Paszenda, Z., Krawczyk, C.: Trends and perspectives in modification of zirconium oxide for a dental prosthetic applications - a review. Biocybern. Biomed. Eng. **37**(3), 453–465 (2017)
13. Bączkowski, B., Ziębowicz, A., Ziębowicz, B., Wojtyńska, E.: Laboratory evaluation of the fit of anti-rotational elements at the hybrid implant abutment used in prosthetic dentistry. In: Innovations in Biomedical Engineering. Advances in Intelligent System and Computing, vol. 526, pp. 11-16, Springer, Heidelberg (2017)
14. Łagodzińska, P., Dejak, B.: Wpływ obróbki strumieniowo-ściernej powierzchni tlenku cyrkonu na wytrzymałość połączenia z cementem kompozytowym, Protetyka stomatologiczna, Warszawa, LXIV 3, str. 195–203 (2014)
15. Łapińska, B., Domarecka, M., Sokołowski, G., Sokołowski, J.: Wpływ sposobu przygotowania powierzchni ceramiki krzemionkowej na wytrzymałość jej połączenia z materiałem kompozytowym za pomocą systemu Tender, Protetyka stomatologiczna, Warszawa, LXIII 4, str. 301–306 (2013)
16. Dubiel, M., Łagan, S.: Ocena kąta zwilżania oraz swobodnej energii powierzchniowej różnych typów materiałów stomatologicznych, Aktualne Problemy Biomechaniki, Kraków, nr 7 (2013)
17. Preoteasa, C.T., et al.: Wettability of some dental materials. Optoelectron. Adv. Mater. Rapid Commun. **5**(8), 874–878 (2011)
18. Sobolewska, E., Frączak, B., Błażewicz, S., Seńko, K., Lipski, M.: Porównanie kąta zwilżalności podstawowych materiałów protetycznych stosowanych w wykonawstwie protez ruchomych w badaniach in vitro, Protetyka stomatologiczna, Warszawa, LIX 6, str. 401–406 (2009)

Stabilization of Synthetic Materials with Silver Particles

Roman Major[1](✉), Gabriela Imbir[1], Aldona Mzyk[1], Piotr Wilczek[2], Marek Sanak[3], and Jürgen M. Lackner[4]

[1] Institute of Metallurgy and Materials Science, Polish Academy of Sciences, Reymonta St. 25, Cracow, Poland
R.Major@imim.pl

[2] Foundation for Cardiac Surgery Development, 354a Wolności Street, Zabrze, Poland

[3] Department of Medicine, Jagiellonian University Medical College, 8 Skawińska Street, Cracow, Poland

[4] Joanneum Research Forschungs-GmbH, Materials - Functional Surfaces, Leoben, Austria

Abstract. The work was related to study of the impact of material on the tissue. A synthetic material of the type of tissue analogue was developed. A disadvantage of synthetic materials is their instability, in particularly in the environmental fluids. Within the framework of the statutory test silver nanoparticles deposited by magnetron sputtering were conducted delivering its stabilization. Silver nanoparticles were introduced into the polyelectrolyte structure, in order to reduce the risk of the bacterial biofilm formation. The introduction of Ag nanoparticles was followed by deposition under high vacuum by magnetron sputtering. The analysis of the blood-material interactions was conducted using commercially available tester, Impact-R (Diamed). The assessment of silver ions assessment into plasma was tested based on prothrombin time (PT) and activated partial thromboplastin time (APTT). The unmodified surface of the polyelectrolyte is a potential activator of blood elements. The introduction of silver nanoparticles reduced the likelihood of clotting. Extrinsic coagulation pathway determined on the basis of internal PT and APTT coagulation pathways did not indicate hazards beyond the acceptable range.

Keywords: Synthetic materials · Polyelectrolytes · Layer by layer · Nanoparticle · Deposition · Blood-material interaction

1 Introduction

Extracellular matrix (ECM) structure can be easily mimic by the synthetic, degradable polymers [1,2]. Degradable scaffold materials are advantageous in tissue engineering applications because they provide initial biomechanical integrity as cells adhere, mature to the desired phenotype, and secrete natural ECM, and,

E. Tkacz et al. (Eds.): IBE 2018, AISC 925, pp. 321–328, 2019.
https://doi.org/10.1007/978-3-030-15472-1_34

most important, they are designed to be replaced by the neo-tissue. Polymers that have been extensively used in tissue engineering applications are poly glycolid acid (PGA), polylactic acid (PLA), and poly-E-caprolactone (PCL) [3–5]. They are widely used because of their biocompatible degradation properties. Hydrolysis of ester bonds in situ is thought to release nontoxic monomers that are excreted or absorbed by surrounding tissues. A release of acidic degradation products can be a source of toxicity that causes adverse local and systemic reactions that could result in implant rejection. An introduction of the nanoparticles of precious metals created an antibacterial function [6]. Binding anions and proteins in biologic systems, silver ion (Ag+) avidly binds to cell surface receptors of bacteria, yeasts and fungi [7]. Silver cations also strongly bind to electron donor groups of biologic molecules containing sulphur, oxygen and nitrogen. The binding of silver ion to sulfhydryl groups and proteins on cell membranes appears to be critical to the antimicrobial action. The ionizing capacity of various silver compounds is critical in comparing the antimicrobial activities. Nanotechnologies allow an efficient exploitation of the antimicrobial properties of silver by using silver in the form of nanoparticles [8]. These are used in applications such as preservatives in cosmetics, textiles, water purification systems, coatings in catheters and wound dressings. The anticipated widespread exposure to silver nanoparticles in the near future has prompted governmental bodies and the public to raise questions about the safety of such applications. An understanding of protein adsorption and macromolecular interactions is crucial for the control of surface initiated coagulation, the humoral immune response, cell binding, activation of inflammation, and wound healing [9–11]. Such processes are of large interest for currently expanding technological areas in biosensors and the controlled drug release, but may be of even larger significance for the in vivo behaviour of artificial surfaces in the cardiovascular system during short to medium long contact times (minutes to hours) [12,13]. The dynamics of plasma protein adsorption, exchange and degradation at solid interfaces are still obscure phenomena during physiological conditions. The time course of adsorption and displacement of fibrinogen in plasma is called the Vroman effect [14]. Its kinetics depends on surface chemical properties, where negatively charged hydrophilic surfaces are suggested to facilitate the displacement of adsorbed proteins and hydrophobic surfaces to retain them over extended times in plasma [15]. Positively charged hydrophilic surfaces (with point of zero charge (p.z.c.) on the other hand, are not at all understood in this context. The research activity was focused on studies of displacement of pre-adsorbed human serum albumin (HSA) by buffer and blood plasma, and the activation of the intrinsic pathway of coagulation and complement on silver nanoparticles introduces.

2 Materials and Methods

Polyelectrolyte multilayer coatings were deposited using so-called "layer by layer" method [16]. The substrate was activated with 10 M NaOH and washed with clean Milli-Q water. Poly-L-lysine (PLL) and hyaluronic acid (HA) and in the second combination chitosan (Chi) and chondroitin sulfate (CS) were dissolved in 400 mM HEPES/0.15 M NaCl solutions at concentrations of 0.5 to 1 mg/ml. The pH of the solutions was determined to 7.4 by the addition of 0.5 M NaOH. The coatings were made cyclically dipping by an automatic system that was developed. The deposition process was performed by submerging the substrates to the polycation and polyanion, building up the porous structure. The thickness of the coating depended on the number of bilayers. The polyelectrolytes deposition and diffusion process is based on electrostatic interactions. After each deposition step, a washing step was carried out in 0.15 M NaCl buffered solution pH 7.4 to remove excess polyelectrolyte. The process is repeated until the desired number of boulders is reached. Then, the samples were subjected to the final rinsing steps and stored at 4 °C in 400 mM HEPES/0.15 M NaCl buffered at pH 7.4. One of the methods to stabilize the synthetic materials is to apply an ultra-thin layer of metals or ceramic with atomic thickness. In this case, the research was carried out by introducing silver nanoparticles. The aim to introduce silver nanoparticles was to stabilize the porous structure and to protect against the formation of bacterial biofilms. The introduction of Ag nanoparticles was accomplished by deposition under high vacuum using a magnetron sputtering process, from pure Ag material (99.9%) at very high gas pressure. Silver was deposited to the dried substrate with a polyelectrolyte synthetic coating. The substrate was introduced into the vacuum chamber, then evacuated to 3×10^{-5} mbar. The deposition process was carried out at a pressure of 1×10^{-2} mbar Ar at a distance of 150 mm between the Ag disk and the substrate. The deposition was stopped after 30 s. In this case, the vacuum evaporation time determined the thickness of the coating.

2.1 Blood-Material Interaction

Analyzes of blood- material interaction were performed using a commercial Impact-R (Diamed) tester [17]. Platelet activation was assessed by glycoproteins IIb/IIIa, using CD62P for P-selectin. The IIb/IIIa integrin receptor is characteristic for the evaluation of the activation level. IIb/IIIa is the receptor for fibrinogen and von Willebrand factor and helps in their activation. Complex IIb/IIIa is produced in natural conditions by the increase of calcium ions resulting from the direct contact with a foreign factor, in this case a biomaterial. The most potent blood activator is adenosine-di-phosphate (ADP). It leads to a strong change in the conformation of platelet IIb/IIIa receptors and induces fibrinogen formation, a protein of blood plasma involved in the final phase of the blood coagulation process. The second analyzed platelet membrane receptor is selectin-P. P-selectin is a human protein encoded by the SELP gene. In inactivated plates, this protein is stored in granules of cells (in this case platelets) called Weibel-Palade. If the plates are activated, a membrane flip occurs. The receptors are released from granulation and exposure on the outer surface of the cell. After direct contact with the material, active plates and plate aggregates are analyzed. Platelet aggregates are analyzed by mixing 25 µl of blood with 0.4 ml of FLS5 and subsequent stabilization by adding 3.5 ml of 1% paraformaldehyde dissolved in PBS4. The cell material was recovered by centrifugation (1000 g, 7 min) and stained (25 µl aliquots) with the following antibodies CD14-PerCP and FITC-CD61 at room temperature for 30 min. The samples were then washed with PBS4 and analyzed using a flow cytometer. Activation of the coagulation system and the nature of the immune response on the surface of the materials was analyzed using a laser confocal scanning microscope, Carl Zeiss Exciter 5. For this purpose, anti-CD62P and anti-CD45 monoclonal antibodies were used, respectively, to analyze the degree of activation of the coagulation system and immune response (defense against foreign tissue body). The antibody finds a self-corresponding membrane receptor on the surface of cells. Monoclonal antibodies were conjugated with two different fluorochromes (anti-CD62P with fluorescein isothiocyanate (FITC), anti-CD45 with PE PE-Texas Red) which allowed distinguishing active platelets (green excitation) from active white blood cells (excitation red). The results of the cytometric analysis are presented in Fig. 1.

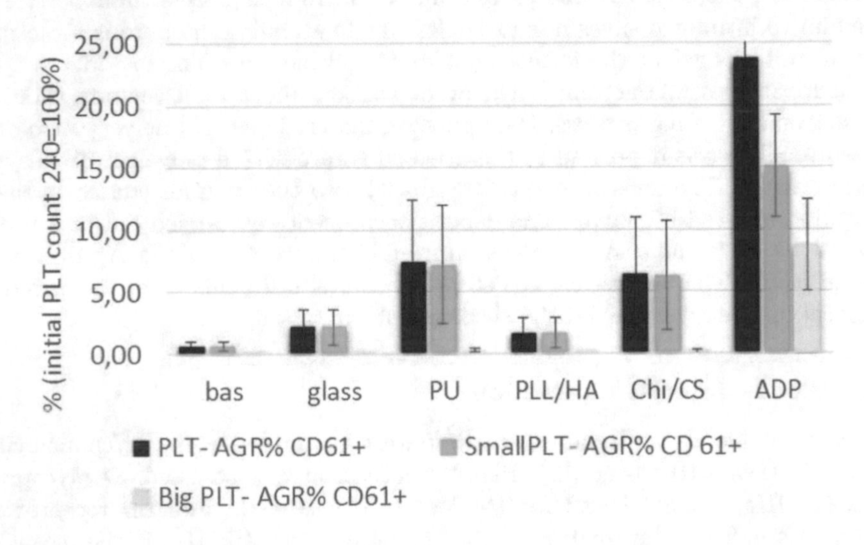

Fig. 1. The likelihood of platelet aggregation (CD61 +) formation depending on the material being tested; bas-negative control, ADP-adenosine-di-phosphate activated platelets - positive control

In the case of plate aggregates, small and large aggregates have been distinguished. Small aggregates were defined as two plates connected together, the large ones were defined as a conglomerate with more than two plates. The reference medium was a clinically used polyurethane. In the case of surface modification with polyelectrolyte modified with silver nanoparticles, in all cases, the modification improved hemocompatibility. Lowering the aggregate formation rate was found for small and large aggregates. Expression of platelet activation markers was measured with anti-CD61 antibodies to determine conformational changes of glycoprotein IIb/IIIa, and with anti-CD62P antibody determining activation of P-selectin. The results are shown in Fig. 2.

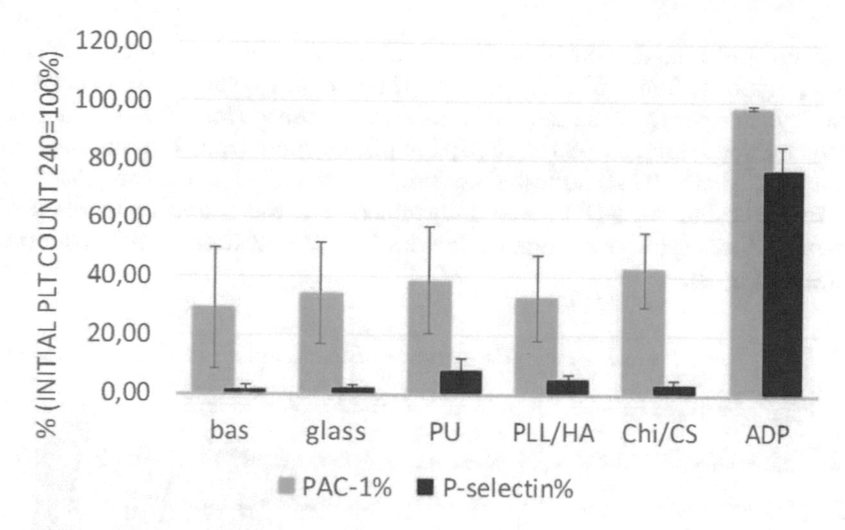

Fig. 2. Number of plates with active IIb/IIIa receptor and selectyin-P respectively in the function of the material being tested; bas-negative control, ADP-adenosine-di-phosphate activated platelets - positive control

2.2 Silver Release to Blood Plasma

Tissue is a cell and a intercellular substance. The specific type of extracellular matrix is plasma. In the evaluation of the release of silver ions prothrombin time (PT) and partial thromboplastin time after activation (APTT) were taken into plasma. PT is used to assess extrinsic coagulation factor. Its value depends on the concentration of coagulation factors such as factor II, factor V, factor VII, factor X and fibrinogen. The tissue factor (TF) is added to the plasma with the addition of citrate to activate Factor VII. The binding of calcium ions under the influence of citric acid causes that activation of factor X does not take place through the activated factor VIIa (VII) [18]. After incubation, the time from the introduction of calcium ions is measured. APTT measures the activity of blood coagulation factors in plasma. Kaolin-activated partial thromboplastin time - one of the indicators of blood coagulation, it is a measure of the activity of plasma coagulation factors such as factors XII, XI, IX and VIII. They form an inseparable prothrombin activation system. 22.5 ml of human blood was collected for analysis, which was collected on citrate (5×4.5 ml). Blood was poured into one Falcon 50 ml tube to ensure the homogeneity of the experiment. 1 ml of blood was transferred to small Eppendorf tubes and centrifuged at 4000 g/8 min/4 °C. After centrifugation, 20 µl of plasma was transferred to a test tube described as (0 bass, i.e. negative control) and frozen at −80 °C to determine the input concentration value of the microparticles. The rest of the plasma (370 µl) was transferred to a new tube labeled (negative-negative control) and the prothrombin time (PT) and time of partially activated thromboplastin (APTT) were measured. The test samples were thoroughly washed with sterile physiological saline. 3.5 ml of whole blood with anticoagulant (sodium citrate) was used for each study and mixed at 37 °C with shaking. The experiments were carried out respectively at 1, 3 and 6 h. Then, 1 ml of blood was collected and centrifuged at 4000 g/8 min/4 °C. Then 20 µl of plasma were taken (for micro-particle measurements) and frozen at −80 °C. 370 µl of plasma were transferred to new tubes for marking APTT PT. The tested coatings do not affect blood coagulation factors. Prothrombin time (PT) was determined. PT was found to increase with decreasing fibrinogen, prothrombin, factors V, VII and X (external coagulation pathway) (Fig. 3).

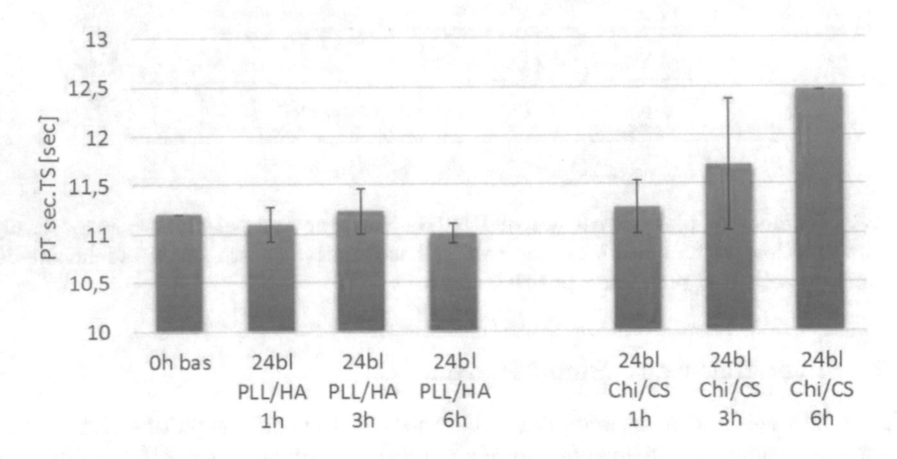

Fig. 3. Prothrombin time (PT), analysis of the extrinsic coagulation path V, VII and X

APTT (time of partial activation of thromboplastin), measures the extrinsic coagulation pathway. This parameter is able to additionally show the consumption of factor VIII, IX, XI and XII (Fig. 4).

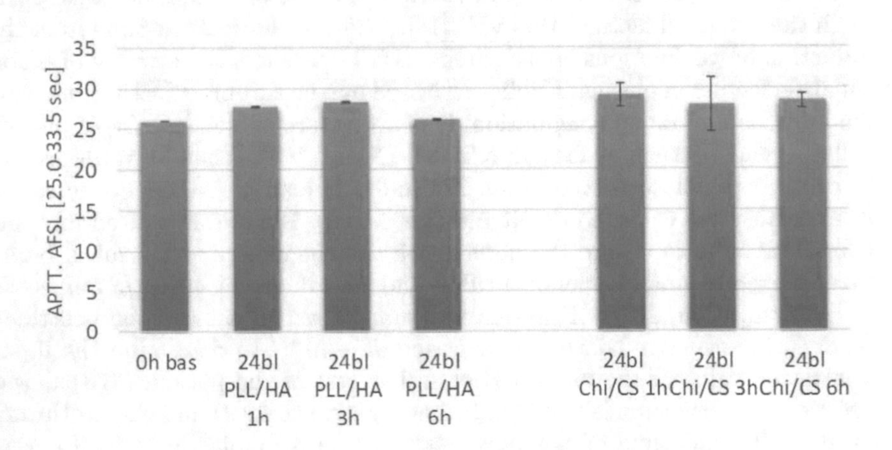

Fig. 4. APTT consumptions of factor VIII, IX, XI and XII

APTT is also dependent on factors involved in the formation of thrombin (prothrombin, factor X and V) and in the conversion of fibrinogen to fibrin. Kaolin-activated partial time should stay in the range of 26 to 36 s. During our experiment, the incubation time had to be reduced to 6 h because the mixed blood showed hemolysis.

3 Discussion and Summary

The best haemocompatibility properties are found in materials that mimic tissue structure and its properties. Because of this, a surface was prepared in the form of a three-dimensional artificial extracellular matrix made from polymeric structures under natural conditions or in favorable conditions. The extracellular matrix is produced by cells, filling the space between them. The clean surface (without cells) of polyelectrolytes causes a strong activation of the coagulation system. A similar phenomenon is observed on the surface of a pure extracellular matrix, i.e. excluding vascular endothelial cells in the blood vessels. Introduction of silver nanoparticles in the structure of the synthetic coating to stabilize the polyelectrolyte layer, reduced the blood activation processes without causing the unwanted release of Ag ions into the body. In this case, significant differences between the modified polyurethane with a polyelectrolyte layer containing silver and unmodified polyurethane was not observed. Polymers in the tissue environment change the characteristics and their physical and mechanical properties. They become stiffer, which can affect the increased risk of coagulation on the surface. Synthetic materials have a major disadvantage. In the tissue environment, they are biodegradable. This is not a serious problem when materials dissolve into simple compounds. A real problem arises when synthetic materials break down into complex compounds. There is a toxic effect on surrounding tissues. Therefore, first of all, it is necessary to choose the right polyelectrolyte. Another important point in the analysis was the impact on the time of biodegradation and the maintenance of control over the process. To this end, an ultra-thin Ag coating was proposed. They were introduced into the porous structure of the synthetic porous coating. Silver is a well-known antibacterial material. It protects completely against the formation of a bacterial biofilm. Another factor that was studied was the effect of Ag on thrombosis processes and the release of Ag ions to blood plasma. It has been found that Ag introduced into the porous structure of the polymer reduces the coagulation process and their release into the plasma is within the limits of acceptability.

Acknowledgements. The research was financially supported by the Project no. 2016 /23/B/ST8/01481 'Interdisciplinary methods of creating and functioning of biomimetic materials based on Animal origin extracellular matrix' of the Polish National Center of Science. Part of the work was co-financed by NCN: 2014/15/B/ST8/00103 and the European Union from resources of the European Social Fund (Project No. WND-POWR.03.02.00-00-I043/16).

References

1. Chung, S., King, M.W.: Biotechnol. Appl. Biochem. **58**, 423–438 (2011)
2. Naderi, H., Matin, M.M., Bahrami, A.R.: J. Biomater. Appl. **26**, 383–417 (2011)
3. Richert, L., Boulmedais, F., Lavalle, P., Mutterer, J., Ferreux, E., Decher, G., Schaaf, P., Voegel, J.C., Picart, C.: Biomacromolecules **5**(2), 284–294 (2004)
4. Major, R.: J. Mater. Sci. Mater. Med. **24**, 725–733 (2013)
5. Mzyk, A., Major, R., Kot, M., Gostek, J., Wilczek, P., Major, B.: Arch. Civ. Mech. Eng. **14**, 262–268 (2014)
6. Marx, D.E., Barillo, D.J.: Burns **40S**, 9–18 (2014)
7. Lansdown, G.: Curr. Probl. Dermatol. **33**, 17–34 (2006)
8. Parka, M., Neigh, A.M., Vermeulen, J.P., de la Fonteyne, L., Verharen, H.W., Briedé, J.J., van Loveren, H., de Jong, W.H.: Biomaterials **32**, 9810–9817 (2011)
9. Arvidsson, S., Askendal, A., Tengvall, P.: Biomaterials **28**, 1346–1354 (2007)
10. Horbett, T.A.: The role of adsorbed proteins in tissue response to biomaterials. In: Ratner, B.D., Hoffman, A.S., Schoen, F.J., Lemons, J.E. (eds.) Biomaterials Science: An Introduction to Materials in Medicine, 2nd edn, pp. 237–246. Elsevier Academic Press, London (2004)
11. Brash, J.L., Ten Hove, P.: Thromb. Hemost. **51**, 326–330 (1984)
12. Merrill, E.W.: Distinctions and correspondences among surfaces contacting blood. In: Leonard, E.F., Turitto, V.T., Vroman, L. (eds.) Blood in Contact with Natural and Artificial Surfaces, vol. 516, pp. 196–203. Annals of the New York Academy of Sciences, New York (1987)
13. Hoffman, A.S.: Modification of material surfaces to affect how they interact with blood. In: Leonard, E.F., Turitto, V.T., Vroman, L. (eds.) Blood in Contact with Natural and Artificial Surfaces, vol. 516, pp. 96–101. Annals of the New York Academy of Sciences, New York (1987)
14. Wojciechowski, P., Ten Hove, P., Brash, J.L.: J. Colloid Interface Sci. **111**, 455–465 (1986)
15. Elwing, H., Askendal, A., Lundstrom, I.: J. Biomed. Mater. Res. **21**, 1023–1028 (1987)
16. Picart, C., Lavalle, P., Hubert, P., Cuisinier, F.J.G., Decher, G., Schaaf, P.: Langmuir **17**, 12531–12535 (2001)
17. Major, L., Lackner, J.M., Major, B.: RSC Adv. **4**, 21108–21114 (2014)
18. Wein, M., Sterbinsky, S.A., Bickel, C.A., Schleimer, R.P., Bochner, B.S.: Am. J. Respir. Cell Mol. Biol. **12**(3), 315–319 (1995)

Tests of Threaded Connections Made by Additive Manufacturing Technologies

Wojciech Kajzer[1(✉)], Katarzyna Gieracka[2], Mateusz Pawlik[3], Marcin Basiaga[1], Anita Kajzer[1], and Janusz Szewczenko[1]

[1] Department of Biomaterials and Medical Devices Engineering, Faculty of Biomedical Engineering, Silesian University of Technology, ul. Roosevelta 40, 41-800 Zabrze, Poland
wojciech.kajzer@polsl.pl
[2] Science Club "SYNERGIA", Department of Biomaterials and Medical Devices Engineering, Faculty of Biomedical Engineering, Silesian University of Technology, ul. Roosevelta 40, 41-800 Zabrze, Poland
[3] CABIOMEDE Sp. z o.o., K. Olszewskiego 6, 25-663 Kielce, Poland

Abstract. The aim of the work was to determine breaking strength of the threaded connection between produced in additive manufacturing technology threaded component (fitting) and steel screw. Samples made in Selective Laser Sintering (SLS) and Multi Jet Fusion (MJF) technologies were used. Depending on method of production, in this research two kinds of materials were applied. Polyamide PA12 and composite material made of polyamide with glass particles PA-GF in volume proportion 70% to 30% were used. The results give a base of the statement, that the printed threaded connection allows to obtain stable and durable connection, enabling practical application of that type of connection.

Keywords: Additive manufacturing · SLS - Selective Laser Sintering · MJF - Multi Jet Fusion ·
Mechanical research, strength of threaded connections

1 Introduction

The additive manufacturing is a dynamically developing technology of manufacturing, which application is showed for example in biomedical engineering and scientific research. Use of 3D printing allows to create a part of the medical devices or prosthesis either all prototyping each part of body, for example hand or leg prostheses. Very often to design and manufacture personalized surgical implants, tissue scaffolding and preparing, planning and testing very complicated reconstructive procedures 3D printing technology are used [1]. The use of additive manufacturing methods gives the possibility to create a personalized model, both anatomically and biomechanically tailored to the specific patient needs [2].

E. Tkacz et al. (Eds.): IBE 2018, AISC 925, pp. 329–337, 2019.
https://doi.org/10.1007/978-3-030-15472-1_35

In order to ensure the best properties of products made in additive manufacturing technology, on the printed models the different kind of tests are performed [4–6]. For example: test of dimensional accuracy, tensile strength, water absorption test, hardness test i.e. Vicers or Shore tests, density test, surface roughness and microscopic analysis [7–11]. To fully analyze the effect of the printing direction on both mechanical and physical properties, samples printed in three directions are used to carry out these tests. The horizontal, side and vertical samples were used [11,12].

According to the 3D printing method, several types of materials which models can be made of are available. For both SLS and MJF technologies the main materials are polyamide PA12. This is thermoplastic material, in the powder form, which is melting under the influence of laser beam [13]. The prototypes made of PA12 are characterized by very good durability and flexibility [14]. That gives also thermal resistance and resistance of most chemicals [15]. Beside polyamide PA12 there are available a lot of different materials, including PA-GF composite used in SLS technology. PA-GF is a polyamide powder with addition of glass participles with very good mechanical properties [3].

However, the main drawback of 3D printing is limited strength, which lead to lower durability of physical objects [16]. One on the places susceptible for damage in printed elements are fittings, in particular threaded connection. They are exposed to excessive stress which may lead to connection rapture [17].

Threaded connection are type of releasable connection including screw or bolts. Each of the thread is described by some parameters, such as: thread diameter and thread pitch. Currently, to create threaded connection between fitting (steel screw) and 3D printed hole, a threaded insert was used. After a slight heating, insert was placed into prepared hole. Thanks to the heat, the bore becomes more plastic and the material was flowing around knurls from the insert and blocking movements (put out). After that stage, it was possible to obtain threaded connection.

Nowadays it is intended to replace such a solution with a direct threaded connection between the printed element and the connecting element (steel screw). Therefore, the purpose of the presented work was to evaluate the strength the threaded joint between the fitting (printed in SLS or MJF technologies), and the steel screw.

2 Materials and Methods

Samples consisting of the threaded printed element (fitting) and the steel screw were used for the tests. The samples were made in two kind of additive manufacturing technologies: SLS and MJF. The PA12 material was used for half of the samples, while second part of sorts where made of PA-GF (70/30) composite. In the MJF method, only polyamide PA12 was applied. Two kinds of threaded connection were analyzed: thread pitch of 0.5 mm and diameter 3 mm and also which pitch of 0.7 mm and diameter 4 mm. For each of the two threads variants, three type of printed fittings (depending on the thickness wall g, which was respectively: g = 1 mm, g = 1.5 mm and g = 2 mm) were applied (Fig. 1). Three samples were prepared for each tested variant of the threaded connection.

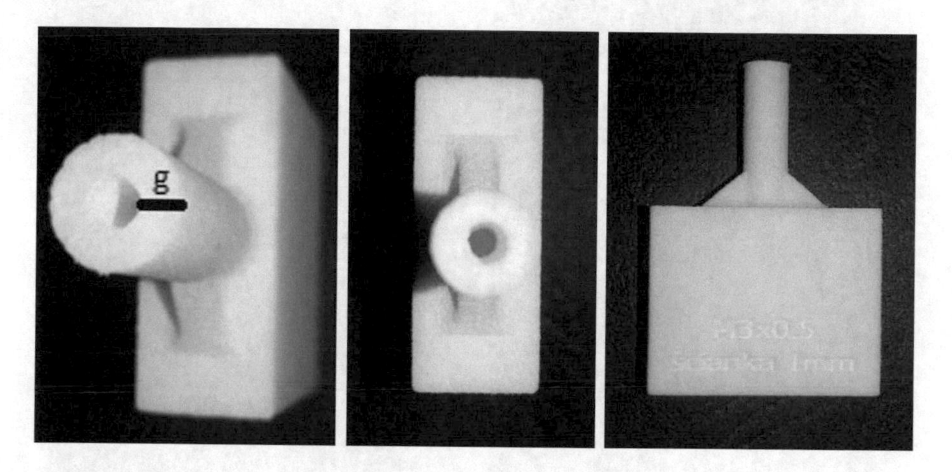

Fig. 1. Examples of fittings sampled to strength tests of a threaded connection made in SLS technology: material PA12, thread: M3x0.5, thickness wall respectively: g = 1.5 mm, g = 2 mm, g = 1 mm

2.1 Tensile Strength Test

To specify the breaking force of the threaded connection, tensile strength test was carried out. The MTS Criterion 45 testing machine was used. The test rate was equal to V = 1 mm/min. The end of the test was defined as the breakage of the threaded connection break or damage of the sample. For the samples made of polyamide PA12 and PA-GF the screw was mounted in the way that part of it (2/5 length of the bolt) was above the sample's neck (Fig. 2a). Next the sample was placed in the handle of the machine (Fig. 2b). The machine was connected to the computer with TestRun Suit software by MTS.

2.2 Microscopic Observation

Both, before and after the tests, the samples were observed with the use of stereo microscope Zeiss SteREO Discovery.V8 at a magnification of 3x, 3.75x, 4.8x, 6x, 12x.

3 Results and Discussion

3.1 The Results of the Tensile Strength Test

Examples of graphs presenting the force as a function of elongation for the prepared variants of samples were shown in (Figs. 3, 4 and 5). The obtained

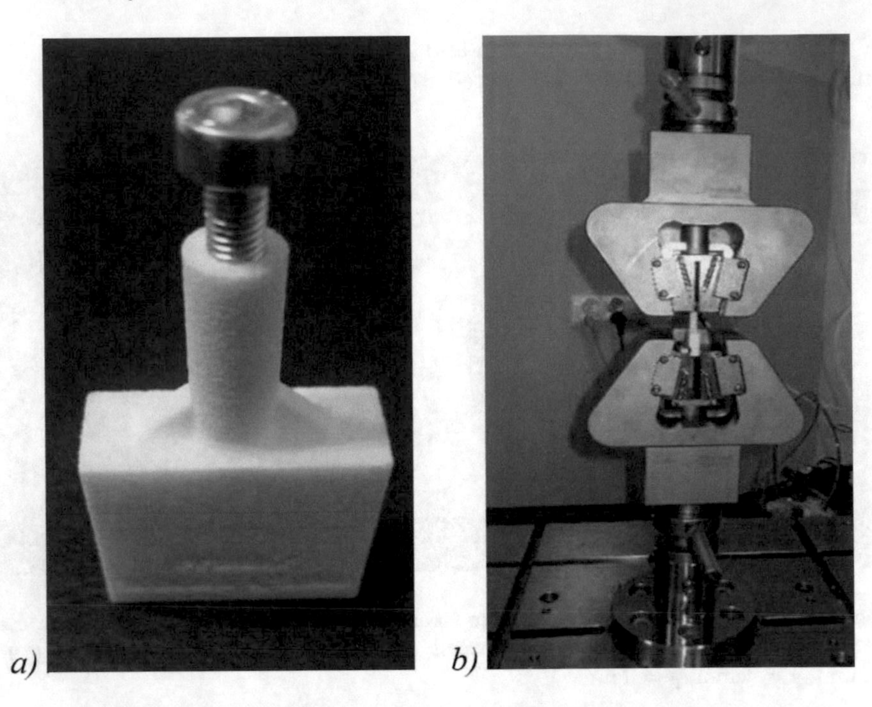

Fig. 2. Examples of: (a) threaded connection used in tensile strength test; (b) test stand

maximum values of forces F_m, at which the samples were broken are presented in Tables 1 and 2.

The analysis of the obtained results showed the influence of wall thickness "g" (printed in Selective Laser Sintering (SLS)) on the maximum force F_m Table 2, Fig. 3 – the thicker wall, the higher value of the force was recorded. This relationship found for two analyzed variants of threaded connection i.e. M3x0.5 and M4x0.7. However, for the samples made in SLS technology with using composite PA–GF, such clear relationship was not found. That was observed that for both connection types (M3x0.5 and M4x0.7), the maximum force F_m, at which the samples were broken, occur for the wall thickness: g = 2 mm (Table 1, Fig. 4). In

Table 1. The values of maximum force values of F_m, obtained for all kind of force variants samples made in SLS technology

Technology	SLS											
Material	PA12						PA-GF					
Kind of thread	M3x0.5			M4x0.7			M3x0.5			M4x0.7		
Wall thickness, "g", mm	1	1.5	2	1	1.5	2	1	1.5	2	1	1.5	2
Average of F_m [N]	219	521	983	396	707	1277	226	297	453	110	152	378
Standard deviation [N]	66	82	100	94	155	77	-	306	258	-	108	256

Table 2. The values of maximum force values F_m, obtained for all kind of prepared variants made in MJF technology

Technology	MJF					
Material	PA12					
Kind of thread	M3x0.5			M4x0.7		
Wall thickness, "g", mm	1	1.5	2	1	1.5	2
Average of F_m [N]	212	344	289	331	929	624
Standard deviation [N]	69	207	6	52	78	203

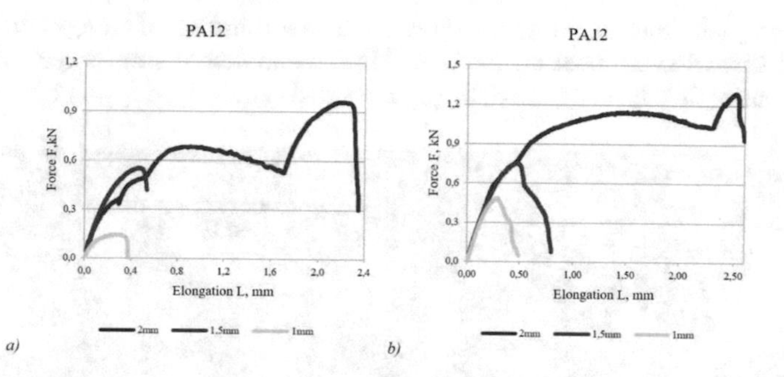

Fig. 3. Examples illustration which show force as a function of elongation for analysed variants of threaded connection made in SLS technology using polyamide PA12: (a) threaded connection type of M3x0.5, (b) threaded connection type of M4x0.7

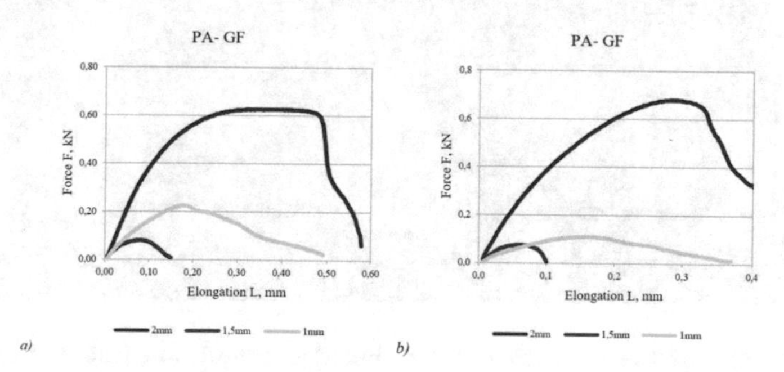

Fig. 4. Examples illustration which show force as a function of elongation for analysed variants of threaded connection made in SLS technology using polyamide PA-GF: (a) threaded connection type of M3x0.5, (b) threaded connection type of M4x0.7

contrast, the values of maximum force determined for wall thickness g = 1 mm and g = 1.5 mm are definitely lower and similar to each other (Table 1). Additionally it was noted that the fitting made of PA–GF were characterized by high brittleness, which in case of samples with wall thickness of g = 1 mm led to their damage while screwing the screw before starting the main test. For the samples made in MJF technology, no relation between wall thickness of the fitting "g"

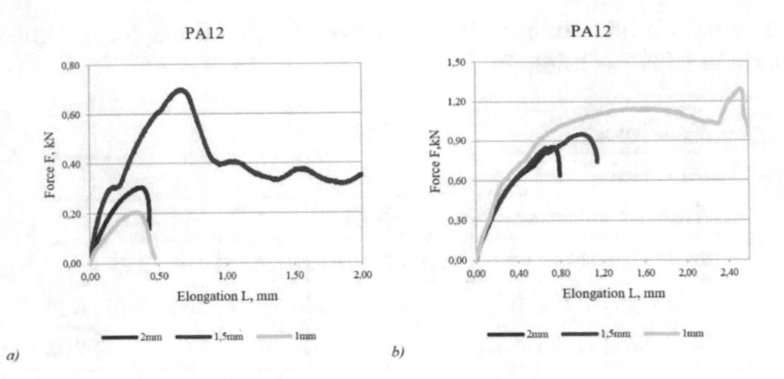

Fig. 5. Examples illustration which show force as a function of elongation for analysed variants of threaded connection made in MJF technology using polyamide PA12: (a) threaded connection type of M3x0.5, (b) threaded connection type of M4x0.7

Fig. 6. Example of the strength test for threaded connection (method MJF, material: PA12, thread: M3x0.5, wall thickness: g = 2 mm), for which the brake threaded connection was observed)

and maximum force F_m was registered. However, for all the tested samples made of PA12 polyamide, the registered maximum forces F_m had got higher values for M4x0.7 thread than M3x0.5 – Table 2, Fig. 5. In addition, observation of the samples showed, that for variant M3x0.5, g = 2 mm the threaded connection was broken – Fig. 6. The samples for which the thread break was observed, characteristic forces curves as a function of elongation were recorded – Fig. 7. The periodic force increase was related to the displacement of the screw with another thread turn.

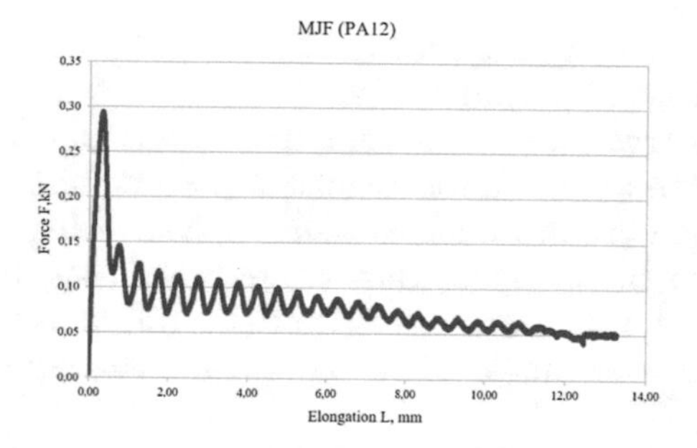

Fig. 7. Example curve of the strength test, for which threaded connection was broken (method MJF, material: PA12, thread: M3x0.5, wall thickness: g = 2 mm)

3.2 Results of Macroscopic Observation

Analysis of the obtained results showed that in the case of the samples printed in SLS technology, in each of the variants, the printed threaded fitting has been damaged. The destruction of the samples had character of a crack just below the thread connection the fitting – steel screw. This provides that threaded

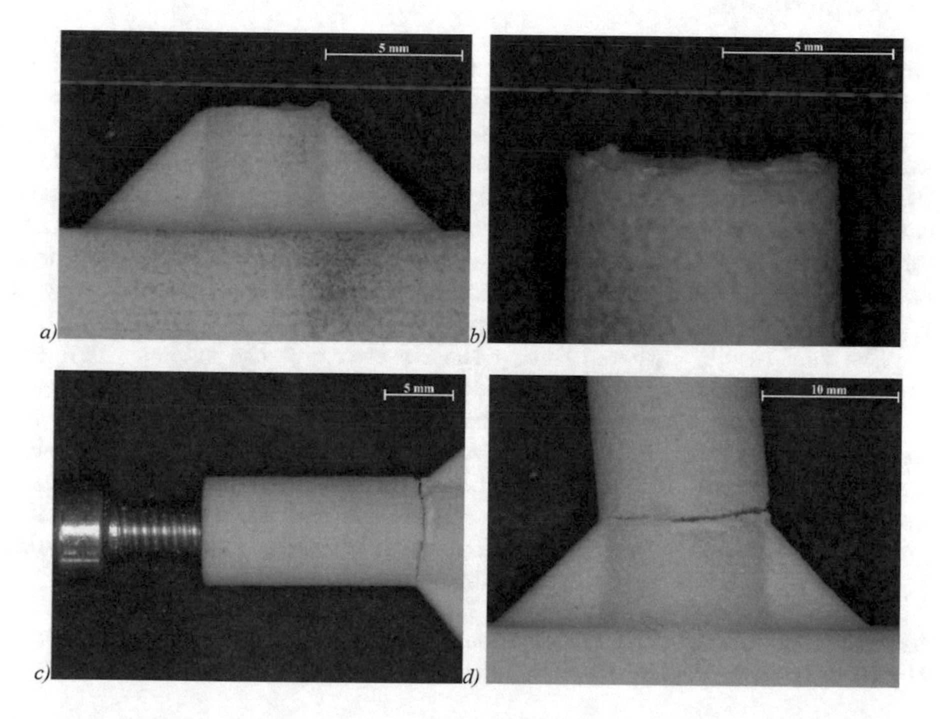

Fig. 8. Example macroscopic picture of breaking samples: (a) SLS, PA12, M4x0.7, g = 1 mm – pow. 4.8x, (b) SLS, PA12, M4x0.7, g = 1 mm – pow. 9.6x; (c) SLS, PA-GF, M4x0.7, g = 2 mm: pow. 3x (d) SLS, PA-GF, M4x0.7, g = 2 mm: pow. 4.8x

connection is more durable compared to printout made of both polyamide PA12 and composite on base of polyamide PA–GF.

Macroscopic observations of the samples' breaksites – Fig. 8 have shown a brittle character with surface development of a characteristic mirror powdered material the fittings were made of. In addition, comparing the fittings made in SLS technology, can be observed, that samples neck was completely raptured (PA12), in comparation to samples made of PA–GF, where the material had a visible fracture of a more malleable without complete rupture of the sample – Fig. 8b.

In the case of samples made in MJF technology with wall thickness g = 2 mm and thread M3x0.5 the break of threaded connection was observed – Fig. 6. For these samples a sliding of the steel screw with a printed fitting and also residues of polyamide PA12 in the screws thread were found. These observations support the results obtained in the static tensile test for this variant – Fig. 7. For other types of samples made in MJF technology made of polyamide PA12, as for the SLS samples, the fitting was raptured just below the threaded connections. The nature of the rapture was brittle in contrast to the raptures obtained for the samples made in SLS characterized by lesser surface development.

4 Conclusion

On the basis of the obtained results it was found that for the sample materials used in the SLS method, the connection has been broken just below the thread connection. It means, that for those samples, the highest maximum force F_m was not registered for the thread connection itself, but for the printout made of the polyamide PA12 or the composite based on PA–GF polyamide. The research carried out on the samples made with the MJF technology has concluded, that for the thread type M3x0.5 and the g = 2 mm, the connection has been broken exactly at the thread connection. For other analysed samples a similar way of rapture of the printed fitting was observed (similar to the SLS printouts).

The final results gathered during the research give the evidence that the thread connection is stable and resistant, which can find practical applications.

In order to find a better characteristics of the thread connection made with the additive manufacturing technologies, future research needs to be made using a modified construction of the printed fitting, in a way that the weakest element would be the thread connection itself. Making that sort of modification should provide the evidence of how using various materials, technologies and the types of the thread, can influence the durability of the threaded connection.

Acknowledgements. The work has been financed from research project no. $BK-210/RIB2/201807/020/BK\_18/0028$.

References

1. Laska-Leśniewska, A.: Wykorzystanie metod szybkiego prototypowania (rapid prototyping) w nowoczesnej medycynie. Zeszyty naukowe towarzystwa doktorantów UJ Nauki Ścisłe. Nr. 15, 39–48, 2 2017
2. Wyleżoł, M., Ostrowska, B., Wróbel, E., Muzalewska, M., Grabowski, M., Wyszyński, D., Zubrzycki, J., Przech, P., Klepka, T.: Inżynieria biomedyczna Metody przyrostowe w technice medycznej. Monografia. Wydawnictwo Politechniki Lubelskiej, Lublin (2016)
3. Source: http://www.materialise.com/pl/manufacturing/technologie-i-materialy/spiekanie-laserowe-sls
4. Kajzer, W., Krauze, A., Kaczmarek, M., Marciniak, J.: FEM analisys of the expandable intramedullar nail. In: Information Technologies in Biomedicine. Advances in Intelligent and Soft Computing, vol. 47, pp. 537–544 (2008)
5. Ziebowicz, A., Kajzer, A., Kajzer, W., Marciniak, J.: Metatarsal osteotomy using double-threaded screws - biomechanical analysis. In: Information Technologies in Biomedicine. Advances in Intelligent and Soft Computing, vol. 69, pp. 465–472 (2010)
6. Basiaga, M., Kajzer, W., Walke, W., Kajzer, A., Kaczmarek, M.: Evaluation of physicochemical properties of surface modified Ti6Al4V and Ti6Al7Nb alloys used for orthopedic implants. Mater. Sci. Eng. C **68**, 851–860 (2016)
7. Joshi, M., Shetty, N., Shetty, S.D., Bharath, N.L.S., Varma, C.S.: Mechanical characterization of additive manufacturing processes. Indian J. Sci. Technol. **9**(36), 1–15 (2016)
8. Spyra, M., Kajzer, W., Czyrnia, R.: Wytrzymałościwa analiza z wykorzystaniem metody elementów skończonych zmodyfikowanych ortez Grafo. Aktualne Problemy Biomechaniki **11**, 127–132 (2016)
9. Basiaga, M., Paszenda, Z., Walke, W.: Study of electrochemical properties of carbon coatings used in medical devices. Electrochem. Rev. **87**(12B), 12–15 (2011)
10. Walke, W., Paszenda, Z., Pustelny, T., Opilski, Z., Drewniak, S., Kościelniak-Ziemniak, M., Basiaga, M.: Evaluation of physicochemical properties of SiO2-coated stainless steel after sterilization. Mater. Sci. Eng. C Mater. Biol. Appl. **63**, 155–163 (2016)
11. Loncierz, D., Kajzer, W.: Wpływ parametrów druku 3D w technologii FDM na własności mechaniczne i użytkowe obiektów wykonanych z PLA. Aktualne Problemy Biomechaniki **10**, 43–48 (2016)
12. Liu, C.-Y.: A comparative study of rapid prototyping systems. Thesis, University of Missouri (2013)
13. Mierzejewska, Ż.A., Markowicz, W.: Selective laser sintering - binding mechanism and assistance in medical applications. Adv. Mater. Sci. **15**(3(45)), 5–16 (2015)
14. Lewandowski, G., Milchert, E., Rytwińska, E.: Właściwości fizyczne i zastosowanie poliamidu 12. Source. http://www.ichp.pl/polimery-lewandowski-rytwinska-milchert-wlasciwosci-poliamidu
15. Podhora, P., Madaj, R., Poljak, S.: Verification of construction properties materials for rapid prototyping using SLS technology. In: 58th ICMD, pp. 306–313 (2017)
16. Cichoń, K., Brykalski, A.: Zastosowanie druku 3D w przemyśle. Przegląd Elektrotechniczny **93**(3), 156–158 (2017)
17. Kromka-Szydek, M., Wrona, M., Jędrusik-Pawłowska, M.: Analiza wytrzymałościowa systemu UNILOCK 2,4 stosowanego w chirurgii szczkowotwarzowej. Modelowanie Inżynierskie **16**(47), 117–122 (2013)

Author Index

E. Tkacz et al. (Eds.): IBE 2018, AISC 925, pp. 339–340, 2019.
https://doi.org/10.1007/978-3-030-15472-1

Printed in the United States
By Bookmasters